Fluid Flow
and
Heat Transfer

Fluid Flow
and
Heat Transfer

AKSEL L. LYDERSEN

*Professor of Chemical Engineering,
The University of Trondheim, Norway*

A Wiley–Interscience Publication

JOHN WILEY & SONS

Chichester · New York · Brisbane · Toronto

Copyright © 1979 by John Wiley & Sons Ltd.

All rights reserved. No part of this book may be reproduced by any means, nor transmitted, nor translated into a machine language without the written permission of the publisher.

Library of Congress Cataloging in Publication Data:
Lydersen, Aksel.
    Fluid flow and heat transfer.

    'A Wiley–Interscience publication.'
    Includes bibliographical references and indexes.
    1. Fluid dynamics.   2. Heat–Transmission.
I. Title.
TP156.F6L93   620.1'064   78-18467

ISBN 0 471 99697 1(Cloth)
ISBN 0 471 99696 3 (Pbk.)

Typeset in IBM Press Roman by Preface Ltd., Salisbury, Wilts and printed by Unwin Brothers Ltd., The Gresham Press, Old Woking, Surrey

Preface

As early as the first half of this century it was impossible for one person to have detailed knowledge of the whole of the chemical process industry. As a result, chemical engineering as a special field of study was introduced into the universities of America and in Europe.

Processes are always built up of unit operations that can be studied and treated separately, whatever the process in question. As an example, the principles of filtration are the same, whether this unit operation is being used in mineral dressing, in the production of fertilizers or synthetic dyestuffs, in the bleaching of oil, or in the pharmaceutical industry. Therefore the study of chemical processes can be rationalized by treatment of a limited number of unit operations which provide the foundations for understanding the complex chemical and allied process industries of today.

In the second half of this century the increase in the number of unit operations, the more elaborate treatments, and the accumulation of empirical data have made it impossible for one person to be a specialist in all fields of unit operations. This resembles the situation with industrial chemistry earlier in this century. The author's conclusion is:

Today's practising chemical engineer needs a short refresher text containing the basic equations, empirical data for common cases, and references to suitable literature where more details can be found. This text should also contain worked examples from engineering practice, with comments on important facts that may easily be overlooked by anyone who has not worked with the same type of problem.

It would also be useful if students of chemical engineering had a condensed text giving important fundamentals and examples that they can work out — and get guidance and a check on their results.

This text is an attempt to cover these needs, both for the practising engineer and for the student in the fields of mechanical unit operations and heat transfer in chemical engineering. The text itself has been kept short, and a high proportion of the examples come directly from industrial work. Numerous other books are available for the student who wants detailed descriptions of equipment and specialist calculation methods that have to be omitted from a concentrated text. In addition, the most comprehensive single source of information in this field, R. H. Perry and C. H. Chilton, *Chemical Engineers' Handbook*, should always be at hand.

The goal of a chemical engineer should be that his work results in profitable production. This depends on product quality and price. Therefore the quality of the final product and the relative costs of production should decide the choice between competing unit operations. These factors can only be judged from knowledge of the main dimensions of the equipment and its efficiency — which in some cases can only be obtained after pilot plant runs. The results must be quantitative. Lord Kelvin's words are valid here, as in other branches of science and technology: 'it is only what can be measured and expressed in figures that one really knows'.

This text is concerned with calculations of the major dimensions of equipment and of the consumption of energy. Stress calculations and many details of design in general will be the task of mechanical engineers.

The engineer must economize with his own time. This means that he will use the methods which give the necessary accuracy with minimum work. It does not make sense to spend days on complex calculations, if the variations during operation are greater than the limits of error of a rapid approximate calculation. It is also wise to use common sense, which includes the habit of checking results. A good guess may be all that is needed to reveal an error in the calculations carried out.

The manuscript was first written for use at the University of Dar es Salaam, Tanzania.

Trondheim, January 1978 AKSEL LYDERSEN

Contents

CHAPTER 1   PRESSURE DROP IN PIPES, CHANNELS, FITTINGS, AND EQUIPMENT 1
    Laminar flow 2
    Turbulent flow 3
    Bernoulli's equation 4
    Fittings 7
    Helices 8
    Tube banks 8
    Granular beds 11
    Two-phase flow 12
    Examples 14
    Problems 34
    References 35

CHAPTER 2   DIMENSIONAL ANALYSIS AND DIMENSIONLESS GROUPS 37
    Dimensional analysis 37
    Model laws 38
    Examples 39
    Problems 53
    References 53

CHAPTER 3   FLOW MEASUREMENTS 55
    Nozzles and orifices 55
    Venturi meters 58
    Pitot tubes 58
    Rotameters 60
    Magnetic flowmeters 61
    Examples 61
    Problems 67
    References 68

CHAPTER 4   PUMPING, COMPRESSION, AND EXPANSION   69
    Pumping 69
    Compression 74
    Vacuum pumps 81
    Expansion 85

	Steam ejectors	87
	Examples	90
	Problems	114
	References	116
CHAPTER 5	**AGITATION**	**117**
	Agitators	117
	Power consumption	119
	Effects of agitation	119
	Examples	121
	Problems	126
	References	126
CHAPTER 6	**PARTICLE AND DROP MECHANICS**	**128**
	Terminal velocity	129
	Classification	132
	Hindered settling	133
	Fluidization	134
	Pneumatic conveying	136
	Drops and bubbles	137
	Decanting	137
	Centrifuges, cyclones, and hydrocyclones	137
	Impingement separators	148
	Wire mesh demisters	150
	Fibre mist eliminators	152
	Bacteria eliminating filters	153
	Bag filters	153
	Scrubbers	155
	Electrostatic precipitators	161
	Examples	167
	Problems	190
	References	191
CHAPTER 7	**LIQUID FILTRATION AND FLOTATION**	**193**
	Filtration	193
	Flotation	205
	Examples	208
	Problems	217
	References	217
CHAPTER 8	**ATOMIZATION, DISPERSION, HOMOGENIZATION, CRUSHING, AND GRINDING**	**219**
	Atomization of a liquid in a gas	219
	Dispersion of a gas in a liquid	219
	Dispersion of a liquid in a liquid	221

	Homogenization	221
	Crushing and grinding	223
	Examples	227
	Problems	230
	References	230

CHAPTER 9 STEADY STATE HEAT TRANSFER 231

Heat conduction	231
Radiation	235
Convection	238
Tube banks	243
Tube coils	243
Agitated vessels	244
Scraped tubes	244
Granular beds	244
Liquid films	245
Condensation	246
Boiling	247
Overall heat transfer	250
Temperature difference	251
Heat exchangers	254
Evaporation	263
Examples	266
Problems	290
References	292

CHAPTER 10 UNSTEADY STATE HEAT TRANSFER 294

With phase change	294
Without phase change	297
Schmidt's graphical method	300
Examples	305
Problems	319
References	320

CHAPTER 11 ENERGY ECONOMY 321

Minimum energy requirement	321
Recovery of waste heat	323
Operation at reduced temperature	323
Multiple-effect evaporation	324
Heat pumps	327
Other processes	328
Examples	328
Problems	333
References	334

APPENDIX 1 UNITS, CONVERSION FACTORS, AND SYMBOLS 335
  Units 335
  Prefixes 336
  Conversion factors 336
  Symbols 340

APPENDIX 2 PHYSICAL PROPERTIES OF GASES AND LIQUIDS . . . 344

INDEX 353

CHAPTER 1
Pressure Drop in Pipes, Channels, Fittings, and Equipment

It is assumed that the fundamentals of fluid statics and fluid mechanics are known by the reader. Hence, this chapter is limited mainly to a collection of equations and empirical data for common types of chemical engineering calculations. In addition, discussion is limited to *Newtonian fluids*, fluids in which the viscosity is independent both of time and of velocity gradient du/dy, and is defined by the equation

$$\tau = \mu \frac{du}{dy} \tag{1.1}$$

where τ is the shear stress and du/dy is the velocity gradient perpendicular to the direction of flow as shown in Figure 1.1.

The viscosity μ has dimensions g/(cm s) or poise in the cgs-system and N s/m^2 in the SI-system.

Pastes and suspensions containing fibres are usually *non-Newtonian fluids*. In Newtonian fluids the velocity gradient du/dy is a linear function of the shear stress as shown by curve b in Figure 1.2. The curves a, c, and d are examples of the same relationship for different types of non-Newtonian fluid. For many non-Newtonian fluids, the viscosity is also a function of the prehistory of the flow, i.e. is a function both of shear stress and of time.

Figure 1.3 shows the flow of a Bingham plastic fluid through a tube of length L. The force acting on a cylinder with radius r is $F = p\pi r^2$ and the surface area of the cylinder is $A = 2\pi rL$, giving a shear stress

$$\tau = p\pi r^2 / 2\pi rL = \frac{p}{2L} r, \tag{1.2}$$

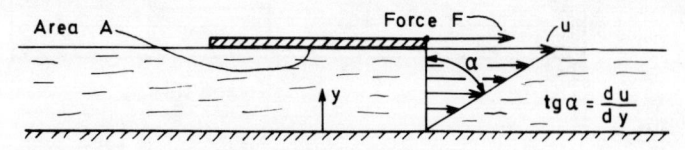

Figure 1.1 Velocity profile for laminar flow between a plate pulled by a force F and a wall at rest

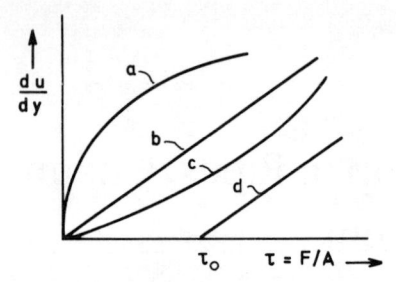

Figure 1.2 Velocity gradient du/dy as a function of the shear stress $\tau = F/A$. a, dilatant fluid; b, Newtonian fluid; c, pseudo-plastic fluid; d, Bingham plastic fluid

i.e. the shear stress increases linearly with the radius r. The threshold shear stress, τ_0 (Figure 1.2), is obtained at a radius r_0. For $r < r_0$ the shear stress is less than the threshold value τ_0, and this part of the fluid will move as a solid rod. For $r > r_0$ the velocity gradient increases with increasing radius as shown in Figure 1.3.

For calculation of non-Newtonian flow, the reader is referred to literature on rheology.[1,2]

Laminar flow

At low flow rates through a straight tube all fluid elements move parallel to the tube axis. Also all fluid elements the same distance from the axis move at the same velocity. The velocity profile is parabolic as shown in Figure 1.4, and the average velocity V is half the velocity in the centre, u_{max}. This type of motion, where the velocity at a certain point is constant and independent of time, is termed laminar flow. It is sometimes called *streamline flow* or *viscous flow*.

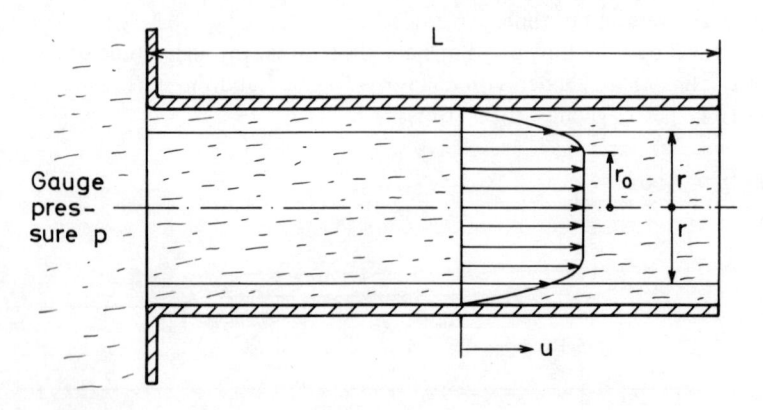

Figure 1.3 Flow of a Bingham plastic fluid through a tube of length L and pressure drop p

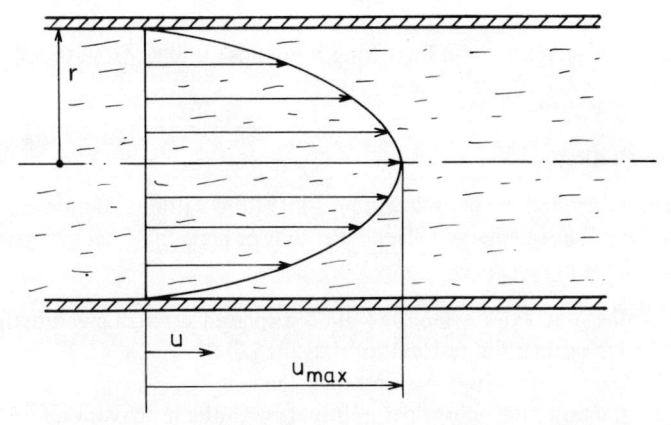

Figure 1.4 Laminar flow: parabolic velocity profile with average velocity $V = u_{max}/2$

In unit operations, plain laminar flow is mainly limited to flow in the tiny channels of a filter cake, the flow of some non-Newtonian fluids, and the flow of extremely viscous Newtonian fluids.

Turbulent flow

At high velocities where the ratio of the inertial force of an eddy to the viscous force, i.e. Reynolds number Re = $\rho VD/\mu$, exceeds approximately 2300 in a straight tube, there will be, superimposed on the net flow, eddies that are born and die away irregularly. In general, however, flow in the layer close to the wall will be laminar with a transition zone between the laminar layer and the turbulent core.

The average velocity in a straight tube, V, equals $u_{max}/1.2$ in turbulent flow (Figure 1.5). The laminar layer near the wall is important as it offers resistance to

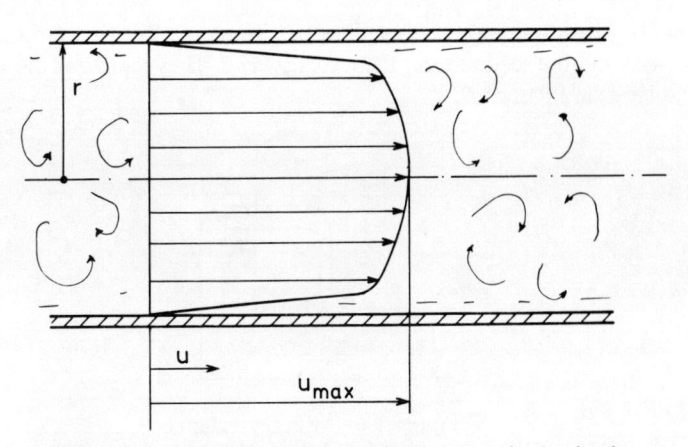

Figure 1.5 Turbulent flow with average velocity in the direction of flow $V = u_{max}/1.2$

heat and mass transfer between the bulk of the fluid and the wall. The thickness of the layer decreases with increasing Reynolds number, from 0.0043 r at Re = 10,000 to 0.00055 r at Re = 100,000.

Bernoulli's equation

The Bernoulli equation for an incompressible fluid without friction can be derived by a mechanical-energy balance. Per unit of mass the total mechanical-energy in a point is the sum of

    p/ρ the flow work when the fluid is displaced through a volumetric space $1/\rho$ against the restraint of pressure p,
    $V^2/2$ the kinetic energy per unit mass, and
    zg the potential energy per unit mass with the horizontal reference level 0–0.

Hence, and energy balance per unit mass for the flow between point 1 and point 2 in Figure 1.6 without accumulation of energy or fluid gives Bernoulli's equation for frictionless flow,

$$\frac{p_1}{\rho} + \frac{V_1^2}{2} + z_1 g = \frac{p_2}{\rho} + \frac{V_2^2}{2} + z_2 g$$

or

$$p_1 + \rho \frac{V_1^2}{2} + z_1 \rho g = p_2 + \rho \frac{V_2^2}{2} + z_2 \rho g. \tag{1.3}$$

This equation has dimensions of force per unit area, Pa or N/m². Dividing by ρg gives

$$\frac{p_1}{\rho g} + \frac{V_1^2}{2g} + z_1 = \frac{p_2}{\rho g} + \frac{V_2^2}{2g} + z_2. \tag{1.4}$$

Equation (1.4) has dimensions of length or height, metre. The first term is called the *pressure head* and the second term the *velocity head*. They represent the heights of the liquid levels in Figure 1.7.

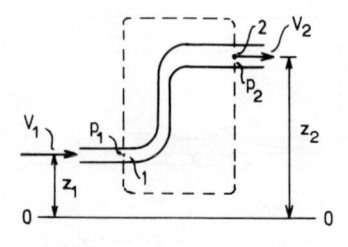

Figure 1.6   Energy balance for the flow from point 1 to point 2

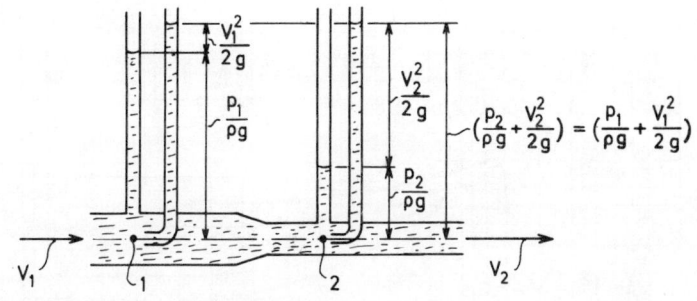

Figure 1.7 Pressure head $(p/\rho g)$ and velocity head $(V^2/2g)$ for frictionless flow from point 1 to point 2

In real fluid flow some of the mechanical energy is converted into heat due to friction in the fluid itself. This loss of mechanical energy can be expressed as a loss in static pressure due to friction, Δp_f. This term added to the right hand side of equation (1.3) gives the mechanical energy balance for flow with friction,

$$p_1 + \rho \frac{V_1^2}{2} + z_1 \rho g = p_2 + \rho \frac{V_2^2}{2} + z_2 \rho g + \Delta p_f \tag{1.5}$$

or

$$\frac{p_1}{\rho g} + \frac{V_1^2}{2g} + z_1 = \frac{p_2}{\rho g} + \frac{V_2^2}{2g} + z_2 + \frac{\Delta p_f}{\rho g} \tag{1.6}$$

where equation (1.5) has dimensions of pressure (Pa or N/m²) and (1.6) dimensions of height (m).

The total pressure drop between two points with separation L in a horizontal tube with constant diameter, D, and constant rate of flow, is due only to friction. This gives a force $\Delta p_f (\pi/4) D^2$ and the shear stress between the tube wall and the fluid [equation (1.1)] is

$$\tau_w = \frac{\Delta p_f \frac{\pi}{4} D^2}{\pi D L} = \frac{\Delta p_f D}{4L} = \mu \left(\frac{du}{dy} \right)_w$$

or

$$\Delta p_f = 4 \frac{L}{D} \mu \left(\frac{du}{dy} \right)_w . \tag{1.7}$$

In turbulent flow the thickness of the laminar layer is almost proportional to the reciprocal of the kinetic energy of the flow, and the velocity gradient at the wall $(du/dy)_w$, is nearly proportional to the kinetic energy, $\rho V^2/2$. Hence, equation (1.7) can be written as

$$\Delta p_f \approx 4 f \rho \frac{L}{D} \frac{V^2}{2} \tag{1.8}$$

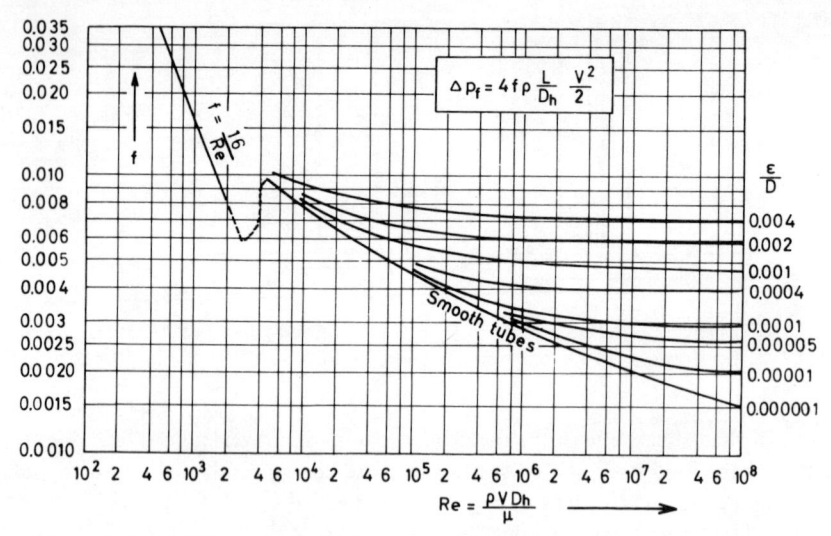

Figure 1.8 Friction factor f plotted as a function of Reynolds number $Re = \rho VD/\mu$ in the laminar and turbulent regions.[3] In turbulent flow (Re > ~2300) the limits of error are about ±10%

where f is a friction factor, which is dependent on both the ratio of inertial forces to viscous forces and the surface roughness of the tube,

$$f = f(Re, \epsilon/D) \tag{1.9}$$

where Re is Reynolds number and ϵ/D is the ratio of the height of unevenness on the surface to the tube diameter. The friction factor for laminar flow (Re < 2300) is $4f = 64/Re$. The friction factor for laminar flow and empirical values of the friction factor in turbulent flow are given in Figure 1.8. Values of surface roughness are given in Table 1.1.

Equation (1.8) and Figure 1.8 are for flow in pipes and tubes. The same values, however, can be used for ducts and channels with other cross sections if the diameter D is replaced by the hydraulic diameter D_h, defined as (see Figure 1.9)

$$D_h = \frac{4(\text{fluid flow cross section})}{\text{wetted circumference}}. \tag{1.10}$$

Reynolds number for *liquids* is usually in the range 30,000–100,000 and $4f \approx 1/40$, i.e. one liquid head is lost for a pipe length $L_e = 40D$. This gives the approximation

$$\Delta p_f \approx \frac{L}{40 D_h} \rho \frac{V^2}{2}. \tag{1.11}$$

Table 1.1 Surface roughness of pipes[4]

Material	ϵ mm
Drawn tubing (brass, lead, glass)	0.0015
Drawn steel pipes	
new	0.04–0.15
used	0.10–0.20
slightly rusted	0.2–0.5
Galvanized iron	0.15
Welded steel pipes	
new	0.05–0.1
used	0.15–0.25
with even rust layer	0.4
Acid resistant steel	0.01
Smooth concrete	0.3–0.8
Wood, new, planed	0.2
Polyethylene, seamless, diameter <200 mm	0.1
Polyester, glassfibre-reinforced, diameter >200 mm	0.05–0.085

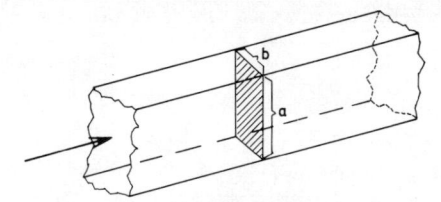

Figure 1.9 Rectangular duct, cross sectional area = ab, wetted circumference = $2(a + b)$, and hydraulic diameter, $D_h = 4\,ab/2(a + b)$

Reynolds number for gases is usually in the range 100,000 to 500,000 and $4f \approx 1/50$ or

$$\Delta p_f \approx \frac{L}{50 D_h} \rho \frac{V^2}{2}. \tag{1.12}$$

The *optimum economical diameter* is that at which the sum of operating costs, interest, and depreciation is a minimum. Economical diameters for pumps and compressors on the pressure side are usually in the velocity range given in Figure 1.10. Liquid velocities over 4.5 to 6 m/s should be avoided to reduce erosion.

Fittings

Pressure drop in fittings is usually given either as the equivalent length of straight pipe, L_e, which gives the same pressure drop as the piece of fittings in question, or as a factor, n, by which $\rho V^2/2$ has to be multiplied to give the pressure drop. Table 1.2 gives L_e and n for some of the more common fittings, valves, and cross section changes.

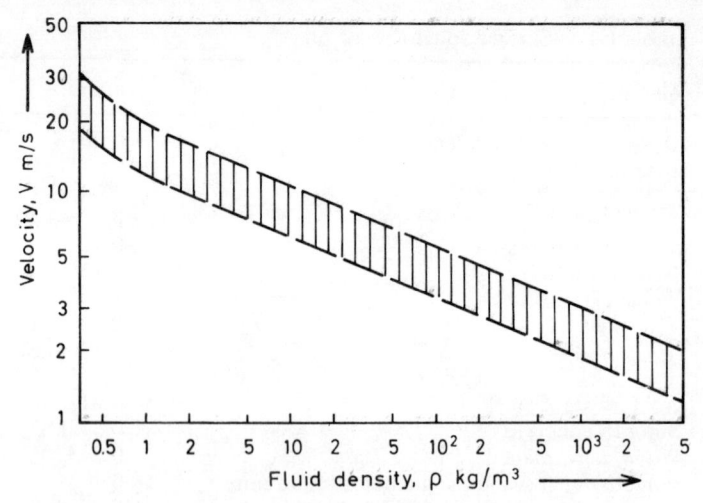

Figure 1.10 Common velocities found on the pressure side of pumps and compressors depending on the fluid density. Somewhat lower velocities are used on the suction side

Helices

Helices as shown in Figure 1.11 with inner tube diameter D and helix diameter D_c have friction factors f [equation (1.8)],[5]

$$f_c = 0.076\, \text{Re}^{-1/4} + 0.0073(D/D_c)^{1/2} \pm 10\% \tag{1.13}$$

where $0.34 < \text{Re}(D/D_c)^2 < 300$. Corresponding data for both cylindrical and flat spirals in the laminar and turbulent regions are given in the literature.[7]

Tube banks

The pressure drop in tube banks (Figures 1.12 and 1.13) is given approximately by the equation[8]

$$\Delta p_f = 4 f_1 N \rho\, V_{max}^2 / 2 \tag{1.14}$$

where N is the number of tube rows in the direction of flow and V_{max} is the velocity in the tube of smallest cross section, m/s.

For *tubes in line* with $\text{Re}_e > 100$

$$f_1 = 0.33\, \text{Re}_e^{-0.2}. \tag{1.15}$$

For *staggered tubes* with $\text{Re}_e > 40$

$$f_1 = 0.75\, \text{Re}_e^{-0.2} \tag{1.16}$$

where $\text{Re}_e = \rho V_{max} D_1 / \mu$ and D_1 is the width at the smallest cross section, m.

Table 1.2 Pressure drop Δp due to friction in fittings, valves, etc., given as length of straight pipe, L_e, with the same pressure drop, or as factor n in the equation $\Delta p = n\rho V^2/2$. For further details the reader is referred to the literature.[6]

Sudden contraction	Sudden enlargement
D/D_0 = 0.0 0.5 0.75 L_e/D = 25 20 14 n = 0.5 0.4 0.3	$\Delta p_f = \rho \dfrac{V^2 - V_0^2}{2}$

Standard Tee	Standard elbow	Sharp elbow
L_e/D = 20, 60, 70, 46 n = 0.4, 1.3, 1.5, 1.0	45°: L_e/D = 15, n = 0.3 ; 90°: 32, 0.74	60, 1.3

90° Bend	180° Bend
R/D = 0.5 1.0 2.0 4.0 8.0 L_e/D = 36 16.5 10 10 14.5 R/D = 0.5 1.0 2.0 4.0 8.0 L_e/D = 50 23 20 26 35	Small radius: L_e/D = 75, n = 1.7 Large radius: L_e/D = 50, n = 1.2

Gate valve	Globe valve
Opening L_e/D n Full 7 0.13 3/4 40 0.8 1/2 200 3.8 1/4 800 15	Opening L_e/D n Full 330 6 Half 470 8.5 Fully open: L_e/D = 170, n = 3

Diaphragm valve	Plug cock
Opening L_e/D n Full 125 2.3 3/4 140 2.6 1/2 235 4.3 1/4 1,140 21	α L_e/D n 5 2.7 0.05 10 16 0.29 20 85 1.56 40 950 17.3 60 11,200 206

Check valve fully open			Water meter			
	L_e/D	n			L_e/D	n
Hinged	110	2		Wheel	300	6
Disk	500	10		Disk	400	8
Ball	3,500	65		Piston	600	12

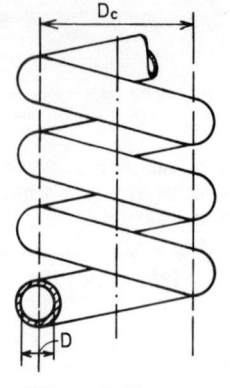

Figure 1.11 Dimensions of a helix

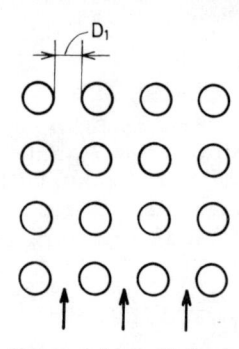

Figure 1.12 Tube bank with $N = 4$ tubes in line

The friction factor for tube banks with *staggered, finned tubes* with 360 to 430 fins per m, fin thickness 0.4 to 0.6 mm, and outer fin diameter 1.5 to 2.1 times the outer tube diameter, is given by the equation[9]

$$f = 9.44 \, \text{Re}_y^{-0.316} \left(\frac{P_t}{D_y}\right)^{-0.937} \pm 8\%. \tag{1.17}$$

where P_t is the distance between centres perpendicular to the direction of flow as shown in Figure 1.13,

$$\text{Re}_y = \rho V_{\max} D_y / \mu$$

where $V_{\max}$ is the maximum flow velocity between tubes in a section vertical to the main direction of flow, m/s, and D_y is the outer tube diameter which equals the inner diameter of the fins, m. Pressure drop by laminar flow past different shapes of fins is given in the literature.[10]

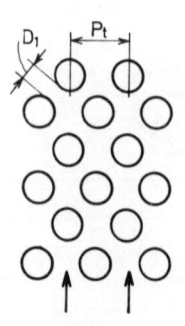

Figure 1.13 Tube bank with $N = 6$ staggered rows

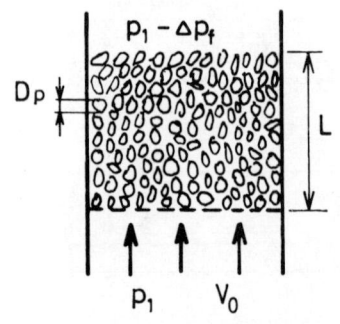

Figure 1.14 Flow through granular bed

Granular beds

Pressure drop through granular beds (Figure 1.14) and packed beds is given by Leva,[11]

$$\Delta p_f = \frac{2f_m \rho V_0^2 L (1-\epsilon)^n}{D_p \Phi_s^n \epsilon^3} \tag{1.18}$$

where V_0 is the superficial velocity referred to the total cross section, m/s, L is the depth of bed in direction of flow, m, ϵ is the void fraction = (void volume)/(total volume), D_p is the average particle diameter, defined as diameter of a sphere with the same volume as the particle, and Φ_s is the shape factor

$$\Phi_s = \frac{\text{(surface area of a sphere with the volume of the particle)}}{\text{(surface area of particle)}} \tag{1.19}$$

≈ 0.2 for spirals of metal wire
≈ 0.3 for Berl saddles and Raschig rings
≈ 0.65 for crushed glass and fine coal <10 mm
≈ 0.75 for crushed coal and sand
= 0.87 for cylinders with length equal to their diameter.

The exponent n and the friction factor f are given in Figure 1.15 as a function of the superficial Reynolds number

$$Re_e = \rho V_0 D_p / \mu. \tag{1.20}$$

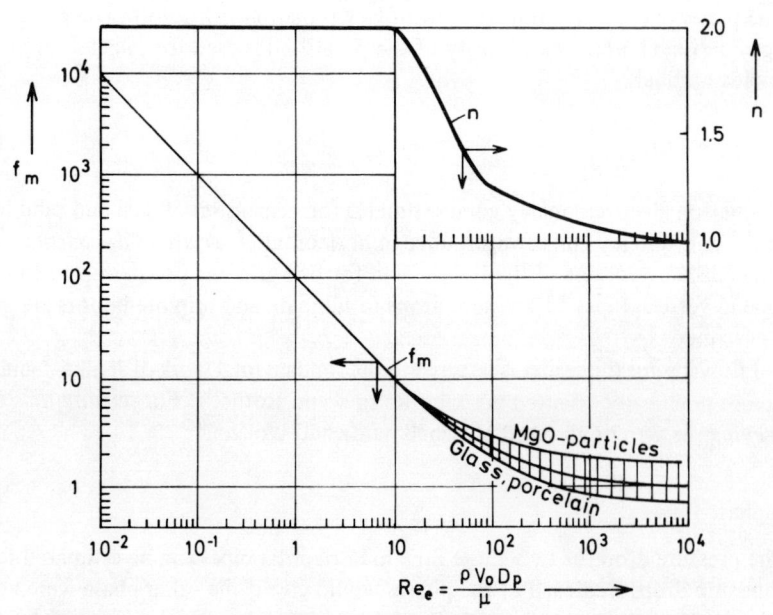

Figure 1.15 Friction factor f_m and exponent n in equation (1.18) as a function of the superficial Reynolds number Re_e. Reproduced by permission of McGraw-Hill Book Co. from Leva, M.: Fluidization, © McGraw-Hill Inc. 1959

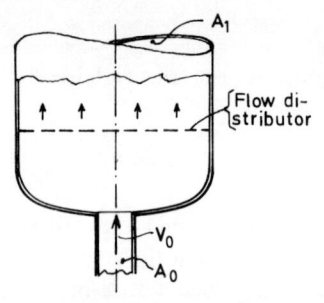

Figure 1.16 A flow distributor plate

The pressure drop in air filters is given in chapter 6, for liquid filtration in chapter 7, and for various heat exchangers in chapter 9.

Flow distributors (Figure 1.16) must have a considerable pressure drop to give an even velocity distribution over the cross sectional area A_1. Richardson[12] recommends a pressure drop of at least 100 times the velocity head at the inlet, i.e.

$$\frac{\Delta p}{\rho g} \geqslant 100 \frac{V_0^2}{2g} \qquad (1.21)$$

where V_0 is the velocity in the inlet (Figure 1.16).

The pressure drop in turbulent flow of Newtonian *suspensions* can be estimated using equation (1.8), if the viscosity of the pure fluid is used for calculation of the Reynolds number,

$$\Delta p_f = 4 f \rho_s \frac{V^2}{2} \frac{L}{D}. \qquad (1.22)$$

This equation gives reasonably good estimates for suspension of coal and sand in water,[13] and for clay and sewage sludge in horizontal pipes where the particles are less than 30% by volume of the fluid[14] and for fine grinding powder up to 11% by volume in vertical pipes.[15] Pressure drops in sulphate and sulphite liquors are given by Leuschner.[16]

For flow in *silos* the reader is referred to the fundamental work of Jenike[17] and to the more practically oriented book by Reisner and Rothe.[18] For *pneumatic conveying* see Gerchow[19] and Muschelknautz and Wojahn.[20]

Two-phase flow

The pressure drop for two-phase flow in horizontal pipes can be estimated from the pressure drops that each of the phases would give if the other phase were not present.[21] The pressure drop Δp_{fG} for the gas phase alone, and pressure drop Δp_{fL} for the liquid phase alone, give a pressure drop for the two-phase flow,

$$\Delta p_{f\,tot} = Y_G (\Delta p_{fG}) = Y_L (\Delta p_{fL}) \pm 50\% \qquad (1.23)$$

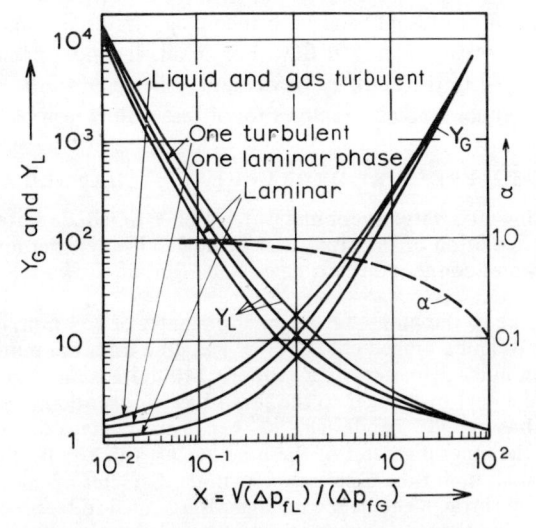

Figure 1.17 Factors Y_G and Y_L in equation (1.23) plotted as function of the square root of the ratio between the pressure drop for the liquid alone, Δp_{fL}, and the pressure drop for the gas alone, Δp_{fG}. α is the gas fraction of the volume

where Y_G and Y_L are functions of $(\Delta p_{fL}/\Delta p_{fG})$ as shown in Figure 1.17. In up-hill sections liquid will collect and give a higher pressure drop than in corresponding length of horizontal pipe. The pressure drop for a pipe with a mixture of natural gas and condensate is today usually calculated by means of the elevation factor introduced by Flanigan,[22]

$$E_H = \frac{\text{pressure drop in the up-hill section}}{\text{pressure drop for liquid filled pipe}} \qquad (1.24)$$

Figure 1.18 gives the elevation factor as a function of the gas velocity to the total cross section.

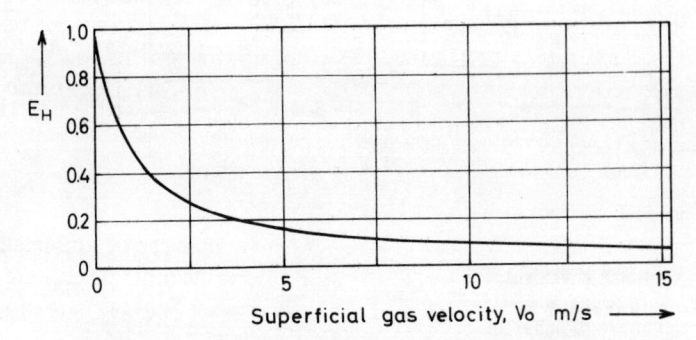

Figure 1.18 Flanigan's elevation factor E_H in equation (1.24) as a function of the gas velocity V_0 without condensate

The two-phase flow of liquid and gas can be plug, stratified, slug, bubble or froth, spray or dispersed, wave, or annular flow. For details and more precise calculations the reader is referred to the book by Govier and Aziz[23] and to the publication by Kern[24] which includes worked examples for different flow regions.

EXAMPLE 1.1. PRESSURE DROP IN HEAT EXCHANGER AND PIPES

A factory using city water as cooling water wishes to reduce water consumption by means of recirculation of the city water through a heat exchanger cooled by sea water. The sea water connections and heat exchanger are shown schematically in Figure 1.19.

The suction line of the pump has an inner diameter of 154 mm, is 22 m long, and has two 90° bends and a hinged check valve. The pipe from the pump to the heat exchanger has an inner diameter of 127 mm, is 140 m long, and has six 90° bends. The pipe from the heat exchanger to the sewer has an inner diameter of 127 mm, is 15 m long, and has two 90° bends. The 90° bends are all made of steel with a radius equal to the inner diameter of the pipe, $R/D = 1.0$. The heat exchanger has 62 tubes in parallel, each tube 6 m long. The inner diameter of the tubes is 18 mm. The pressure drop through the heat exchanger is assumed to be proportional to the square of the liquid velocity. The supplier has stated that the pressure drop is 3.2 m liquid head at a flow of 120 m³/h. All pipes are made of drawn mild steel.

    Density of sea water   1030 kg/m³.
    Viscosity of sea water in front of the heat exchanger (15°C)   1.2 cP.
    Viscosity of sea water after heat exchanger (20°)   1.05 cP.

(*a*)   Check the figure stated by the supplier.

(*b*)   What will the total head of the sea water pump have to be if the flow rate at low tide is 120 m³/h?

(*c*)   What will the amount of sea water (m³/h) be at low tide for a centrifugal pump with the characteristics given in Figure 1.20?

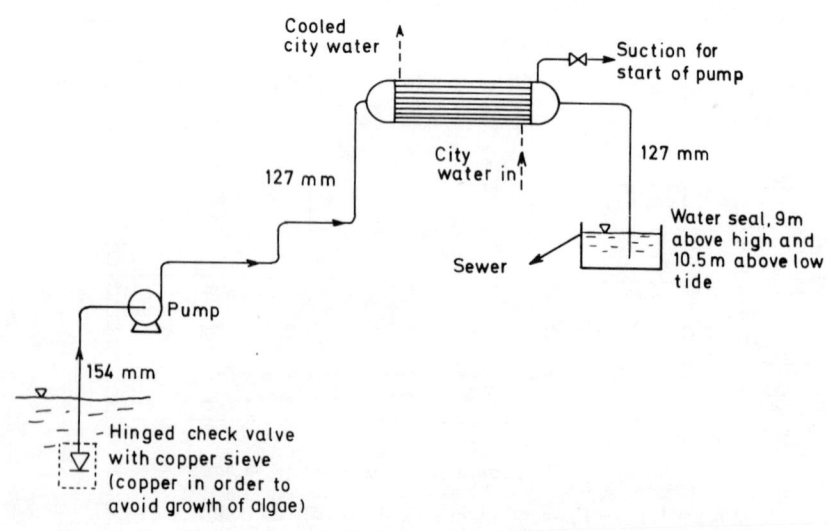

Figure 1.19   Sea water connections to and from heat exchanger in Example 1.1

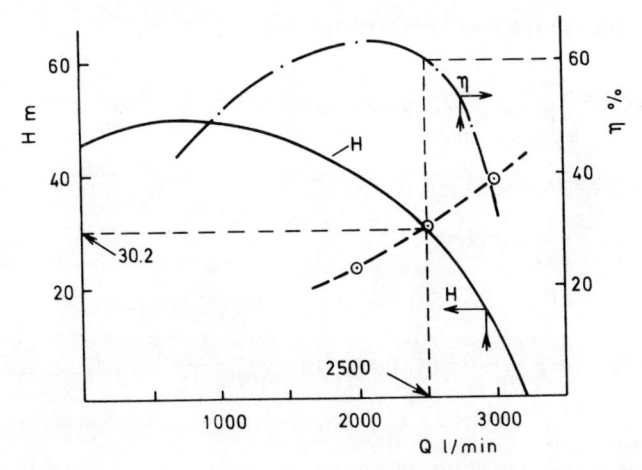

Figure 1.20 Characteristics of a centrifugal pump the factory has in stock

(d) What will the lowest pressure inside the heat exchanger tubes be if the flow rate is 120 m³/h when the top of the heat exchanger is 18.5 m above low tide?

Solution

(a) Velocity in the nozzles of the heat exchanger,

$$V_r = \frac{120}{3600 \frac{\pi}{4} 0.127^2}$$

$$= 2.63 \text{ m/s}.$$

According to Table 1.2, one velocity head is lost at the inlet and 0.5 velocity head at the outlet of the heat exchanger,

$$\Delta p_1 = (1 + 0.5)\rho \frac{V_r^2}{2} = 1.5\rho \frac{2.63^2}{2}$$

$$= 5.19\rho \text{ N/m}^2.$$

Velocity in the heat exchanger tubes is given by,

$$V_v = \frac{120}{(3600)(62)\frac{\pi}{4} 0.018^2}$$

$$= 2.11 \text{ m/s}.$$

Pressure loss by inlet to and outlet from the heat exchanger tubes is given by,

$$\Delta p_2 = (0.5 + 1)\rho \frac{V_v^2}{2} = 1.5\rho \frac{2.11^2}{2}$$

$$= 3.34\rho \text{ N/m}^2.$$

Reynolds number in the heat exchanger tubes ($\mu \approx 1.13$ cP),

$$\text{Re} = \frac{\rho V_v D}{\mu} = \frac{(1030)(2.11)(0.018)}{(1.13)(10^{-3})}$$
$$= 34{,}600.$$

Table 1.1 gives $\epsilon = 0.0445$ mm, i.e.

$$\frac{\epsilon}{D} = \frac{0.045}{18} = 0.0025.$$

Figure 1.8 gives $f = 0.0075$ corresponding to [equation (1.8)]

$$\Delta p_f = 4f\rho \frac{L}{D} \frac{V_v^2}{2} = (4)(0.0075)\rho \frac{6}{0.018} \frac{(2.11)^2}{2}$$
$$= 22.26\rho \text{ N/m}^2.$$

Total pressure drop in the heat exchanger,

$$\Delta p_v = \Delta p_1 + \Delta p_2 + \Delta p_f = (5.19 + 3.34 + 22.26)\rho$$
$$= 30.8\rho \text{ N/m}^2$$

which corresponds to

$$\frac{30.8\rho}{g\rho} = \frac{30.8}{9.82}$$
$$= 3.14 \text{ m liquid head.}$$

The tolerance limits for f indicated in Figure 1.8 are ±10%, hence the tolerance limits for Δp_f,

$$\pm \frac{(0.1)(22.26)\rho}{9.82\rho} = \pm 0.23 \text{ m liquid head}$$

or

$$2.91 < \frac{\Delta p_v}{g\rho} < 3.37 \text{ m liquid head,}$$

i.e. the figure 3.2 m liquid head given by the supplier is within the tolerance limits given by the calculations above.

(b) Length of suction line   22 m straight pipe
     Hinged check valve corresponds to (Table 1.2)
        (110)(0.154) =   17 m straight pipe
     Two 90° bends correspond to (Table 1.2)
        (2)(16.5)(0.154) =   5 m straight pipe

     Suction line with valves and bends corresponds to   44 m straight pipe

Liquid velocity,

$$V_s = \frac{120}{3600 \frac{\pi}{4} 0.154^2}$$
$$= 1.79 \text{ m/s.}$$

Reynolds number,
$$\mathrm{Re} = \frac{\rho V_s D}{\mu} = \frac{(1030)(1.79)(0.154)}{(1.2)(10^{-3})}$$
$$= 237{,}000.$$

Surface roughness,
$$\epsilon/D = 0.045/154 = 0.0003.$$

Corresponding friction factor (Figure 1.8),
$$f = 0.0043.$$

Equation (1.8) gives
$$\Delta p_{fs} = \frac{(4)(0.0043)(44)\rho}{0.154}\frac{1.79^2}{2}$$
$$= 7.9\rho \text{ N/m}^2.$$

For pipes after the pump:

Pipe length, 140 + 15 =	155 m straight pipe
Eight 90° bends correspond to (Table 1.2), (8)(16.5)(0.127) =	17 m straight pipe
Sum	172 m straight pipe

$$\mathrm{Re} = \frac{(1030)(2.63)(0.127)}{0.0012} = 287{,}000$$

and
$$\frac{\epsilon}{D} = \frac{0.045}{127} = 0.00035$$

which gives (Figure 1.8) $f = 0.0043$. Equation (1.8) gives
$$\Delta p_{fr} = \frac{(4)(0.0043)(172)\rho}{0.127}\frac{2.63^2}{2}$$
$$= 80.6\rho \text{ N/m}^2.$$

The pump must cover the following pressure differences:

(1)	Elevation difference	10.5 m liquid head
(2)	Pressure drop in heat exchanger	3.2 m liquid head
(3)	Friction loss, $\dfrac{\Delta p_{fs} + \Delta p_{fr}}{g\rho} = \dfrac{(7.9 + 80.6)\rho}{9.82\rho}$	9.0 m liquid head
(4)	Outlet loss, $\dfrac{V_r^2}{2g} = \dfrac{2.63^2}{2 \times 9.82}$	0.4 m liquid head
Sum		23.1 m liquid head

or
$$23.1\rho g = (23.1)(1030)(9.82) = 234{,}000 \text{ N/m}^2$$
$$= 2.3 \text{ bar}.$$

(c) The volumetric flow rate, 120 m³/h = 2000 litre/min, gives a friction loss 23.1 − 10.5 = 12.6 m liquid head. The friction loss is practically proportional to the square of the flow rate, giving a total liquid head of $H = (\dot{Q}/2000)^2 \, 12.6 + 10.5$ which have been calculated below for three values of $\dot{Q}$.

$\dot{Q}$ litre/min	2000	2500	3000
H m	23.1	30.2	38.9

These points are plotted in Figure 1.20 and the dotted curve through the points intercepts the characteristics of the pump at $\dot{Q} = 2500$ litre/min,

or $\quad \dfrac{(2500)(60)}{1000} = 150 \text{ m}^3/\text{h}$

and $H = 30.2$ m liquid head.

(d) The pressure drop from point 1 to point 2 in Figure 1.21 consists of

(1) Discharge loss from the heat exchanger tubes,

$$\frac{V_1^2}{2g} = \frac{2.11^2}{(2)(9.82)} = \qquad 0.23 \text{ m liquid head}$$

(2) Inlet loss in the heat exchanger tubes (Table 1.2),

$$0.5 \frac{V_r^2}{2g} = 0.5 \frac{2.63^2}{(2)(9.82)} = \qquad 0.18 \text{ m liquid head}$$

(3) Exit loss in the water seal,

$$\frac{V_r^2}{2g} = \frac{2.63^2}{(2)(9.82)} = \qquad 0.35 \text{ m liquid head}$$

(4) Friction loss corresponds to
$15 + (2)(16.5)(0.127) = 19.2$ m straight tube,

i.e. $\quad \dfrac{19.2}{172} \dfrac{80.6 \rho}{\rho g} = 0.1116 \dfrac{80.6}{9.82} = \qquad \underline{0.92 \text{ m liquid head}}$

Pressure drop from point 1 to point 2 $\qquad$ 1.68 m liquid head

which can be inserted into equation (1.6) with the liquid level of the water seal as reference level for z:

$$\frac{p_1}{g\rho} + \frac{2.11^2}{2g} + 8 = 0 + 0 + 0 + 1.68$$

$$\frac{p_1}{g\rho} = 1.68 - 8 - \frac{2.11^2}{(2)(9.82)}$$

$$= -6.55 \text{ m liquid head}$$

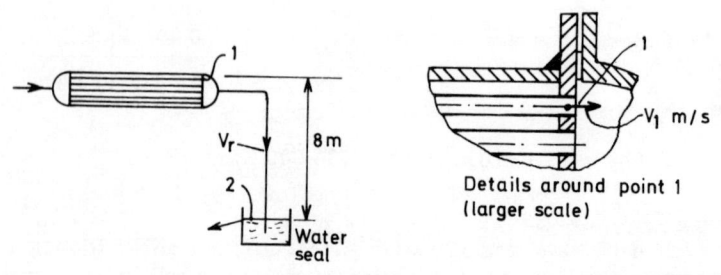

Figure 1.21 The discharge side of the heat exchanger

or

$$(6.55)(1030)(9.82) = 66{,}000 \text{ N/m}^2 \text{ suction pressure.}$$

Comment The suction pressure should be well within the limits required to prevent boiling in the heat exchanger.

EXAMPLE 1.2. PRESSURE DROP IN THE SUCTION LINE FOR A CENTRIFUGAL PUMP

The centrifugal pump in Figure 1.22 pumps hydrocarbons from the bottom of a distillation column. The pressure above the liquid in the column is $p = 0.8$ MN/m². The liquid temperature, 77°C, is the boiling temperature at the pressure mentioned above. Density of the liquid, ρ, is 610 kg/m³ and its viscosity, μ, is 0.00015 N s/m².

The supplier of the pump has stated that it can operate with a net positive suction head of 7.5 m with water at 20°C without danger of cavitation. Calculate the necessary diameter of the suction pipe which is 7.5 m long, has one 90° elbow, and a disk check valve. The liquid level difference z (Figure 1.22) varies between 4.5 and 4.8 m and the flow rate is 12 m³/h

Atmospheric pressure is 0.1 MN/m² and saturation pressure for water vapour at 20°C is 0.002 MN/m².

Solution

The minimum allowable difference between the pressure in front of the pump and the saturation pressure of the liquid is

$$p_2 - p_1 = 0.1 - \frac{(7.5)(1000)(9.82)}{10^6} - 0.002$$
$$= 0.0243 \text{ MN/m}^2,$$

i.e. the static pressure at the pump inlet should be more than

$$p_2 = 0.8 + 0.0243$$
$$= 0.8243 \text{ MN/m}^2.$$

Equation (1.5) gives, for velocity $V_1 = 0$ and the lowest liquid level,

$$(0.8)(10^6) + (4.5)(610)(9.82) = (0.8243)(10^6) + 610\frac{V_2^2}{2} + \Delta p_f$$

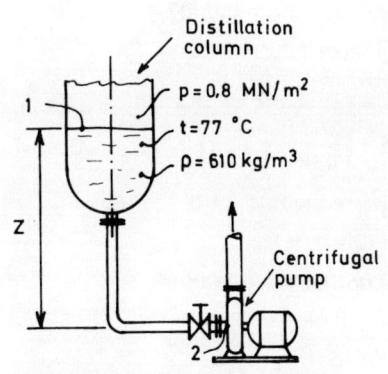

Figure 1.22 Centrifugal pump for distillation column

or

$$2660 = 305 V_2^2 + \Delta p_f. \qquad (a)$$

Table 1.2 gives the equivalent pipe length for the calculation of Δp_f,

inlet in the bottom of the column	$L_e = 25\ D$
disc check valve	$L_e = 330\ D$
elbow	$L_e = 32\ D$
Sum	$L_e = 387\ D$.

The friction loss corresponds to $(7 + 387\ D)$ m straight pipe. Equation (1.8) gives

$$\Delta p_f = 4f 610 \left(\frac{7}{D} + 387\right) \frac{V_2^2}{2}$$

$$= \left(\frac{8540}{D} + 472{,}000\right) f V_2^2$$

which is inserted in equation (a);

$$2660 = 305 V_2^2 + \left(\frac{8540}{D} + 472{,}000\right) f V_2^2$$

or

$$\left[0.115 + \left(\frac{3.21}{D} + 177\right) f\right] V_2^2 = 1.0. \qquad (b)$$

In addition, from the equation of continuity,

$$\frac{12}{3600} = \frac{\pi}{4} D^2 V_2$$

or

$$D = 0.065/\sqrt{V_2}. \qquad (c)$$

The equations (b) and (c) together with Figure 1.8 provide three equations with three unknowns. Table 1.1 gives $\epsilon = 0.045$ mm. The equations are solved by iteration. First trial: $V_2 = 1.0$ m/s inserted in equation (c) gives

$$D = 0.065/\sqrt{1.0} = 0.065 \text{ m},$$

$$\text{Re} = \frac{(610)(1.0)(0.065)}{0.00015} = 264{,}000$$

$\epsilon/D = 0.045/(0.065)(10^3) = 0.0007$,

$f = 0.0047$ (Figure 1.8)

$$V_2 = \frac{1}{\sqrt{0.115 + \left(\frac{3.21}{0.065} + 177\right) 0.0047}} = 0.92 \text{ m/s} \qquad [\text{equation } (b)]$$

The calculation is repeated with $V_2 = 0.92$ m/s,

$$D = 0.065/\sqrt{0.92} = 0.068,$$

$\epsilon/D = 0.045/(0.068)(10^3) = 0.00066$

$$\text{Re} = \frac{(610)(0.92)(0.068)}{0.00015} = 254{,}000,$$

$f = 0.0047$,

i.e. the friction factor f remains unchanged.

Diameter of the suction pipe,

$D \geqslant 68$ mm.

Comment Figure 1.8 gives an approximate error of 10% in the friction factor f. Taking this into account gives $D \geqslant 70$ mm as a safe value.

EXAMPLE 1.3. PRESSURE DROP IN SPIRALS. HYDRAULIC DIAMETER

The pressure tank in Figure 1.23 has a cooling spiral that consists of 6 windings of a half tube welded to the outside of the tank as shown in the detail drawing. The cooling medium is circulating hot water. The water density, ρ, is 1000 kg/m³ and viscosity is 0.7 cP.

(*a*)   Calculate the hydraulic diameter of the cooling spiral.

(*b*)   Calculate the water flow in litres per hour when the pressure drop is a maximum 7 m liquid head.

(*c*)   Estimate the error limits in the answer to (*b*).

Solution

(*a*)   From equation (1.10),

$$D_h = \frac{(4)(\tfrac{1}{2})\tfrac{\pi}{4} 0.052^2}{\pi 0.026 + 0.052}$$

$$= 0.0318 \text{ m}.$$

(*b*)   From equation (1.13),

$$f_c = 0.076 \left(\frac{0.0007}{1000 \, V \, 0.0318}\right)^{1/4} + 0.0073 \left(\frac{0.0318}{1.100}\right)^{1/2}$$

$$= 0.0052 V^{-1/4} + 0.00124.$$

Six windings correspond to $6\pi 1.1 = 21$ m in equation (1.8),

$$\Delta p_f = 4(0.0052 V^{-1/4} + 0.00124)\rho \frac{21}{0.0318} \frac{V^2}{2}$$

$$= 7\rho 9.82$$

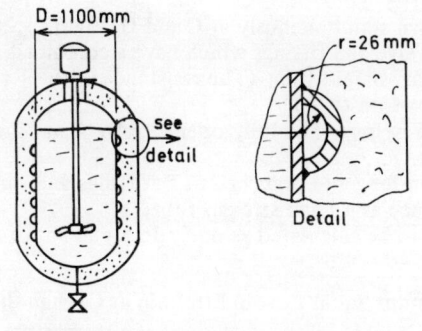

Figure 1.23   Tank with external cooling spiral

or
$$V = \sqrt{\frac{68.7}{1.64 + 6.86 V^{-1/4}}}.$$

This equation can be solved by iteration, giving $V = 3.18$ m/s.

$$\dot{Q} = (3600)(3.18)\frac{1}{2}\frac{\pi}{4}0.052^2$$
$$= 12.2 \text{ m}^3/\text{h}.$$

Checking the product $\text{Re}(D/D_c)^2$,

$$\text{Re}\left(\frac{D}{D_c}\right)^2 = \frac{(1000)(3.18)(0.0318)}{0.0007}\left(\frac{0.0318}{1.100}\right)^2 = 121,$$

i.e.,

$$0.034 < \text{Re}\left(\frac{D}{D_c}\right)^2 < 300$$

which is within the limits of validity of equation (1.13).

(c) Equation (1.13) is stated to have limits of error $\pm 10\%$. Furthermore $f_c V^2$ is constant. The error limit would have been $\pm\sqrt{10} = \pm 3.2\%$ if f_c had been independent of V. But f_c is dependent on V which gives the limiting equations

$$f_c = 1.1\left[0.076\left(\frac{\mu}{\rho VD}\right)^{1/4} + 0.0073\left(\frac{0.0318}{1.100}\right)^{1/2}\right]$$

and

$$f_c = 0.9\left[0.076\left(\frac{\mu}{\rho VD}\right)^{1/4} + 0.0073\left(\frac{0.0318}{1.100}\right)^{1/2}\right].$$

This calculation gives the velocity $V = 3.01$ and 3.37 m/s, respectively, corresponding to

$$11.5 \leqslant \dot{Q} \leqslant 12.8 \text{ m}^3/\text{h}.$$

EXAMPLE 1.4. PRESSURE DROP IN BRANCHED PIPE LINE. VELOCITY HEADS

Liquid can be drawn simultaneously at C and D, Figure 1.24. The tube AB is 44 m long and has a valve and fittings which have a combined resistance to flow corresponding to 14 m straight tube. (The resistance in the T at B can be neglected for the liquid that flows on to C.)

The tube BC is 1.5 m long. The fully opened valve C has the same resistance as an 11 m straight tube.

The tube BD is 4 m long and branched at B; fittings and valve at D fully open have the same resistance as a 13 m straight tube.

The friction loss can be calculated as one velocity head for a straight pipe length $L_e = 40D$.

(a) Find the maximum liquid flow in litre/min at C when the valve at D is closed.

(b) Find the liquid flow in litre/min at C and at D when all valves are fully open.

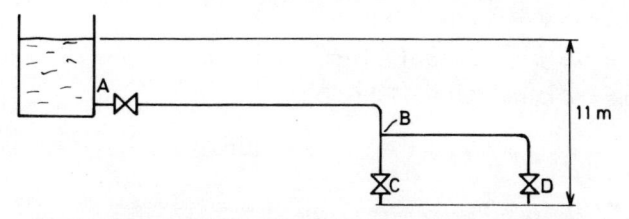

Figure 1.24 Branched pipe line. Internal diameters: Distance AB, D_{AB} = 38 mm; BC, D_{BC} = 32 mm; BD, D_{BD} = 26 mm

Solution

(a) The friction losses can be expressed in terms of the velocity V_C in the tube BC. These figures inserted in equation (1.11) give

$$\Delta p_{fAB} = \frac{44 + 14}{(40)(0.038)} \rho \frac{\left[\left(\frac{32}{38}\right)^2 V_C\right]^2}{2} = 13.53 \rho V_C^2 \text{ N/m}^2$$

$$\Delta p_{fBC} = \frac{1.5 + 11}{(40)(0.032)} \rho \frac{V_C^2}{2} = 4.88 \rho V_C^2 \text{ N/m}^2$$

Exit loss at C,

$$\rho \frac{V_C^2}{2} = 0.50 \rho V_C^2 \text{ N/m}^2$$

Inlet loss at A (Table 1.1),

$$0.5 \rho \frac{\left[\left(\frac{32}{38}\right)^2 V_C^2\right]^2}{2} = 0.18 \rho V_C^2 \text{ N/m}^2$$

therefore total loss,

$$\Delta p = 19.09 \rho V_C^2 \text{ N/m}^2$$

or

$$\frac{\Delta p}{g \rho} = \frac{19.09 \rho V_C^2}{9.82 \rho}$$

$$= 1.94 V_C^2 \text{ m liquid head}$$

$$11 = 1.94 V_C^2,$$

$$V_C = \sqrt{\frac{11}{1.94}} = 2.38 \text{ m/s}$$

$$\dot{Q}_C = (60)(2.38) \frac{\pi}{4} 0.032^2$$

$$= 0.115 \text{ m}^3/\text{min}$$

or

115 litre/min.

(b) Friction loss and outlet loss from B to C,
$$(4.88 + 0.5)\rho V_C^2 = 5.38\rho V_C^2 \text{ N/m}^2.$$

Friction loss and outlet loss from B to D,
$$\frac{4 + 13}{(40)(0.026)} \rho \frac{V_D^2}{2} + \rho \frac{V_D^2}{2}$$
$$= 8.67\rho V_D^2 \text{ N/m}^2.$$

Both branches are at the same level, and the pressure drop has to be the same through both of them:
$$5.38\rho V_C^2 = 8.67\rho V_D^2$$

or
$$V_D = 0.788 V_C. \tag{a}$$

Material balance gives
$$\frac{\pi}{4} 0.038^2 V_A = \frac{\pi}{4} 0.032^2 V_C + \frac{\pi}{4} 0.026^2 V_D. \tag{b}$$

Equation (a) and equation (b),
$$V_A = 1.078 V_C.$$

Inlet at A,
$$0.5\rho \frac{(1.078 V_C)^2}{2} = \qquad 0.29\rho V_C^2 \text{ N/m}^2$$

$$\Delta p_{fAB} = \frac{44 + 14}{(40)(0.038)} \rho \frac{(1.078 V_C)^2}{2} = 22.17\rho V_C^2 \text{ N/m}^2$$

$$\Delta p_{fBC} = \qquad 4.88\rho V_C^2 \text{ N/m}^2$$

Outlet loss at C = $\qquad 0.50\rho V_C^2 \text{ N/m}^2$

Sum $\qquad 27.84\rho V_C^2 \text{ N/m}^2$

or
$$27.84\rho V_C^2 / 9.82\rho = 2.83 V_C^2 \text{ m liquid head.}$$
$$11 = 2.83 V_C^2,$$
$$\therefore \quad V_C = 1.97 \text{ m/s}$$

$$\dot{Q}_C = (60)(1.97) \frac{\pi}{4} 0.032^2$$
$$= 0.095 \text{ m}^3/\text{min} \quad \text{or} \quad 95 \text{ litre/min.}$$

$$\dot{Q}_D = (60)(0.788)(1.97) \frac{\pi}{4} 0.026^2$$
$$= 0.049 \text{ m}^3/\text{min} \quad \text{or} \quad 49 \text{ litre/min.}$$

EXAMPLE 1.5. DISTRIBUTOR DUCT WITH OVERFLOW

A water distributor for a cooling tower consists of a main duct with downcomers and distributor ducts with notches where the water flows out as shown in Figure 1.25.

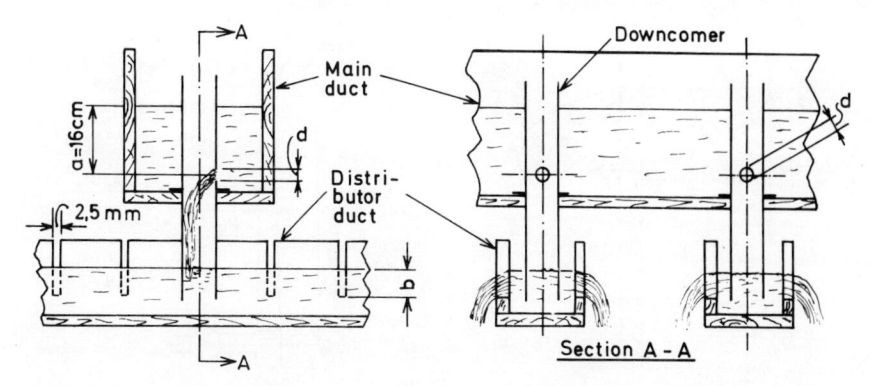

Figure 1.25 Water distributor for cooling tower. (The holes in the downcomers are placed a little above the bottom of the main ducts in order to give space for sludge)

(a) Calculate the hole diameter d in the downcomers for a liquid flow of 350 litre/h through each downcomer at liquid level $a = 16$ cm above the holes. Friction and contraction of the liquid jet reduces the liquid flow to 65% of what it would be for frictionless flow through holes with rounded inlets.

(b) Find the depth b from the liquid level in the distributor duct to the bottom of the 2.5 mm wide notches when the water from each downcomer flows out through 24 notches. Friction and contraction is believed to reduce the water flow to 55% of what it would be for ideal flow.

Solution

(a) Equation (1.6) gives for ideal flow

$$0.16 = \frac{V_2^2}{2g},$$

$$V_2 = \sqrt{(2)(9.82)(0.16)} = 1.77 \text{ m/s}.$$

Real flow of water through the hole,

$$\dot{Q} = 0.35 \text{ m}^3/\text{h}$$

∴ $0.35 = (0.65)(3600)(1.77)\frac{\pi}{4}d^2,$

$d = 0.0104$ m   or   10.4 mm.

(b) For ideal flow and with notation as in Figure 1.26 the velocity in depth y is

$$V = \sqrt{2gy}$$

and the rate of flow with nozzle width B and depth dy,

$$d\dot{Q} = 0.55B \, dy \sqrt{2gy} = 0.55B(2g)^{1/2}y^{1/2} \, dy.$$

Integration with $B = 0.0025$ m gives,

$$\dot{Q} = \frac{2}{3}(0.55)(0.0025)(2 \times 9.82)^{1/2}b^{3/2} = 0.00406b^{3/2}.$$

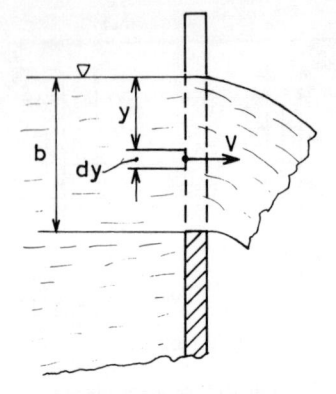

Figure 1.26 Overflow in notch

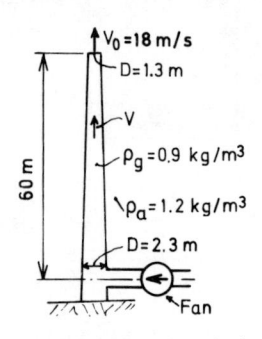

Figure 1.27 A conical stack

With 24 notches, the rate of flow per notch is

$$0.35/(3600)(24) = 0.00406 b^{3/2},$$

$b = 0.01$ m or 10 mm.

EXAMPLE 1.6. PRESSURE DROP IN A STACK

To reduce pollution at ground level, 86,000 m³/h hot flue gases are to be exhausted 60 m above the ground at a velocity $V_0 = 18$ m/s, through a conical stack with inner diameter 2.3 m at ground level and diameter 1.3 m at the top (Figure 1.27). Density of flue gas, ρ_g, is 0.9 kg/m³ and density of ambient air, ρ_a, is 1.2 kg/m³. The pressure drop due to friction is assumed to be one velocity head per length $L_e = 50D$.

(a) Calculate the exit loss in mm water gauge.

(b) Calculate the friction loss in the stack in mm water gauge.

(c) Calculate the total head (static head and velocity head) after the fan. The pressure drop from the fan to the bottom of the stack is estimated to be 6 mm water gauge.

Solution

(a) The exit loss is $\rho \dfrac{V^2}{2} = 0.9 \dfrac{18^2}{2}$

$$= 146 \text{ N/m}^2$$

or

$146/g = 146/9.82 = 15$ mm water gauge.

(b) The velocity, $V = V_0 \left(\dfrac{D_0}{D}\right)^2 = 18 \left(\dfrac{1.3}{D}\right)^2 = \dfrac{30.4}{D^2}$ m/s.

The stack diameter,

$$D = 2.3 - (2.3 - 1.3)L/60,$$

$$L = 138 - 60D.$$

These values are inserted in equation (1.12),

$$\Delta p_f = \frac{138 - 60D}{50D} \, 0.9 \, \frac{30.4^2}{2D^4} = 1148 D^{-5} - 499 D^{-4}$$

$$\frac{d\Delta p_f}{dD} = (-5)(1148)D^{-6} - (-4)(499)D^{-5}$$

$$\Delta p_f = -5740 \int_{2.3}^{1.3} D^{-6}\, dD + 1996 \int_{2.3}^{1.3} D^{-5}\, dD = 134 \text{ N/m}^2$$

134/9.82 = 14 mm water gauge.

(c)   Total loss, 15 + 14 + 6 =   35 mm water gauge
     Natural draft, $60(1.2 - 0.9)g/g$ =   18 mm water gauge

     Total head of the fan   27 mm water gauge.

EXAMPLE 1.7. PRESSURE DROP IN A GRANULAR BED (CATALYST)

Calculate the pressure drop in a 0.5 m thick catalyst bed consisting of cylinders, 6 mm diameter and 8 mm long. A sample of the catalyst gives a void fraction $\epsilon = 0.32$ and gas velocity referred to the total cross section, $V_0 = 0.8$ m/s. The gas density is assumed to be a constant, $\rho = 0.49$ kg/m³, and the gas viscosity, $\mu = 0.034$ cP.

Solution

Calculate the shape factor Φ_s [equation (1.19)]. A sphere with the same volume has the diameter D_p,

$$\frac{1}{6}\pi D_p^3 = \frac{\pi}{4}(0.6)^2(0.8),$$

$D_p = 0.756$ cm.

Shape factor,

$$\Phi_s = \frac{\pi\, 0.756^2}{2\left(\frac{\pi}{4} 0.6^2\right) + \pi(0.6)(0.8)}$$

$$= 0.870.$$

Artificial Reynolds number according to equation (1.20),

$$\text{Re}_e = \frac{(0.49)(0.8)(0.00756)}{(0.034)(10^{-3})}$$

$$= 87$$

which gives (Figure 1.15) an exponent $n = 1.28$ and friction factor $f_m = 2.0$–$3\,0$. These values inserted in equation (1.18) give

$$\Delta p_f = \frac{2 f_m (0.49)(0.8^2)(0.5)(1 - 0.32)^{1.28}}{(0.00756)(0.870^{1.28})(0.32^3)}$$

$$= 923 f_m \text{ N/m}^2.$$

With $f_m = 2.0$,
$$\Delta p_s = (923)(2.0) = 1846 \text{ N/m}^2.$$
With $f_m = 3.0$,
$$\Delta p_s = (923)(3.0) = 2770 \text{ N/m}^2,$$
i.e. $1846 \leq \Delta p_s \leq 2770 \text{ N/m}^2$,

or with one mm water gauge = 9.82 N/m^2, $188 \leq \Delta p_s \leq 282$ mm water gauge.

EXAMPLE 1.8. PRESSURE DROP THROUGH FINNED TUBE HEAT EXCHANGER

Calculate the pressure drop for air that flows past the finned tubes in Figure 1.28 when the air velocity V_0 is 5.5 m/s.

The tubes have 360 fins per metre and the fin thickness is 0.6 mm. The other measurements are given in mm on the figure. The air can be considered as an ideal gas with molecular weight 29; temperature is 30 °C, pressure 1.0 bar, and viscosity $\mu = 0.018$ cP.

Solution

Free cross section vertical to the air flow,
$$[0.055 - 0.025 - (360)(0.0006)(0.05 - 0.025)]/0.055$$
$$= 0.447 \text{ m}^2/\text{m}^2 \text{ total cross section.}$$

Air velocity,
$$V_{max} = 5.5/0.447 = 12.3 \text{ m/s.}$$

Air density,
$$\rho = \frac{M}{R}\frac{p}{T} = \frac{(29)(0.1)(10^6)}{(8314)(273+30)} = 1.15 \text{ kg/m}^3.$$

Reynolds number,
$$Re_y = \frac{(1.15)(12.3)(0.025)}{(0.018)(10^{-3})} = 19{,}646.$$

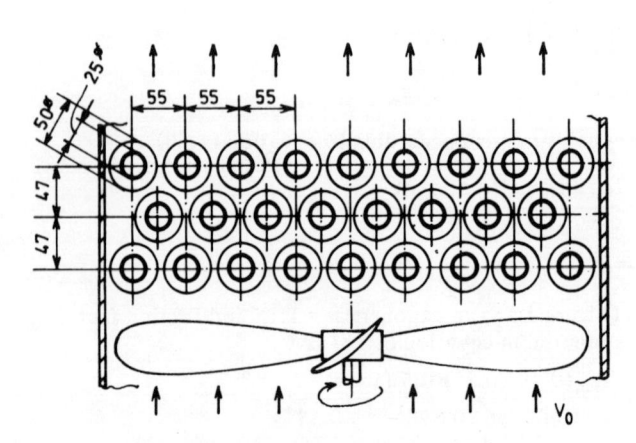

Figure 1.28 A finned tube heat exchanger

Equation (1.17) gives the friction factor

$$f_1 = (9.44)(19{,}646^{-0.316})\left(\frac{55}{25}\right)^{-0.937} \pm 8\%$$

$$= 0.198 \pm 8\%,$$

which is inserted into equation (1.14).

$$f_1 = (9.44)(19{,}646^{-0.316})\left(\frac{55}{25}\right)^{-0.937} \pm 8\% = 0.198 \pm 8\%$$

or $207/9.82 = 21$ mm water gauge $\pm (0.08)(21) = \pm 2$ mm,

i.e. $19 < \Delta p_f < 23$ mm water gauge.

EXAMPLE 1.9. PRESSURE DROP FOR A LIQUID IN A CAPILLARY TUBE

The laboratory test assembly in Figure 1.29 is constructed to give a constant flow rate of 90 cm^3/h of isopropyl alcohol at 20°C.

Determine the length of the capillary tube when the inner diameter is 0.6 mm, the liquid density $\rho = 978$ kg/m^3, the liquid viscosity $\mu = 2.3$ cP, and both inlet loss and friction loss in the tube in front of the capillary, in valves, and in the filter can be neglected.

Solution

$$V = \frac{0.00009}{3600 \frac{\pi}{4} 0.0006^2} = 0.088 \text{ m/s},$$

$$\text{Re} = \frac{(978)(0.088)(0.0006)}{0.0023} = 22.5,$$

i.e. laminar flow (Figure 1.8),

$$f = \frac{16}{\text{Re}} = \frac{16}{22.5} = 0.711.$$

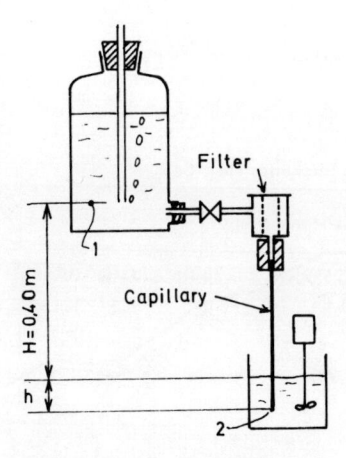

Figure 1.29 Test assembly with a capillary tube to give a constant liquid feed

Equation (1.8),

$$\Delta p_f = (4)(0.711)\rho \frac{L}{0.0006} \frac{0.088^2}{2}$$

$$= 18.4 \rho L.$$

This value inserted into equation (1.6) for the flow from point 1 to point 2 (Figure 1.29) and with the bottom of the capillary as reference level, gives

$$0 + 0 + (0.4 + h) = \frac{hg\rho}{g\rho} + \frac{0.088^2}{2(9.82)} + 0 + \frac{18.4\rho L}{9.82\rho}$$

$$L = (0.4 - 0.0004)(9.82)/18.4$$

$$= 0.21 \text{ m} \quad \text{or} \quad 21 \text{ cm}.$$

EXAMPLE 1.10. PRESSURE DROP DURING FLOW OF SUSPENSION

Calcium fluoride suspended in water shall be pumped at 135 kg/h through a 700 m long plastic tubing with 20 mm inner diameter down to the bottom of a fjord as shown schematically in Figure 1.30. The suspension contains 4% by volume calcium fluoride with density $\rho = 3180$ kg/m^3 and 96% by volume water with density $\rho = 1000$ kg/m^3 and viscosity $\mu = 1.2$ cP. The density of the sea water is $\rho = 1025$ kg/m^3.

Calculate the pressure difference through the pump.

Solution

One m^3 suspension contains $(0.04)(3180) = 127$ kg CaF$_2$ and $(0.96)(1000) = 960$ kg water.

Density of suspension,

$$\rho_s = 127 + 960 = 1087 \text{ kg/m}^3.$$

Volume of suspension,

$$135/127 = 1.063 \text{ m}^3/\text{h}.$$

Flow velocity,

$$V = \frac{1.063}{3600 \frac{\pi}{4} 0.02^2} = 0.94 \text{ m/s}.$$

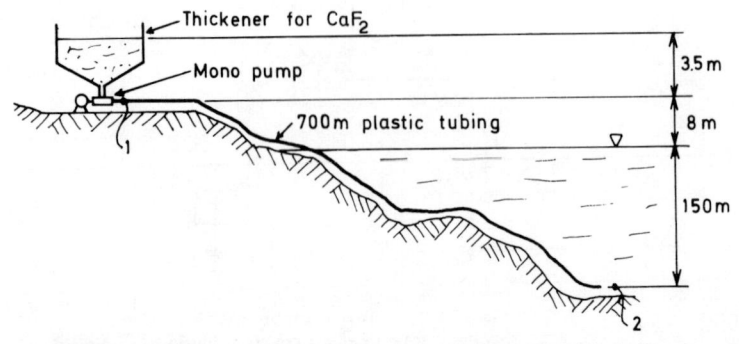

Figure 1.30   Plastic tubing arrangement for calcium fluoride sludge

Reynolds number,

$$\text{Re} = \frac{(1087)(0.94)(0.02)}{(1.2)(10^{-3})} = 17{,}030.$$

Table 1.1 gives $\epsilon = 0.1$ mm. Re = 17,030, and $\epsilon/D = 0.1/20 = 0.005$ inserted in Figure 1.8 gives the friction factor $f \approx 0.0092$. From equation (1.22),

$$\Delta p_f = (4)(0.0092)(1087)\frac{0.94^2}{2}\frac{700}{0.02}$$

$$= 618{,}500 \text{ N/m}^2.$$

Equation (1.5) gives for the flow between point 1 after the pump and point 2 outside the outlet,

$$p_1 + 1087\frac{0.94^2}{2} + (8 + 150)(1087)(9.82)$$

$$= (1025)(150)(9.82) + 0 + 0 + 618{,}500$$

$p_1 = 441{,}300$ N/m^2 or 4.41 bar.

Static pressure in front of the pump,

$$p_0 = (3.5)(1000)(9.82)$$

$$= 34{,}400 \text{ N/m}^2.$$

Pressure difference through the pump,

$$p_1 - p_0 = 441{,}300 - 34{,}400$$

$$= 406{,}900 \text{ N/m}^2$$

$$= 4.1 \text{ bar}.$$

Note Due to uncertainty in the calculation, the pump should be designed for a pressure at least 20% more than calculated, i.e. $(1.2)(4.1) = 4.9$ bar.
The calculations are carried out with the assumption that the velocity 0.94 m/s is sufficient to keep the particles in suspension.

EXAMPLE 1.11. PRESSURE DROP IN TWO-PHASE FLOW

350,000 m^3 natural gas (at normal temperature and pressure) and 15 m^3 condensate per day flows in a steel pipe-line with inner diameter $D = 104$ mm and length $L = 2400$ m from the bore-hole to the purification facilities for removal of hydrogen sulphide, carbon dioxide, water vapour and mercaptans. Pressure and temperature at the bore-hole (after pressure reduction and heating to prevent formation of hydrates) are 81.4 bar and 65°C. The flow is almost isothermal and the gas density proportional to the reciprocal of the pressure.
At pressure 81.4 bar and temperature 65°C:

    average molecular weight of the gas, $M = 22$
    gas density, $\rho_1 = 66$ kg/m^3
    gas viscosity, $\mu = 0.015$ cP
    density of condensate, $\rho_L = 665$ kg/m^3
    viscosity of condensate, $\mu_L = 0.3$ cP.

(*a*) Calculate the pressure drop from the reduction valve after the bore hole for the gas alone, with the assumption that the condensate flows through another pipe-line.

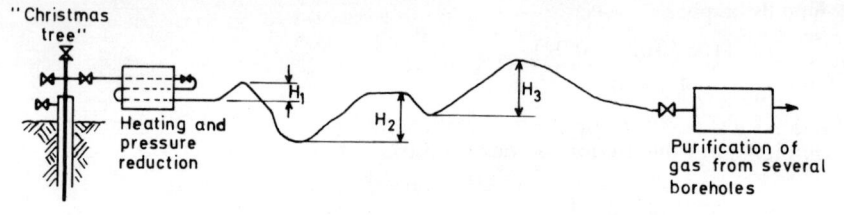

Figure 1.31 Pipe line for gas and condensate from the bore hole to the purification plant, $\Sigma H = H_1 + H_2 + H_3$

(b) Calculate the pressure drop for two-phase flow with the assumption that the pipe-line is horizontal.

(c) Find the pressure drop for two-phase flow when the pipe-line traverses rolling country where the sum of the uphill sections measured vertically (ΣH Figure 1.31) is 280 m.

Solution

(a) 350,000 m³/day (at normal temperature and pressure) corresponds to

$$Mpv/RT = \frac{(22)(10^5)(350,000)}{(24)(3600)(8314)(273)} = 3.93 \text{ kg/s}.$$

Gas velocity at the inlet,

$$V_1 = \frac{3.93}{66 \, \frac{\pi}{4} \, 0.104^2} = 7.0 \text{ m/s}.$$

Reynolds number,

$$Re = \frac{\rho_1 V_1 D}{\mu} = \frac{(66)(7.0)(0.104)}{0.000015}$$
$$= (3.20)(10^6).$$

Surface roughness (Table 1.1),

$$\epsilon = 0.045, \quad \frac{\epsilon}{D} = \frac{0.045}{104}$$
$$= 0.00043.$$

Figure 1.8 gives the corresponding friction factor,

$$f = 0.004.$$

With the approximations of isothermal flow and $\rho \approx \rho_1 p/p_1$, the velocity

$$V = V_1 p_1/p,$$

where V_1 and ρ_1 are the values at the inlet. These values inserted in equation (1.8) give

$$\Delta p_f = 4f\rho_1 \frac{p}{p_1} \frac{L}{D} \frac{V_1^2}{2} \left(\frac{p_1}{p}\right)^2$$

$$= 4fp_1\rho_1 \frac{L}{D} \frac{V_1^2}{2p},$$

where the pressure p varies with the length L. Differentiation of this equation with respect to L gives,

$$p\,d(\Delta p_f) = -2\frac{f}{D}p_1\rho_1 V_1^2\,dL$$

or integration from pressure p_1 at the inlet to pressure p_2 at the outlet,

$$p_1 - \frac{p_2^2}{p_1} = \frac{4f}{D}\rho_1 V_1^2 L$$

$$(8.14)(10^6) - \frac{p_2^2}{(8.14)(10^6)} = \frac{(4)(0.004)}{0.104}(66)(7.0)^2(2400)$$

$$p_2 = (7.5)(10^6)\text{ N/m}^2,$$
$$\Delta p = p_1 - p_2 = 0.64\text{ MN/m}^2$$
$$= 6.4\text{ bar}.$$

The additional pressure drop required to accelerate the gas from the inlet to the outlet velocity is negligible compared to the pressure drop due to friction.

(b)   The liquid alone gives

$$V_L = \frac{15}{(24)(3600)\frac{\pi}{4}(0.104^2)} = 0.0204\text{ m/s}.$$

Reynolds number,

$$\text{Re}_L = \frac{(665)(0.0204)(0.104)}{0.0003} = 4700.$$

$\text{Re}_L = 4700$ and $\epsilon/D = 0.00043$ give (Figure 1.8) $f_L = 0.0094$, and equation (1.8) gives

$$\Delta p_{fL} = (4)(0.0094)(665)\frac{2400}{0.104}\frac{0.0204^2}{2}$$
$$= 120\text{ N/m}^2$$

i.e. in Figure 1.17

$$X = \sqrt{120/(0.64)(10^6)} = 0.014.$$

The corresponding value of Y_G (Figure 1.17) for turbulent flow is 1.7, and equation (1.23) gives,

$$\Delta p_{f\text{ tot}} = (1.7)(0.64)(10^6) = (1.09)(10^6)\text{ N/m}^2$$
$$= 11\text{ bar}$$

(c)   Superficial gas velocity $V_1 = 7.1$ m/s inserted in Figure 1.18 gives $E_H = 0.113$ or an additional pressure drop due to the uphill sections of the pipe,

$$\Delta p_H = (0.113)(665)(9.82)(280) = (0.21)(10^6)\text{ N/m}^2$$
$$= 2.1\text{ bar}.$$

Total pressure drop

$$\Delta p_{\text{tot}} = 11 + 2.1 \approx 13\text{ bar}.$$

Problems

1.1 The suction pipe for an ammonia compressor is 25 m long and made of commercial steel. The inner diameter is 52 mm. It has five 90° standard elbows and one fully open straight globe valve.

Estimate the pressure drop through the pipe with valve and elbows when the flow rate of ammonia gas, the ammonia temperature, and the ammonia pressure all at the inlet of the compressor are 200 m³/h, −25°C, and 1 bar = 0.1 MN/m², respectively. Ammonia can be assumed to be an ideal gas with molecular weight $M = 17$ kg/kmol and viscosity $\mu = 0.0085$ cP.

1.2 Small quantities of a chemical are to be added to the liquid flowing in the pipe A, Figure 1.32. To obtain easily measurable concentrations, the chemical is added to the branch, pipe B, which has an inner diameter of 27 mm.

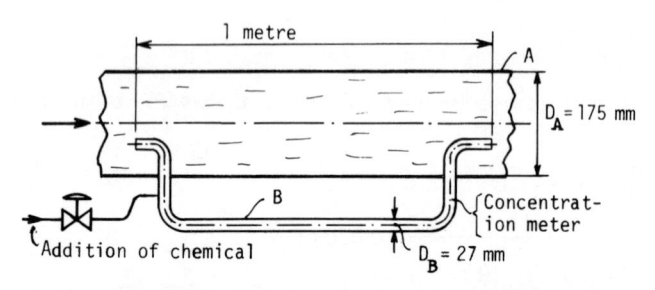

Figure 1.32   Arrangement for addition of a chemical

The pressure drop due to friction in the pipes is estimated to be one velocity head per pipe length $L_e = 40D$. The pipe B is 1.55 m long and has four 90° bends, each giving an additional resistance equal to 0.75 m straight pipe.

Estimate the ratio of mass flow in the two pipes, w_A/w_B, when no chemical is added.

1.3. Figure 1.33 shows an arrangement for charge-wise feed of a liquid into a piece of test equipment. The inner diameter of the vessel A is 125 mm and of the plastic tubing B 6 mm. Liquid is added continuously to the vessel A at a flow rate of Q litre/h, and flows out discontinuously through the siphon B which has sharp-edged inlet and outlet. The pressure drop due to friction in the straight sections and in the bends is

$$\Delta p_f = 4f\rho \frac{L_e}{D} \frac{V^2}{2} = 6.9\rho \frac{V^2}{2}.$$

The meniscus in the plastic tubing will pull the liquid over the top of the tubing at the moment when the liquid level in the vessel reaches C, and the tubing will then act as a siphon until it is emptied at liquid level D.

All dimensions in Figure 1.33 are in mm. The acceleration due to gravity is $g = 9.82$ m²/s.

(a)   Calculate the flow rate through the plastic tubing at liquid levels C and D respectively.

(b)   Find the time t_0 for the liquid level to drop from C to D when $\dot{Q}_1 = 16$ litre/h.

1.4. Air is dried by passing through a vertical glass cylinder with inner diameter $D = 100$ mm. In the bottom of the cylinder is a distributor plate with 224 holes of 1 mm diameter; 1.5 litres of concentrated sulphuric acid has been introduced after the fixed air flow of 24 m³/h was started.

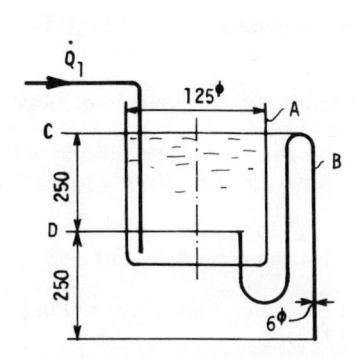

Figure 1.33 Arrangement for discontinuous addition of a liquid

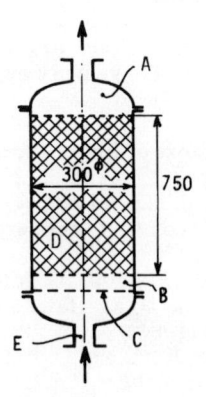

Figure 1.34 The air dryer described in problem 1.5

Density and viscosity of sulphuric acid are 1840 kg/m³ and 15 cP, respectively. Density and viscosity of the air are 1.10 kg/m³ and 0.019 cP, respectively. The water vapour content of the air is negligible.

(*a*) What is the pressure drop through the sulphuric acid?

(*b*) What is the pressure drop through the distributor plate when contraction of the air jets reduces the cross section by 40%?

1.5. The dryer in Figure 1.34 is designed for a mass flow rate of 130 kg/h of air at a pressure of 5 bar (= 0.5 MN/m²) and temperature 25°C after the alumina bed (point A). The bed is 300 mm in diameter, 750 mm deep, and has activated alumina particles with diameter of about 5 mm and void fraction $\epsilon = 0.58$.

(*a*) Calculate the necessary pressure under the bed (point B) to obtain 0.5 MN/m² at A.

(*b*) Calculate the number of holes with diameter 1.5 mm in the distributor plate C when the pressure drop through the holes should be 100 times the velocity head at the inlet E which has a diameter of 25 mm. The air flow through the holes is about 60% of what it would be for ideal flow without contraction or friction.

Air can be assumed to be an ideal gas with molecular weight $M = 29$ and viscosity $\mu = 0.018$ cP.

References

1 Wilkinson, W. L.: *Non-Newtonian Fluids*, Pergamon Press, Oxford, 1960 (single-phase non-Newtonian flow).
2 Oliver, D. R., and A. Y. Hoon: Two-phase non-Newtonian flow, *Trans. Instn. Chem. Engrs.*, **46**, 106–115 (1968).
3 Moody, L. W.: Friction factor for pipe flow, *Trans. Am. Soc. Mech. Engrs.*, **66**, 671–684 (1944).
4 Augland, O.: Pressure drop in pipes. Roughness and choice of friction loss (in Norwegian); *Tidsskr. Kjemi Bergv. Metallurgi*, **32** (No. 9), 11 (1972).
5 Ito, H.: Friction factors for turbulent flow in curved pipes, *Trans. Am. Soc. Mech. Engrs.*, **82**, series D, 131 (1960).
6 Perry, R. H., and C. H. Chilton: *Chemical Engineers' Handbook*, McGraw-Hill, New York, 5th Edn., 1973.

7. Srinivasan, P. S., S. S. Nandapurkar and F. A. Holland: Friction factors for coils, *Trans. Instn. Chem. Engrs.*, **48**, T 156–T 161 (1970).
8. Lapple, C. E.: Velocity head simplifies flow computation, *Chem. Eng.*, 96–104 (May 1949).
9. Robinson, K. K., and D. E. Briggs: Pressure drop of air flowing across triangular pitch banks of finned tubes, *Chem. Eng. Progress, Symposium Series,* Heat Transfer – Los Angeles, No. 64, **62**, 177–184 (1966).
10. Kayes, M. M., and A. L. London: Heat transfer and flow-friction characteristics of some compact heat-exchanger surfaces, *Trans. Am. Soc. Mech. Engrs.*, **72**, 1075–1079 (1950).
11. Leva, M.: *Fluidization*, McGraw-Hill, New York, 1959.
12. Richarson, D. R.: How to design fluid-flow distributors, *Chem. Eng.*, **68** No. 9, 83–86 (1961).
13. Worster, R. C., and D. F. Denny: Hydraulic transport of solid material in pipes, *Proc. Instn. Mech. Engrs.*, **169**, 563–573 (1955).
14. Caldwell, D. H., and H. E. Babbitt: Flow of muds, sludges, and suspensions in circular pipe, *Ind. Eng. Chem.*, **33**, 249–256 (1941).
15. Maude, A. D., and R. L. Whitmore: The turbulent flow of suspensions in tubes, *Trans. Instn. Chem. Engrs.*, **36**, 296–304 (1958).
16. Leuschner, G.: *Kleines Pumpenhandbuch für Chemie und Technik,* Verlag Chemie, Weinheim, 1967.
17. Jenike, A. W.: Storage and flow of solids, *Utah Eng. Expt. Station Bull. 123,* University of Utah, 1964.
18. Reisner, W., and M. v. Eisenhart Rothe: Bins and bunkers for handling bulk materials – practical design and techniques, *Trans. Tech. Publications,* Cleveland, Ohio, 1971.
19. Gerschow, F. J.: How to select a pneumatic-conveying system, *Chem. Eng.*, Feb. 17, 72–86 (1975) and Specified components of pneumatic conveying systems, *Chem. Eng.*, March 31, 88–96 (1975).
20. Muschelknautz, E., and H. Wojahn: Auslegung pneumatischer Förderanlagen, *Chem. Ing. Techn.*, **46**, No. 6. 223–235 (1974).
21. Lockhart, R. W., and R. C. Martinelli: Proposed correlation of data for isothermal two-phase flow in pipes, *Chem. Eng. Progress,* **45**, 39–48 (1949).
22. Flanigan, O.: Effect of uphill flow on pressure drop in design of two-phase gathering systems, *Oil and Gas J.*, 132–141 (March 1958).
23. Govier, G. M., and K. Aziz: *The flow of complex mixtures in pipes,* Van Nostrand-Reinold, New York, 1972.
24. Kern, R.: How to size process piping for two-phase flow, *Hydrocarbon Processing,* 105–116 (Oct. 1969).

CHAPTER 2
Dimensional Analysis and Dimensionless Groups

Dimensional analysis

Dimensional analysis provides a method of reducing the number of measurements necessary to find the relationship between different variable properties. Let us assume that the quantity A is a function of the quantities B and C, and we wish to find a relationship between the three quantities.

A conventional method is to keep B constant, vary C, and measure A, and repeat this for different values of B. These measurements are plotted as shown in Figure 2.1. This figure can be used to find A for intermediate values of B by interpolation of B between the experimental curves at given values of C.

Dimensional analysis may show that A is a function of a product of B and C to powers b and c, i.e.

$$A = f(B^b C^c).$$

In this case values of $B^b C^c$ can be calculated for each measurement, and the result plotted as a single curve as shown in Figure 2.2.

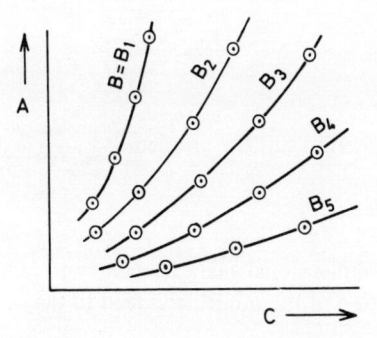

Figure 2.1 Quantity A determined experimentally as a function of B and C.

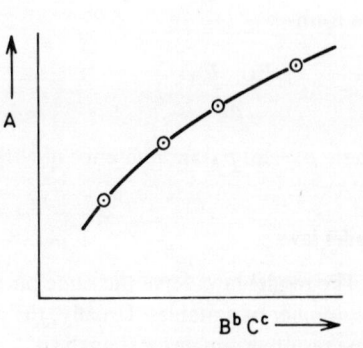

Figure 2.2 Quantity A as a function of $B^b C^c$.

Thus the relationship between A, B, and C can be determined by a small fraction of the measurements required by the conventional method. For example in compiling Figure 2.1, 19 measurements were required. In Figure 2.2 only four were required. The method of Figure 2.2 also has the advantage that interpolation is not involved.

The quantity in question may be a function of more than two variables. For example, p_1 may be a function of $p_2, p_3 \ldots p_n$. If p_1 is a continuous function of these quantities, it may be expressed as an infinite series,

$$p_1 = \sum_{x=1}^{\infty} a_x p_2^{b_x} p_3^{c_x} \ldots p_n^{n_x} \tag{2.1}$$

a_x is a dimensionless quantity so that the right hand side of equation (2.1) will have the same dimensions as p_1. Mass M, length L, time T, and temperature θ are basic quantities in the SI-system. If $p_1, p_2 \ldots p_n$ contain m basic quantities, equation (2.1) results in m equations with n unknowns. Thus, $n - m$ remain unknown. This means instead of n variables $n - m$ variables are needed. These can be arranged in dimensional groups, $P_1, P_2 \ldots P_{n-m}$. These variables are independent if none of them can be obtained as the product of two or more of the other variables to suitable powers.

In some cases it is possible to know in advance how one or more of the variables will affect the quantity p_1. If, for instance, p_1 is the pressure loss due to friction in a straight pipe, where the influence of the inlet and the outlet may be neglected, and p_2 is the pipe length, then p_1 will be proportional to p_2. This means that one can introduce a new variable, $p_1' = p_1/p_2$ on the left hand side of equation (2.1), and the term $p_2^{b_x}$ on the right may be omitted. It may also be useful to substitute one or more of the dimensionless variables found by means of equation (2.1) by products of powers of two or more of the original dimensionless groups. For instance p_1 can be substituted by

$$p_1' = p_1^x p_2^y \tag{2.2}$$

where x and y can be whole numbers or fractions.

The result may be simplified further by introduction of dimensionless groups of the form

$$\pi_1 = \frac{p_1 - p_A}{p_B} \tag{2.3}$$

where p_A and p_B are reference quantities chosen to suit the problem.[1,2]

Model laws

The model laws serve the same purpose as dimensional analysis, namely to reduce the number of variables. Usually, the application of the model laws lead to the same result as dimensional analysis.

Models are often used for scale-up in geometrically and dynamically similar systems.[3]

Dynamic similarity in two systems means that the paths described by corresponding particles in corresponding times are geometrically similar. This occurs when the ratios between the forces acting in both systems are the same.

Examples of such forces are:

pressure	pL^2
inertia	$\rho L^3 \dfrac{V^2}{L} = \rho L^2 V^2$
viscous forces	$\mu L^2 \dfrac{V}{L} = \mu VL$
gravity forces	$\rho L^3 g$
surface forces	γL
elastic forces	EL^2

p is pressure, L length, ρ density, V velocity, μ viscosity, g acceleration due to gravity, γ surface tension, and E modulus of elasticity.

Examples of force ratios are:

$$\frac{\text{inertia forces}}{\text{viscous forces}} = \frac{\rho L^2 V^2}{\mu VL} = \frac{\rho VL}{\mu} = \text{Re} = \text{Reynolds number} \qquad (2.4)$$

$$\frac{\text{inertia forces}}{\text{gravity forces}} = \frac{\rho L^2 V^2}{\rho L^3 g} = \frac{V^2}{Lg} = \text{Fr} = \text{Froude number} \qquad (2.5)$$

$$\frac{\text{inertia forces}}{\text{elasticity forces}} = \frac{\rho L^2 V^2}{EL^2} = \frac{\rho V^2}{E} = \text{Ca} = \text{Cauchy number} \qquad (2.6)$$

$$\frac{\text{inertia forces}}{\text{surface forces}} = \frac{\rho L^2 V^2}{\gamma L} = \frac{\rho L V^2}{\gamma} = \text{We} = \text{Weber number} \qquad (2.7)$$

$$\frac{\text{pressure forces}}{\text{inertia forces}} = \frac{pL^2}{\rho L^2 V^2} = \frac{p}{\rho V^2} = \text{Eu} = \text{Euler number.} \qquad (2.8)$$

(Some publications use similar notations for the reverse force ratios, for example Fr = gravity forces/inertia forces. The nomenclature N_{Re}, N_{Fr}, etc., is also commonly used.)

The ratio of the forces acting in systems must be unchanged from model to prototype in order to give dynamic similarity. It is in general impossible to keep all force ratios constant in scale-up, but in most cases only one ratio is of deciding importance, and other ratios may be neglected.

EXAMPLE 2.1. DIMENSIONAL ANALYSIS. PRESSURE DROP

(*a*)   Upon what properties does the pressure drop Δp depend in an apparatus, a valve, a nozzle, etc.?

(*b*)   Upon which dimensionless groups does the pressure drop depend? Arrange these groups in such a way that the viscosity μ is included in one of them only.

(c) Give an equation for geometrically similar apparatuses, where all dimensions are known as products of one length or diameter D in the special case where the influence of viscosity can be neglected.

Solution

(a) The pressure drop Δp depends on the velocity V in a characteristic cross section, size characterized by a length D_1 and a length ratio D_1/D_2, the density of the fluid ρ, and its viscosity μ, i.e.

$$\Delta p = f\left(\rho, V, \mu, D_1, \frac{D_1}{D_2}\right) \qquad (a)$$

(b) With mass M, length L, and time T as basic quantities, equation (a) has the following dimensions:

Δp N/m² $= M(LT^{-2})/L^2 = ML^{-1}T^{-2}$ $\qquad \mu$ kg/ms $= ML^{-1}T^{-1}$
ρ kg/m³ $= ML^{-3}$ $\qquad\qquad\qquad\qquad\qquad\quad D$ m $\quad = L$
V m/s $\;\; = LT^{-1}$

Equation (a) can now be written in the form of equation (2.1),

$$\Delta p = \sum_{x=1}^{\infty} a_x \rho^{b_x} V^{c_x} \mu^{d_x} D_1^{e_x} \left(\frac{D_1}{D_2}\right)^{f_x}. \qquad (b)$$

The dimensions of the xth term are

$$(M^{b_x} L^{-3b_x})(L^{c_x} T^{-c_x})(M^{d_x} L^{-d_x} T^{-d_x})(L^{e_x})(1^{f_x})$$
$$= M^{(b_x+d_x)} L^{(-3b_x+c_x-d_x+e_x)} T^{(-c_x-d_x)}.$$

This has the same dimensions as Δp, $ML^{-1}T^{-2}$. Hence,

for M: $\quad 1 = b_x + d_x$
for L: $\; -1 = -3b_x + c_x - d_x + e_x$
for T: $\; -2 = -c_x - d_x$.

These three equations with four unknowns are now solved as three of the unknowns are expressed by means of the fourth, $c_x = 1 + b_x$, $d_x = 1 - b_x$, and $e_x = b_x - 1$. These are inserted in equation (b) to give

$$\Delta p = \sum_{x=1}^{\infty} a_x \rho^{b_x} V V^{b_x} \mu \mu^{-b_x} D^{b_x} D^{-1} \left(\frac{D_1}{D_2}\right)^{f_x}$$

$$\Delta p = \sum_{x=1}^{\infty} a_x \left(\frac{V\mu}{D}\right) \left(\frac{\rho V D}{\mu}\right)^{b_x} \left(\frac{D_1}{D_2}\right)^{f_x}.$$

This equation can be divided through by the term without x, $V\mu/D$, to give

$$\frac{\Delta p D}{V\mu} = f\left(\frac{\rho V D}{\mu}, \frac{D_1}{D_2}\right).$$

By application of equation (2.2), a new dimensionless group, which does not contain the viscosity μ, is obtained by multiplication of $\Delta p D / V \mu$ by the reciprocal of $\rho V D / \mu$,

$$\left(\frac{\Delta p D}{V \mu}\right)\left(\frac{\rho V D}{\mu}\right)^{-1} = \frac{\Delta p}{\rho V^2}$$

$$\frac{\Delta p}{\rho V^2} = f\left(\frac{\rho V D}{\mu}, \frac{D_1}{D_2}\right).$$

(c) Equation (a) simplifies to

$$\Delta p = f(\rho, V, D)$$

$$\Delta p = \sum_{x=1}^{\infty} a_x \rho^{b_x} V^{c_x} D^{d_x}$$

$$ML^{-1}T^{-2} = (M^{b_x}L^{-3b_x})(L^{c_x}T^{-c_x})(L^{d_x}) = M^{b_x}L^{(-3b_x+c_x+d_x)}T^{-c_x}.$$

For

$M: \quad 1 = b_x$
$L: \quad -1 = -3b_x + c_x + d_x$
$T: \quad -2 = -c_x$

or

$b_x = 1, \quad c_x = 2, \quad d_x = 0$

$$\Delta p = \sum_{x=1}^{\infty} a_x \rho V^2 = a \rho V^2$$

where a is a constant equal to $\sum_{x=1}^{\infty} a_x$. In this special case the pressure drop is independent of the length D.

Comment This means that calculation of the pressure drop in fittings by means of the equation $\Delta p = n\rho(V^2/2)$ as given in Table 1.2, is based on the simplifying condition that the influence of viscosity is negligible.

EXAMPLE 2.2. DIMENSIONAL ANALYSIS. POWER CONSUMPTION IN AGITATION

The power consumption of an impeller, P, depends on the dimensions of the impeller, expressed for instance by the diameter D, the number of revolutions per unit time n, the density of the liquid ρ, and the viscosity of the liquid μ.

Find the relationship, which makes it possible by means of a few measurements of the power consumption of different revolution rates of the impeller in one liquid, to find the power consumption in other liquids of other densities and viscosities.

Solution

The given quantities are inserted into equation (2.1),

$$P = \sum_{x=1}^{\infty} a_x D^{b_x} n^{c_x} \rho^{d_x} \mu^{e_x}.$$

The dimensions are:

$$P \text{ Nm/s} = (MLT^{-2})L/T = ML^2 T^{-3} \qquad \rho \text{ kg/m}^3 = ML^{-3}$$
$$D \text{ m} = L \qquad \qquad \mu \text{ kg/ms} = ML^{-1} T^{-1}$$
$$n \text{ rev./s} = T^{-1}.$$

This gives the dimensions in equation (a)

$$ML^2 T^{-3} = L^{b_x} T^{-c_x} (M^{d_x} L^{-3d_x})(M^{e_x} L^{-e_x} T^{-e_x}).$$

For

M: $\quad 1 = d_x + e_x$
L: $\quad 2 = b_x - 3d_x - e_x$
T: $\quad -3 = -c_x - e_x.$

Three of the four unknowns in these three equations can be expressed in terms of the fourth unknown, $b_x = 5 - 2e_x$, $c_x = 3 - e_x$, $d_x = 1 - e_x$.
These values can be inserted into equation (a),

$$P = \sum_{x=1}^{\infty} a_x D^5 D^{-2e_x} n^3 n^{-e_x} \rho \rho^{-e_x} \mu^{e_x}$$

$$P = \sum_{x=1}^{\infty} a_x \left(\frac{\mu}{\rho n D^2}\right)^{e_x} (\rho D^5 n^3)$$

$$\frac{P}{\rho D^5 n^3} = f\left(\frac{\rho n D^2}{\mu}\right),$$

i.e. the experimental results can be plotted as $P/\rho D^5 n^3$ against the modified Reynolds number $\rho n D^2/\mu$. A curve through these points can be used to read off $P/\rho D^5 n^3$. This means that P for liquids with different values of ρ and μ can be calculated if $\rho n D^2/\mu$ is within the experimental range.

EXAMPLE 2.3. TERMINAL VELOCITY OF FALLING SPHERES

The terminal velocity of a sphere sinking in a liquid can be used to give a rapid estimate of the liquid viscosity.

(a) What variables will influence the terminal velocity of the sphere?

(b) Find, by means of dimensional analysis, an equation with dimensionless groups, where the terminal velocity V_t is included in only one of the dimensionless groups.

(c) Find the dimensionless groups for a corresponding equation where the liquid viscosity μ is included in only one of the dimensionless groups.

(d) The table below gives six measurements of the terminal velocity of glass spheres falling in a mixture of glycerol and water. Plot a single curve that relates liquid viscosities to the terminal velocities of glass spheres. Give the limits of validity for this correlation.

Experimentally determined terminal velocities for glass spheres with density $\rho = 2720$ kg/m³ in mixtures of glycerol and water

Measurement number	1	2	3	4	5	6
Density of liquid, ρ kg/m³	1260	1260	1200	1200	1145	1145
Viscosity of liquid, μ cP	350	350	115	115	23	23
Sphere diameter, D mm	3	6	3	6	3	6
Terminal velocity, V_t m/s	0.020	0.080	0.065	0.15	0.14	0.30

Solution

(a) The terminal velocity V_t will depend on

ρ = density of the liquid, kg/m^3, $[ML^{-3}]$
D = sphere diameter, m $[L]$,
μ = liquid viscosity, kg/m s $[ML^{-1}T^{-1}]$,
F = resulting force (gravity minus buoyancy), newton $[MLT^{-2}]$,

i.e. $V = f(\rho, D, \mu, F)$.

(b) $V_t = \sum\limits_{x=1}^{\infty} a_x \rho^{b_x} D^{c_x} \mu^{d_x} F^{e_x}$ (a)

$LT^{-1} = M^{b_x} L^{-3b_x} L^{c_x} M^{d_x} L^{-d_x} T^{-d_x} M^{e_x} L^{e_x} T^{-2e_x}$
$LT^{-1} = M^{(b_x+d_x+e_x)} L^{(-3b_x+c_x-d_x+e_x)} T^{(-d_x-2e_x)}$.

For

L: $1 = -3b_x + c_x - d_x + e_x$
T: $-1 = -d_x - 2e_x$
M: $0 = b_x + d_x + e_x$.

This gives three equations with four unknowns,

$b_x = e_x - 1, \quad c_x = -1, \quad d_x = 1 - 2e_x, e_x = e_x$.

From equation (a)

$V_t = \sum\limits_{x=1}^{\infty} a_x \rho^{e_x} \rho^{-1} D^{-1} \mu \mu^{-2e_x} F^{e_x}$

$V_t = \sum\limits_{x=1}^{\infty} a_x \left(\dfrac{\mu}{\rho D}\right) \left(\dfrac{\rho F}{\mu^2}\right)^{e_x} = \left(\dfrac{\mu}{\rho D}\right) \sum\limits_{x=1}^{\infty} a_x \left(\dfrac{\rho F}{\mu^2}\right)^{e_x}$. (b)

Equation (b) multiplied by $\rho D/\mu$ gives

$\dfrac{\rho V_t D}{\mu} = f\left(\dfrac{\rho F}{\mu^2}\right)$. (c)

(c) Equation (2.2) with convenient powers of x and y gives the new dimensionless group

$P' = \left(\dfrac{\rho F}{\mu^2}\right)^{-1} \left(\dfrac{\rho V_t^2 D^2}{\mu}\right) = \dfrac{\rho V_t^2 D^2}{F}$

or

$\dfrac{\rho V_t D}{\mu} = f\left(\dfrac{\rho V_t^2 D^2}{F}\right)$ (d)

(d) The quantities in equation (d) are given in the table below. The resulting force is

$F = \dfrac{\pi}{6} D^3 (\rho_p - \rho) g = 5.14(\rho_p - \rho_g) D^3$.

Measurement number	1	2	3	4	5	6	Notes
$\rho_p - \rho$ kg/m³	1460	1460	1520	1520	1575	1575	$\rho_p - \rho = 2720 - \rho$
$F \times 10^6$ N $\times 10^6$	203	1621	211	1688	219	1749	$F = 5.14(\rho_p - \rho)D^3$
$\dfrac{\rho V_t^2 D^2}{F}$	0.0223	0.179	0.216	0.576	0.922	2.12	
$\dfrac{\rho V_t D}{\mu}$	0.216	1.73	2.03	9.39	20.9	89.6	
$\log\left(\dfrac{\rho V_t^2 D^2}{F}\right)$	0.348−2	0.253−1	0.335−1	0.760−1	0.965−1	0.326	
$\log\left(\dfrac{\rho V_t D}{\mu}\right)$	0.335−1	0.238	0.308	0.973	1.320	1.952	

As seen from the table the dimensionless groups vary over a wide range (∼1--100). This means that it might be convenient to use logarithmic scales as in Figure 2.3. This figure can be used to find the viscosity of a liquid based on measurements of the liquid density ρ, terminal velocity V_t, sphere diameter D, and resulting force F when

$$0.02 < \frac{\rho V_t^2 D^2}{F} < 2$$

which is the range of the experimental measurements.

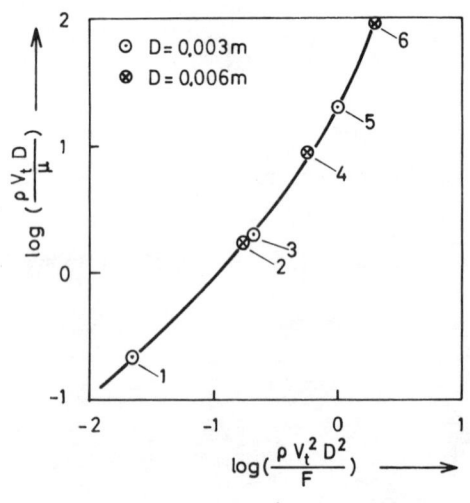

Figure 2.3 $\text{Log}\left(\dfrac{\rho V_t D}{\mu}\right)$ plotted against $\log\left(\dfrac{\rho V_t^2 D^2}{F}\right)$.

The results are least accurate for $\log(\rho V_t^2 D^2)/F > -0.5$ as the curve for $\rho V_t D/\mu$ is steepest in this region.

EXAMPLE 2.4. DIMENSIONAL ANALYSIS. FORMATION OF DROPS

Drops of liquid A in liquid B, with which A is immiscible, are formed as liquid A flows slowly through a tube with inner diameter D as shown in Figure 2.4.

(a) What quantities influence the mass W in drops of liquid A that falls through liquid B?

(b) Give an equation for W in terms of the fewest variables to be measured experimentally.

Solution

(a) The mass of the drops W, will depend on the tube dimeter D, the surface tension of the drop γ, the density difference for the two liquids, $\Delta\rho = \rho_A - \rho_B$, and the acceleration of gravity g.

$$W = f(D, \gamma, \Delta\rho, g)$$

(b) With the same basic quantities as in the previous example the dimensions are

W kg $= M$
D m $= L$
γ N/m $= (MLT^{-2})/L = MT^{-2}$.
$\Delta\rho$ kg/m^3 $= ML^{-3}$
g m/s^2 $= LT^{-2}$

From equation (2.1):

$$W = \sum_{x=1}^{\infty} a_x D^{b_x} \gamma^{c_x} (\Delta\rho)^{d_x} g^{e_x}. \tag{a}$$

The dimensions in this equation are

$$M = (L^{b_x})(M^{c_x} T^{-2c_x})(M^{d_x} L^{-3d_x})(L^{e_x} T^{-2e_x})$$
$$M = L^{(b_x - 3d_x + e_x)} M^{(c_x + d_x)} T^{(-2c_x - 2e_x)}.$$

For

M: $\quad 1 = c_x + d_x$
L: $\quad 0 = b_x - 3d_x + e_x$
T: $\quad 0 = -2c_x - 2e_x$.

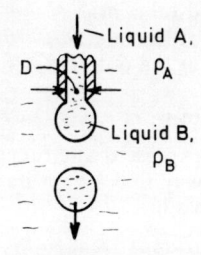

Figure 2.4 Drop formation of liquid A, density ρ_A, in liquid B, density $\rho_B < \rho_A$.

These three equations with four unknowns give
$$b_x = 2e_x + 3, \quad c_x = -e_x, \quad d_x = e_x + 1.$$
Insertion in equation (a) gives
$$W = \sum_{x=1}^{\infty} a_x D^{2e_x} D^3 \gamma^{-e_x} (\Delta\rho)^{e_x} (\Delta\rho) g^{e_x}$$
$$W = \sum_{x=1}^{\infty} a_x (\Delta\rho) D^3 \left(\frac{D^2(\Delta\rho)g}{\gamma}\right)^{e_x}$$
i.e
$$W = (\Delta\rho) D^3 f\left(\frac{D^2(\Delta\rho)g}{\gamma}\right),$$
where $(D^2(\Delta\rho)g)/\gamma$ is the only free variable and
$$y = \frac{W}{(\Delta\rho)D^3} = f\left(\frac{D^2(\Delta\rho)g}{\gamma}\right)$$
is the only function to be determined experimentally.

Note The above derivation is based on the assumption that the velocity in the tube was so small that it had no influence on the size of the drops. At higher velocities both velocity and viscosity become new variables on the right hand side of equation (a).

EXAMPLE 2.5. HEATING OF GRANULAR MATERIAL

Granular material with specific heat capacity c kJ/(kg °C) is heated from temperature θ_1 to temperature θ_2 as it passes through a steam-jacketed tube with constant temperature θ_3 at the inner tube surface (Figure 2.5). Equations (9.42) and (9.43) give the rate of heat transfer,
$$\dot{Q} = h(\pi d l)\frac{\Delta\theta_1 - \Delta\theta_2}{\ln(\Delta\theta_1/\Delta\theta_2)} \quad \text{W or J/s}$$
where
h = heat transfer coefficient in W/(°C m²) or J/(s °C m²) (h is assumed to be a function of w, c, k, l, and d)
w = rate of mass flow, kg/s
k = thermal conductivity of the granular material, W/(m °C) or J/(s m °C)
l = length of heated tube, m
d = inner diameter of heated tube, m
$\Delta\theta_1$ = temperature difference at the inlet, $\theta_3 - \theta_1$ °C
$\Delta\theta_2$ = temperature difference at the outlet, $\theta_3 - \theta_2$ °C.

(a) Express $\Delta\theta_1/\Delta\theta_2$ in terms of w, h, c, l, and d.

(b) The ratio $\Delta\theta_1/\Delta\theta_2$ can be expressed as a function of two dimensionless groups A and B. Find A and B in such a way that k appears only in the numerator of A and A contains as few quantities as possible.

(c) Find a new dimensionless group A' that fulfils the requirements given for A in question (b).

(d) Based on the results for (b) and (c) and the experimental data in the following table, find a suitable correlation for $\log(\Delta\theta_1/\Delta\theta_2)$ for the granular material used in test runs.

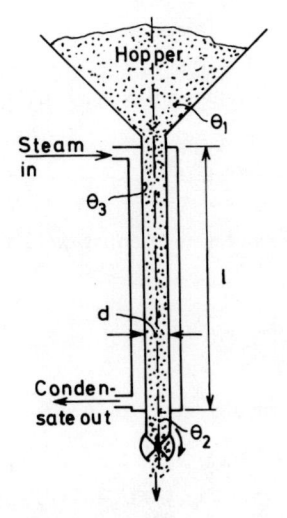

Figure 2.5 Heat exchanger for granular material.

(e) Estimate the tube wall temperature needed to heat 400 kg/h of the same granular material from 16 to 90°C in a 10 m long tube with inner diameter 32 mm.

Run no.	Tube length mm	Tube diameter mm	θ_1 °C	θ_2 °C	θ_3 °C	w kg/h
3	6100	22	18.1	169.8	174.3	43.5
8	6100	22	23.8	135.7	174.6	133
10	6100	22	22.5	155.5	173.5	80.5
11	6100	22	23.7	80.6	101.3	143
12	6100	22	21.4	90.0	100.6	81.5
13	3050	22	23.5	127.5	177.5	82
16	3050	22	21.6	64.0	101.0	135
17	3050	22	21.7	90.3	101.0	38.1
18	3050	15.8	23.0	156.2	176.5	36.3
19	3050	15.8	23.0	125.0	176.0	83.7
21	3050	15.8	26.5	100.0	100.5	15.8
22	3050	15.8	23.5	93.9	100.8	32.0

Solution

(a) The rate of heat transfer is $\dot{Q} = wc(\theta_2 - \theta_1) = wc(\Delta\theta_1 - \Delta\theta_2)$. This value inserted in the given equation yields

$$h\pi dl \frac{\Delta\theta_1 - \Delta\theta_2}{\ln(\Delta\theta_1/\Delta\theta_2)} = wc(\Delta\theta_1 - \Delta\theta_2)$$

$$\ln \frac{\Delta\theta_1}{\Delta\theta_2} = \frac{hld}{wc}.$$

(b) The answer above can be written as

$$\frac{\Delta\theta_1}{\Delta\theta_2} = f(h, l, d, w, c) \quad \text{where } h = \Phi(w, c, k, l, d).$$

Combination of the two equations gives

$$\frac{\Delta\theta_1}{\Delta\theta_2} = F(w, c, k, l, d) \text{ or in terms of equation (2.1),}$$

$$\frac{\Delta\theta_1}{\Delta\theta_2} = \sum_{x=1}^{\infty} a_x w^{b_x} c^{c_x} k^{d_x} l^{e_x} d^{f_x}. \tag{a}$$

The dimensions of the xth term are

$$(M^{b_x} T^{-b_x})(L^{2c_x} T^{-2c_x} \theta^{-c_x})(M^{d_x} L^{d_x} T^{-3d_x} \theta^{-d_x})(L^{e_x})(L^{f_x})$$
$$= M^{(b_x + d_x)} T^{(-b_x - 2c_x - 3d_x)} L^{(2c_x + d_x + e_x + f_x)} \theta^{(-c_x - d_x)}.$$

This product must have the dimensions of the left hand side of equation (a),

$M: 0 = b_x + d_x$
$T: 0 = -b_x - 2c_x - 3d_x$
$L: 0 = 2c_x + d_x + e_x + f_x$
$\theta: 0 = -c_x - d_x.$

The four equations with five unknowns give the exponents in equation (a),

$$c_x = b_x, d_x = -b_x, \quad \text{and} \quad e_x = -b_x - f_x$$

$$\frac{\Delta\theta_1}{\Delta\theta_2} = \sum_{x=1}^{\infty} a_x w^{b_x} c^{b_x} k^{-b_x} l^{-b_x} l^{-f_x} d^{f_x}$$

$$\frac{\Delta\theta_1}{\Delta\theta_2} = \sum_{x=1}^{\infty} a_x \left(\frac{wc}{kl}\right)^{b_x} \left(\frac{d}{l}\right)^{f_x} = f_1\left(\frac{wc}{kl}, \frac{d}{l}\right) = f_2\left(\frac{kl}{wc}, \frac{l}{d}\right) \tag{b}$$

where

$$A = \frac{kl}{wc} \quad \text{and} \quad B = \frac{l}{d}.$$

(c) Equation (2.2) gives the new variable,

$$A' = \left(\frac{kl}{wc}\right)\left(\frac{l}{d}\right)^{-1} = \frac{kd}{wc}$$

and

$$\frac{\Delta\theta_1}{\Delta\theta_2} = f_3\left(\frac{kd}{wc}, \frac{l}{d}\right). \tag{c}$$

(d)   The dimensionless groups in equations (b) and (c) are calculated for the eleven test runs and tabulated below together with $\log(\Delta\theta_1/\Delta\theta_2)$:

Run no.	$\dfrac{l}{d}$	$\Delta\theta_1$ °C	$\Delta\theta_2$ °C	$\dfrac{\Delta\theta_1}{\Delta\theta_2}$	$\log\dfrac{\Delta\theta_1}{\Delta\theta_2}$	$\dfrac{kl}{wc}$	$\dfrac{kd}{wc}$
3	277	156.2	4.5	34.7	1.540	0.140 k/c	0.000506 k/c
8	277	150.8	38.9	3.88	0.590	0.046 k/c	0.000165 k/c
10	277	151.0	18.0	8.39	0.923	0.076 k/c	0.000273 k/c
11	277	77.6	20.7	3.75	0.573	0.043 k/c	0.000154 k/c
12	277	79.2	10.6	7.47	0.873	0.075 k/c	0.000270 k/c
13	139	154.0	50.0	3.08	0.488	0.037 k/c	0.000268 k/c
16	139	79.4	37.0	2.15	0.332	0.023 k/c	0.000163 k/c
17	139	79.3	10.7	7.41	0.670	0.080 k/c	0.000577 k/c
18	193	153.5	20.3	7.56	0.756	0.084 k/c	0.000435 k/c
19	193	153.0	51.0	3.00	0.477	0.0364 k/c	0.000189 k/c
21	193	74.0	0.5	148	2.17	0.193 k/c	0.001000 k/c
22	193	77.3	6.9	11.2	1.05	0.095 k/c	0.000494 k/c

$\log(\Delta\theta_1/\Delta\theta_2)$ is plotted in Figure 2.6 as a function of l/w and in Figure 2.7 as a function of d/w.
Figure 2.6 gives one curve for all values of l/d. The plotted line corresponds to

$$\log(\Delta\theta_1/\Delta\theta_2) = 0.096 + 10.2\,(l/w) \qquad (d)$$

(e) $l/d = 10/0.032 = 313$ which is close to the region of the experimental data, $(139 \leqslant l/d \leqslant 277)$. $l/w = 10/400 = 0.025$ is inserted in equation (d),

$$\log(\Delta\theta_1/\Delta\theta_2) = 0.096 + (10.2)(0.025) = 0.351$$

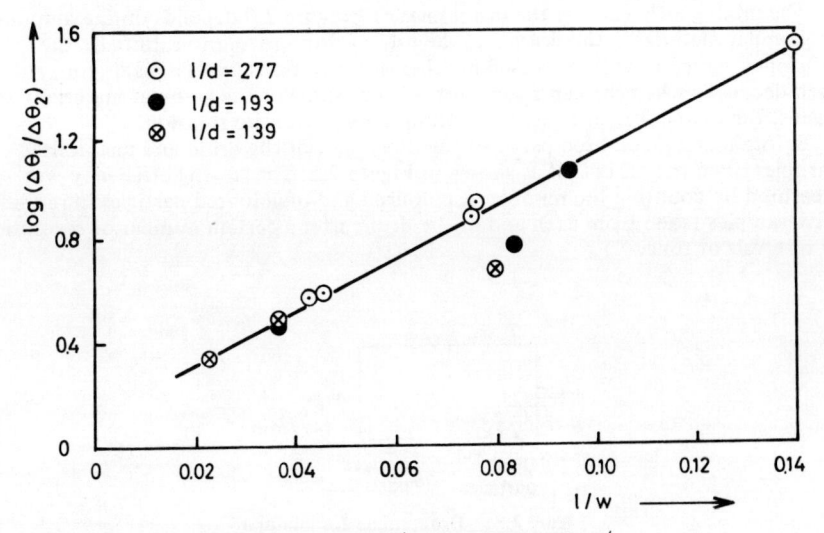

Figure 2.6   Log $(\Delta\theta_1/\Delta\theta_2)$ plotted against l/w.

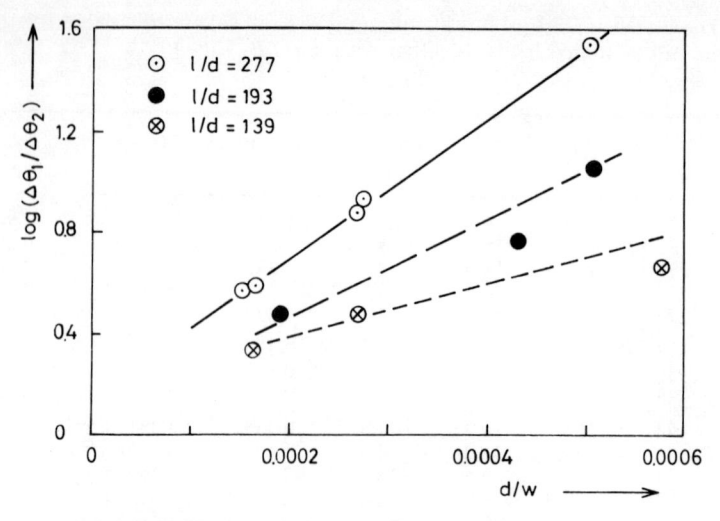

Figure 2.7 Log $(\Delta\theta_1/\Delta\theta_2)$ plotted against d/w.

or $\quad \Delta\theta_1/\Delta\theta_2 = 2.24$

$$\frac{\theta_3 - 16}{\theta_3 - 90} = 2.24,$$

$\theta_3 = 150°C.$

Note The numbers in this example are for w in kg/h.

EXAMPLE 2.6. DRUM MIXER

The mixing efficiency of the drum mixer of Figure 2.8 depends on the amount of granular material in the drum and the rate of drum revolution. Initially the efficiency increases with increased number of revolutions up to a maximum, and then decreases when the centrifugal force keeps some of the granular material fixed against the cylindrical surface of the drum during its entire rotation.

Before mixing, coloured particles filled one half of the drum and uncoloured particles filled the other half as shown in Figure 2.8. The mixing efficiency was measured by counting the number of coloured and uncoloured particles in representative samples taken from each end of the drum after a certain number of revolutions or intervals of time.

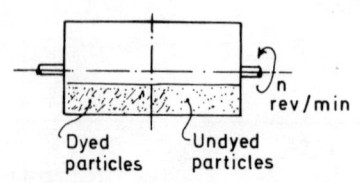

Figure 2.8 Drum mixer for granular material.

Test runs with a three litre mixing drum showed that the maximum amount which could be mixed satisfactorily per unit time was 1.4 kg granular material at $n_m = 73$ rev./min.

(a) Find the most favourable number of revolutions, n_p, for a geometrically similar drum for mixing 300 kg of the same granular material.

(b) Sufficient mixing in the test drum was obtained after nine minutes. Estimate the volume of a geometrically similar drum with the same degree of mixing when the production rate w_p is 1100 kilograms per hour. The total time to charge and discharge the large drum is 15 minutes.

Solution

(a) It is assumed that the same material is handled, and that the particle diameter is very small compared to the diameter and length of both drums. This means that inertia forces and gravitational forces determine the mixing of the particles, i.e. Froude number [equation (2.5)] must be the same for both mixers,

$$\frac{V_m^2}{L_m g} = \frac{V_p^2}{L_p g}. \tag{a}$$

With the same fraction of the volume filled in both cases, the length ratio is

$$\frac{L_p}{L_m} = \sqrt[3]{\frac{300}{1.4}} = 6.0.$$

$V = \pi n L$ inserted into equation (a) gives

$$\frac{\pi^2 n_m^2 L_m^2}{L_m g} = \frac{\pi^2 n_p^2 L_p^2}{L_p g}$$

$$n_p = n_m \sqrt{L_m/L_p} = \frac{73}{\sqrt{6}} = 30 \text{ rev./min.}$$

(b) The velocity V in equation (a) is replaced by $\pi n L$:

$$n_m^2 L_m = n_p^2 L_p, \quad n_p = n_m \sqrt{\frac{L_m}{L_p}} = n_m \sqrt[6]{\frac{v_m}{v_p}},$$

where v is the volume of the drum mixer. With mixing time t and the same number of revolutions for sufficient mixing for each charge,

$$n_p t_p = n_m t_m \quad \text{or} \quad t_p = t_m \sqrt[6]{\frac{v_p}{v_m}}$$

With an additional 15 minutes for charging and discharging, the total time for each charge is

$$t_s = t_m \sqrt[6]{\frac{v_p}{v_m}} + 15 \text{ min.}$$

The number of charges per hour is $60/t_s$. With w_p kg/h, each charge contains

$$\frac{w_p t_s}{60} = \frac{w_p}{60}\left(t_m \sqrt[6]{\frac{v_p}{v_m}} + 15\right). \tag{b}$$

With 1.4 kg granular material in a three litre drum the amount of granular material in the large drum is $1.4\, v_p/v_m$,

$$1.4\frac{v_p}{v_m} = \frac{w_p}{60}\left(t_m \sqrt[6]{\frac{v_p}{v_m}} + 15\right)$$

$$\frac{v_p}{v_m} = \frac{1100}{(1.4)(60)}\left(9\sqrt[6]{\frac{v_p}{v_m}} + 15\right)$$

$$= 118\sqrt[6]{\frac{v_p}{v_m}} + 196. \tag{c}$$

This equation solved by iteration gives $v_p/v_m = 532$, or

$$v_p = (532)(3)$$
$$= 1600 \text{ litre}.$$

EXAMPLE 2.7. MODEL LAWS. PLATE THICKNESS OF A STORAGE TANK

Large storage tanks for liquids may be given the shape of a drop as shown in Figure 2.9. This shape gives the same stress on all parts of the tank and most efficient utilization of the plates. The drop can be used as a model. An enlarged copy of a photo of the drop is used to establish the shape of a geometrically similar tank.

Calculate the necessary plate thickness in a water tank with a volume 1000 m³ when the tank is made geometrically similar in shape to a mercury drop with a volume of 37 mm³.

The plates in the tank should be designed for a tensile stress of $k = 80$ N/mm² plus an additional 1.5 mm to allow for corrosion. The density of mercury is $\rho_m = 13{,}600$ kg/m³ and its surface tension $\gamma_m = 0.54$ N/m (540 dyn/cm).

Solution

Gravity forces and surface forces are involved. The ratio between these two is the same in the model as in the prototype, if

$$\frac{\rho_m L_m^3 g}{\gamma_m L_m} = \frac{\rho_p L_p^3 g}{\gamma_p L_p},$$

or

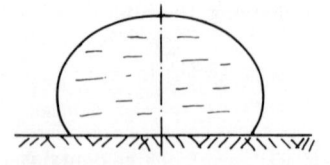

$$\gamma_p = \gamma_m \frac{\rho_p}{\rho_m}\left(\frac{L_p}{L_m}\right)^2.$$

Figure 2.9 Shape of a mercury drop.

Figure 2.10 Cross section through a steel plate. Force $\gamma_p = 0.36$ MN/m perpendicular to the paper.

With $(L_p/L_m)^3 = (1000)(1000^3)/37$, then

$$\gamma_p = 0.54 \frac{1000}{13,600} \left(\frac{1000^4}{37}\right)^{2/3}$$

$$= 357,600 \text{ N/m} \quad \text{or} \quad 0.36 \text{ MN/m}.$$

The force γ_p is the tensile stress of a plate s metre thick (Figure 2.10).

$$k = \frac{\gamma_p}{(1)(s)} = \frac{(0.36)(10^6)}{s} = (80)(10^6) \text{ N/m}^2$$

$s = 0.30/80 = 0.0045$ m = 4.5 mm
Allowance for corrosion   1.5 mm

Plate thickness   6.0 mm

Problems

2.1. The equation $f(P, D, n, \rho, \mu) = 0$ is valid for certain cases of agitation. Here P is the power consumption, D the characteristic length, n the number of revolutions per unit time, ρ the density, and μ the dynamic viscosity. Replace the equation above with an equation relating two independent, dimensionless groups. What requirements must be fulfilled to make the equation valid?

2.2 A model of a submarine (length ratio 1/5) is to be tested in a wind tunnel instead of in water. What air velocity must be used in the wind tunnel in order to simulate the submerged vessel moving at a velocity of 6 knots (3.1 metres per second)?

	Air	Water
Density, kg/m^3	1.23	1000
Viscosity, cP	0.0175	0.95

2.3 A steel bridge weighing 8 tons carries an evenly distributed load of 20 tons. An aluminium model of 1/12 scale is used. What evenly distributed load should be used on the model to simulate the deformation of the bridge?

	Steel	Aluminium
Density, kg/m^3	8,000	2,700
Modulus of elasticity, E MN/m^2	200,000	70,000

The deformations are assumed to be elastic within the validity of Hooke's law,

$$F = EA(\Delta L/L)$$

where A is the area, L the length, ΔL the elongation, and F the force.

2.4 The heat transfer coefficient, h, in a fluidized bed is to be determined experimentally as a function of the properties of the fluid, k, ρ, μ, the superficial velocity V_0 and the particle size D. Write a relation between these quantities in terms of dimensionless groups.

References

1 Hellums, J. D., and S. W. Churchill: Dimensional analysis and natural circulation, *Chem. Eng. Progress Symposium Series*, **57**, Nr. 32, 75–80 (1961).
2 Hellums, J. D., and S. W. Churchill: Simplification of the mathematical description of boundary and initial value problems, *A.I.Ch.E. Journal*, **10**, 110–114 (1964).

3 Johnstone, R. E., and M. W. Thring: *Pilot Plants, Models, and Scale-up in Chemical Engineering*, McGraw-Hill Book Co., New York, 1957. (This reference gives a more detailed treatment of scale-up.)
4 Boucher, D. F., and G. E. Alves: Dimensionless numbers, *Chem. Eng. Progress*, **55**, 55–64 (1959). (This reference contains 59 literature references.)
5 Pawlowski, J.: *Die Ähnlichkeitstheorie der physikalisch-technischen Forschung*, Springer-Verlag, Berlin, 1971.

CHAPTER 3
Flow Measurements

Nozzles and orifices

A nozzle and an orifice are shown in Figures 3.1 and 3.2. These are the most common devices for flow measurements in pipes and tubes. Orifices are the simplest to make, but give more permanent pressure drop and are more sensitive to wear than nozzles. Hence, they are usually used in temporary installations.

For frictionless flow from a point 1 in a pipe in front of a nozzle or orifice in a horizontal pipe to a point 2 behind the measuring plate,

$$\frac{p_1}{\rho_1} + \frac{V_1^2}{2} = \frac{p_2}{\rho_2} + \frac{V_2^2}{2}. \tag{3.1}$$

The velocity of an incompressible fluid ($\rho_1 = \rho_2 = \rho$), is

$$V_1 = mV_2$$

where the cross section ratio is

$$m = \left(\frac{D_2}{D}\right)^2, \tag{3.2}$$

D_2 is the diameter of the opening in the measuring plate, and D is the pipe diameter. From equation (3.1),

$$V_2 = \sqrt{\frac{1}{1-m^2}} \sqrt{2(p_1 - p_2)/\rho} \ \ \text{m/s}. \tag{3.3}$$

This equation neglects the increase in static pressure in front of the measuring plate, contraction of the jet (D_2' Figure 3.2 is smaller than D_2), and friction loss. If the length of straight pipe in front of and behind the measuring plate is large, and the measuring plates are geometrically similar, the corrections for these factors are all functions of the ratio between the inertia forces and the viscous forces and of the geometrical ratios expressed by m. This gives the modified equation

$$w = \alpha S_2 \sqrt{2\rho(p_1 - p_2)} \ \ \text{kg/s} \tag{3.4}$$

$$\alpha = f(\text{Re}, m) \tag{3.5}$$

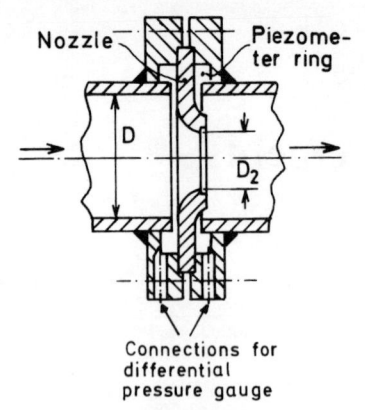

Figure 3.1 Nozzle with piezometer rings and curved inlet.

The coefficient α also includes $(1 - m^2)^{-1/2}$ in equation (3.3). S_2 is the cross section of the opening in the measuring plate, $S_2 = (\pi/4)D_2^2$.

The function described by equation (3.5) is given in Figure 3.3 for nozzles and in Figure 3.4 for orifices with dimensions in accordance with the German standard DIN 1952.[1] The error limits in the coefficient α increase with increased opening ratio m, with increased roughness of the pipe, and with reduced length of straight pipe in front of and after the measuring plate. Recommended minimum lengths of straight pipe between a 90° bend and a measuring plate are given in Table 3.1.

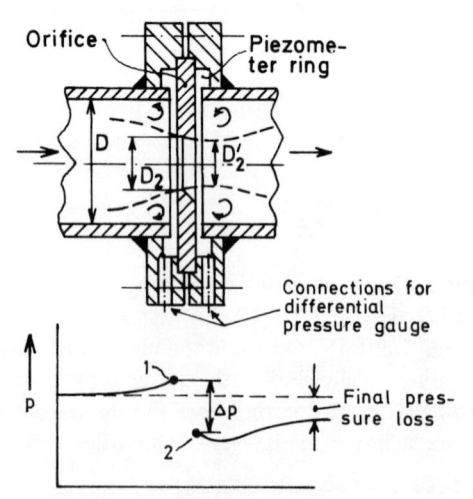

Figure 3.2 Orifice with piezometer rings. Below: static pressure measured in front of, and behind, the orifice. Direction of flow is *towards* the sharp edge.

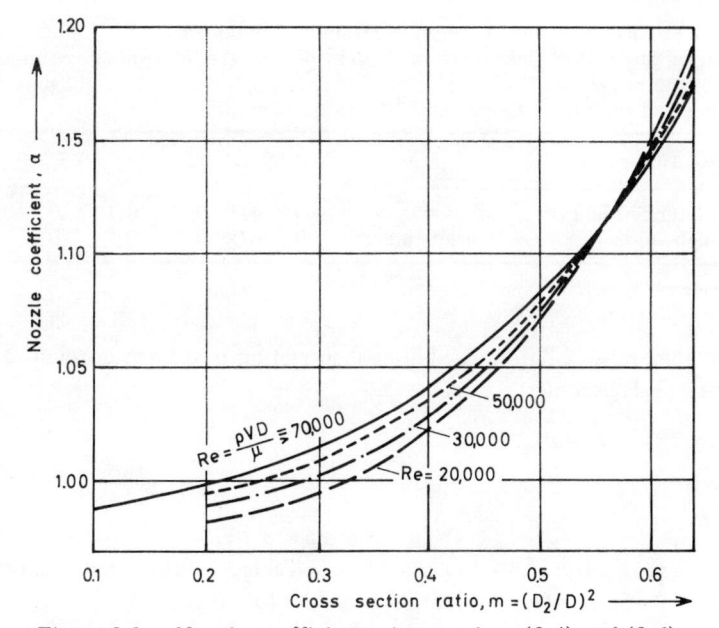

Figure 3.3 Nozzle coefficient α in equations (3.4) and (3.6) as a function of the cross section ratio m according to DIN 1952. Reynolds number refers to the pipe diameter. The diagram is valid for pipe diameters between 50 and 500 mm.
Reproduced by permission of Deutsches Institut für Normung eV.

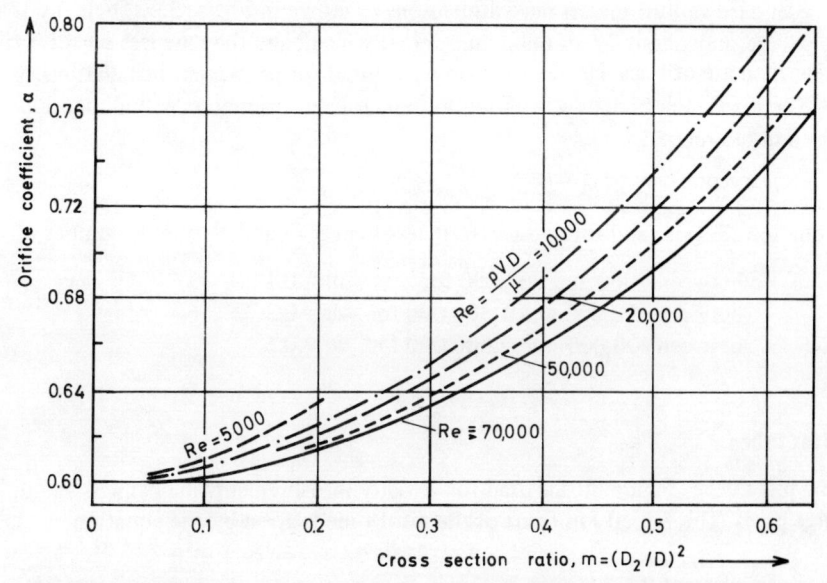

Figure 3.4 Coefficient α in equations (3.4) and (3.6) for orifices as a function of the opening ratio m according to DIN 1952. Reynolds number refers to the pipe diameter. The figure is valid for pipe diameters from 50 to 1000 mm.
Reproduced by permission of Deutsches Institut für Normung eV.

Table 3.1 Recommended length of straight pipe between a 90° bend and a measuring plate. Half this length of straight pipe at the inlet side gives an additional error in the nozzle coefficient α of ±0.5%. Half the length of straight pipe on both the inlet and the outlet side give additional error ±1.0%.*

Cross section ratio m		0.1	0.2	0.3	0.4	0.5	0.6	0.64
Recommended length of straight pipe	inlet side	10D	14D	16D	20D	28D	40D	46D
	outlet side	5D	6D	6D	7D	7D	8D	8D

*Reproduced by permission of Deutsches Institut für Normung eV.

For compressible fluids, an additional correction factor ϵ, is introduced, and equation (3.4) becomes

$$w = \epsilon \alpha S_2 \sqrt{2\rho(p_1 - p_2)} \tag{3.6}$$

where

$$\epsilon = f(p_2/p_1, m, \kappa = c_p/c_v). \tag{3.7}$$

Some values of ϵ from DIN 1952 are given in Table 3.2 for nozzles. For orifices with $m < 0.55$, $(1 - \epsilon)$ is one half of the values for nozzles.

$$(1 - \epsilon)_{\text{orifice}} = 0.5(1 - \epsilon)_{\text{nozzle}} \pm 0.5\% \tag{3.8}$$

Venturi meters

Standard venturi meters have dimensions as shown in Figure 3.5. Their friction losses are only about 2% of maximum velocity head, and they are less sensitive to wear than are orifices. Hence, they are well suited for permanent installations.

The nozzle coefficient α in equation (3.4) for venturi meters with dimensions as given in reference 1 is

$$\alpha = 0.993 + 0.68 \, m^{2.75}. \tag{3.9}$$

Equation (3.9) is valid for $m = (D_2/D)^2$ less than 0.77 and Reynolds number

Re between 7000 and 700,000 for $m = 0.1$
between 130,000 and 1,300,000 for $m = 0.3$
between 200,000 and 2,000,000 for $m = 0.6$.

Pitot tubes

The pitot tube (Figure 3.6) is used for velocity measurements in channels and in large pipes. The velocity in front of the pitot tube is given by the equation

$$V = \alpha \sqrt{2\Delta p/\rho} \tag{3.10}$$

where Δp is the pressure difference and α, the correction factor, is 0.99 ± 0.01 when the length of straight pipe in front of the pitot tube is at least 50D.

Table 3.2 Expansion coefficient ϵ for standard nozzles (DIN 1952). ϵ for other values of p_2/p_1, κ, and m^2 can be found by linear interpolation. For $p_2/p_1 = 1.0$, $\epsilon = 1.0$.*

$p_2/p_1 =$		0.98	0.94	0.90	0.85	0.80	0.75
m	m^2	\multicolumn{6}{c}{ϵ for $\kappa = c_p/c_v = 1.3$}
0	0	0.988	0.965	0.941	0.910	0.878	0.846
0.316	0.1	0.987	0.960	0.933	0.899	0.865	0.829
0.447	0.2	0.985	0.954	0.924	0.886	0.848	0.810
0.548	0.3	0.982	0.947	0.912	0.870	0.828	0.788
0.633	0.4	0.979	0.937	0.897	0.850	0.804	0.760
		\multicolumn{6}{c}{ϵ for $\kappa = c_p/c_v = 1.4$}
0	0	0.989	0.967	0.945	0.916	0.887	0.856
0.316	0.1	0.988	0.963	0.938	0.906	0.873	0.840
0.447	0.2	0.986	0.957	0.929	0.893	0.858	0.822
0.548	0.3	0.983	0.950	0.918	0.878	0.839	0.800
0.633	0.4	0.980	0.941	0.904	0.859	0.815	0.773
		\multicolumn{6}{c}{ϵ for $\kappa = c_p/c_v = 1.66$}
0	0	0.991	0.972	0.953	0.929	0.903	0.877
0.316	0.1	0.990	0.969	0.947	0.920	0.892	0.863
0.447	0.2	0.988	0.964	0.939	0.909	0.878	0.846
0.548	0.3	0.986	0.958	0.930	0.895	0.861	0.827
0.633	0.4	0.983	0.950	0.918	0.878	0.840	0.802

*Reproduced by permission of Deutsches Institut für Normung eV.

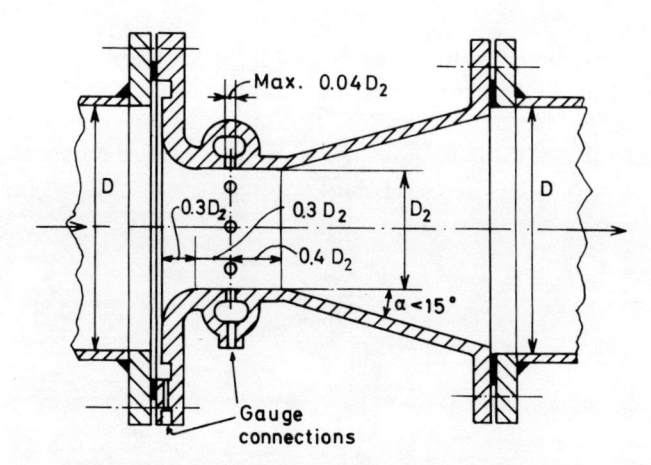

Figure 3.5 Standard venturi meter for pipes with diameters D from 50 to 500 mm for liquids, and from 65 to 500 mm for gases.

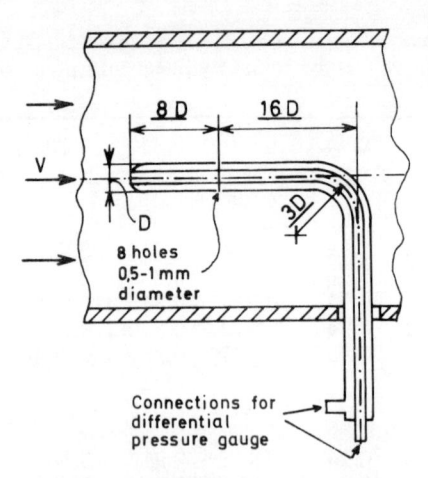

Figure 3.6 Pitot tube with outer diameter D.

For gases with velocities greater than 60 metres per second and less than 0.7 times the velocity of sound, the equation for velocity becomes

$$V = \sqrt{\frac{2\kappa}{\kappa - 1}\left(\frac{p_1}{p_2}\right)\left[\left(\frac{p_1}{p_2}\right)^{(\kappa - 1)/\kappa} - 1\right]} \pm 1\% \qquad (3.11)$$

where $\kappa = c_p/c_v = 1.4$ for air, p_1 is the pressure in the inner tube, p_2 is the static pressure measured in the annular space of the pitot tube, and ρ_2 is the fluid density at pressure p_2.

Rotameters

The rotameter (Figure 3.7) has a float placed in a slightly tapered tube. The float is heavier than the fluid it replaces. It can have grooves that make it rotate in the fluid stream. The elevation of the float depends on the flow velocity and on the density and viscosity of the fluid. Rotameters require individual calibration.

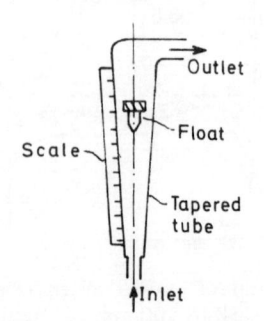

Figure 3.7 A rotameter.

Standard rotameters are delivered for connection to pipes with diameters from 3 to 150 mm. The corresponding capacities for water are from 0.0003 m^3/h to 200 m^3/h. The corresponding values for the flow of air at atmospheric pressure and room temperature are 0.005 m^3/h and 17,000 m^3/h respectively.

Magnetic flowmeters

Magnetic flowmeters are available for pipe sizes from 2.5 mm to 2.5 m for liquids or slurries with some electrical conductivity. Standard systems are rated for liquids with conductivity down to 0.5 mS/m and special systems down to 10 μS/m. (The conductivity of water is usually of the order of millisiemens per metre; siemens = ampere per volt.)

The liquid flows through an electromagnetic field with two electrodes inserted through the walls of the pipe at right angles to the magnetic field. The voltage at the electrodes is proportional to the flow rate. The instrument is obstructionless with accuracy ±1%. It is widely used for water and wastewater applications, and in many areas of the pulp and paper, mining, food, petroleum, and power industries.

Other types of flowmeters are cup anemometers, hot wire anemometers,[2] elbow meters, and gas meters based on volumes trapped between a sealing liquid and vanes. For further details see references 2–4.

EXAMPLE 3.1. MEASURING NOZZLE FOR LIQUID: LIMITS OF ERROR

A refrigeration plant has a condenser cooled by sea water. The volume of sea water flowing through the condenser is measured with a nozzle in the pressure line from the sea water pump, as shown in Figure 3.8. The sea water pipe is of mild steel with an inner diameter D = 102 mm. The rate of flow is calculated to be 60 m^3/h.

Density of sea water, ρ = 1026 kg/m^3.
Viscosity of sea water, μ = 1.7 cP.
Density of mercury, ρ_{Hg} = 13,600 kg/m^3.

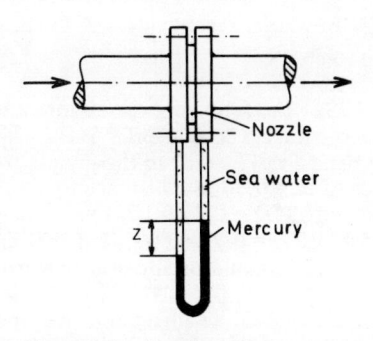

Figure 3.8 Nozzle with mercury differential gauge.

(a) Calculate a suitable nozzle diameter when the difference z measured on the mercury gauge is to lie between 45 and 55 mm for a flow rate 60 m^3/h.

(b) The diameter of the nozzle measured after it was made was 75.15 mm. Find the rate of flow in m^3/h when z equals 42 ± 1 mm.

(c) Estimate the tolerance limits for the calculated rate of flow.

See VDI-Durchflussmessregeln, DIN 1952, and Table 1.1.

Solution

(a) Reynolds number,

$$\text{Re} = \frac{(1026)(0.102)}{(1.7)(10^{-3})} \frac{60}{(3600)(\pi/4)(0.102^2)} = 125{,}600.$$

The required gauge reading, $z \approx 50$ mm, hence

$$\Delta p = 0.05(13{,}600 - 1026)\,9.82 = 6174 \text{ N/m}^2.$$

The diameter D_2 is obtained by a trial and error procedure with values of α obtained from Figure 3.3. Equation (3.4) with $\alpha = 1.1$ gives

$$w = \frac{(60)(1026)}{3600} = 17.1 = 1.1 S_2 \sqrt{(2)(1026)(6174)},$$

hence $S_2 = 0.0044$ m^2 ($D_2 = 0.0748$ m) and $m = (74.8/102)^2 = 0.538$.
From Figure 3.3 at this value of m and Re = 125,600, $\alpha = 1.103$ or
$S_2 = 0.00436$ m^2 ($D_2 = 0.0745$).
The design diameter $D_2 = 74.5$ mm.

(b) The cross section ratio is almost as calculated by the first trial, hence $\alpha = 1.10$.

Multiplication of equation (3.4) by $3600/\rho$ gives the flow rate of sea water $\dot{Q}$ in m^3/h,

$$\dot{Q} = (3600)(1.10)\frac{\pi}{4}(0.07515^2)\sqrt{\frac{(2)(0.042)(13600 - 1026)9.82}{1026}}$$

$$= 55.8 \text{ m}^3/\text{h}.$$

One millimetre error in the reading of z corresponds to $100(1/42) = 2.4\%$ error in z, or approximately 1.2% error in $\sqrt{z}$ and in $\dot{Q}$. This gives an error of $(0.012)(55.8) = 0.7$ m^3/h, or

$$55.1 < \dot{Q} < 56.5 \text{ m}^3/\text{h}.$$

(c) According to DIN 1952[1] the following corrections and tolerance limits should be taken into account for the nozzle coefficient.

(1) Disturbances from bends, valves, etc., in the pipe in front of and behind the nozzle. For instance, if $m = 0.54$, the distance to a 90° bend in front of the nozzle should be at least $34D = (34)(0.102) = 3.5$ metres, and after the nozzle at least $7.5D = (7.5)(0.102) = 0.77$ metre. Half that distance on both sides gives limits of error ±1.0%. This should be added to the error limits calculated below.

(2) The nozzle coefficient α should be adjusted if the pipe roughness, ϵ/D is greater than 1/1400. From Table 1.1, $\epsilon/D = 0.045/102 = 1/2266$. Hence, no correction for pipe roughness need be made in this case.

(3) Without corrections for pipe roughness and with $m = 0.2$ to 0.64, the tolerance limit (reference 1, section 4.5) is

$$\tau_\alpha = \pm 0.5[1 + 3m^2 + (\log \text{Re} - 6)^2 + 0.05/D]\%$$

where τ_α is twice the standard deviation for α, i.e. within 95% probability. With $m = 0.54$,

$D = 0.102$ metre,

$$\text{Re} = \frac{55.8}{60}(125,600) = 116,800.$$

This gives

$$\tau_\alpha = \pm 0.5\left[1 + (3)(0.54^2) + (5.0675 - 6)^2 + \frac{0.05}{0.102}\right]$$

$$= \pm 1.62\%.$$

In addition 1.2% uncertainty enters due to errors in the height z, giving a total tolerance limit of

$$\tau = \pm\sqrt{1.62^2 + 1.2^2}$$

$$= \pm 2.0\%.$$

Assuming no other sources of error (such as air bubbles in the water), the flow rate is

$\dot{Q} = 55.8 \pm (0.02)(55.8)$

$= 55.8 \pm 1.1 \text{ m}^3/\text{h}.$

EXAMPLE 3.2. ORIFICE METER FOR VAPOUR

Figure 3.9 shows the arrangement of an orifice meter with a mercury differential gauge in a steam pipe line where the 'Zeppeliners'[*] ensure the same level of condensate above each leg of the differential gauge. The steam pressure in front of the orifice is 7.5 bar or 0.75 MN/m^2, and the temperature $175°$C. The pipe diameter $D = 89$ mm and the diameter of the orifice $D_2 = 33.3$ mm. The scale on the differential gauge gives the rate of steam flow, w, in kg per hour. Establish the height h in mm for different values of w, where h ranges from 20 to 750 mm.

See DIN 1952, steam tables and Mollier diagram for steam. The viscosity of steam at $175°$C is $\mu = 0.0156$ cP.

Solution

At a pressure of 7.5 bar and a temperature of $175°$C, the vapour density of steam $\rho = 3.85$ kg/m^3. Figure 3.4 shows that the coefficient α in equation (3.6) is independent of velocity for Reynolds number Re greater than 70,000. The velocity V_1 at Re = 70,000 is obtained from

$$70,000 = \frac{3.85 V_1 0.089}{0.0000156},$$

$V_1 = 3.19$ m/s

---

[*]Named after the designer of Zeppelin airships, the German Count Ferdinand von Zeppelin, who made his first balloon ascent as a volunteer in the Federal army during the American Civil War.

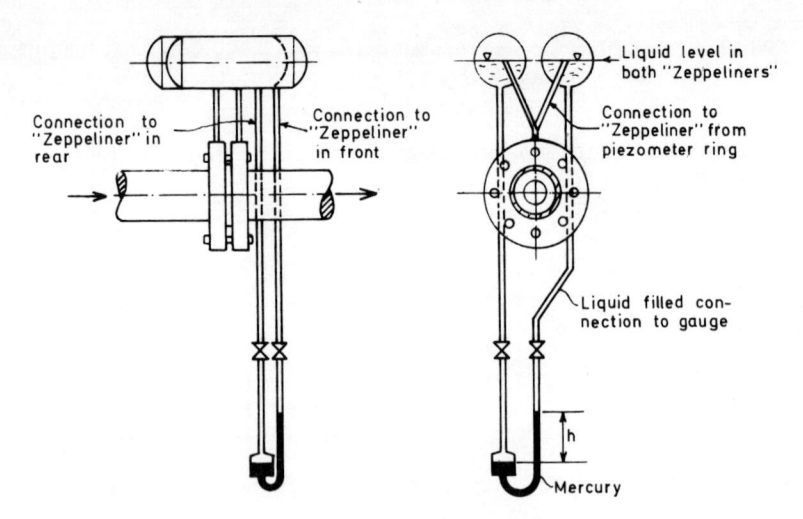

Figure 3.9 An orifice meter in a steam pipe.

and the mass flow rate

$$w = (3600)(3.19)(3.85)(\pi/4)(0.089^2) = 275 \text{ kg/h}.$$

With a cross section ratio $m = (33.3/89)^2 = 0.14$ and $\text{Re} > 70{,}000$ (Figure 3.4), $\alpha = 0.605$.

The expansion coefficient ϵ depends on the cross section ratio m, the pressure ratio p_2/p_1, and the ratio of the heat capacities, $\kappa = c_p/c_v$. In the region of interest, κ is found from equation (4.10) with pressures and temperatures read from a constant entropy line on a Molliere diagram. With $p_1 = 0.75$ MN/m^2 and $\theta_1 = 175°$C as one point and $p_0 = 1.0$ MN/m^2 as the other point, the resultant temperature at constant entropy is $\theta_0 = 206.5°$C:

$$\frac{273.2 + 206.5}{273.2 + 175} = \left(\frac{1.0}{0.75}\right)^{(\kappa-1)/\kappa}$$

or

$$\kappa = 1.31.$$

With h in mm and condensate of density 1000 kg/m^3, the pressure difference in the mercury gauge is

$$p_1 - p_2 = \frac{h}{1000}(13{,}600 - 1000)9.82$$
$$= 123.7h \text{ N/m}^2.$$

This means that for $w > 275$ kg/h [equation (3.6)], then

$$w = 3600\epsilon 0.605(\pi/4)0.0333^2\sqrt{(2)(3.85)123.7h}$$
$$= 58.5\epsilon\sqrt{h} \text{ kg/h} \qquad (a)$$

Table 3.3

h mm	20	100	300	500	750	Notes
$\dfrac{p_1 - p_2}{p_1}$	0.0033	0.0165	0.0495	0.0825	0.1237	$\dfrac{p_1 - p_2}{p_1} = \dfrac{123.7h}{750,000}$
$\dfrac{p_2}{p_1}$	0.9967	0.9835	0.9505	0.9175	0.8763	$\dfrac{p_2}{p_1} = 1 - \dfrac{p_1 - p_2}{p_1}$
ϵ	0.999	0.994	0.982	0.971	0.953	DIN 1952 or equation (3.8) and Table 3.2.
w kg/h	261	581	995	1270	1527	$w = 58.5\epsilon\sqrt{h}$
$(w/100)^2$	6.81	33.8	99	161	233	

The mass flow rates w calculated from equation (a) are given in Table 3.3. It shows that ϵ decreases almost linearly with h or with w^2 (Figure 3.10). Table 3.4 gives the height h in mm for w in the range from 300 to 1500 kg/h.

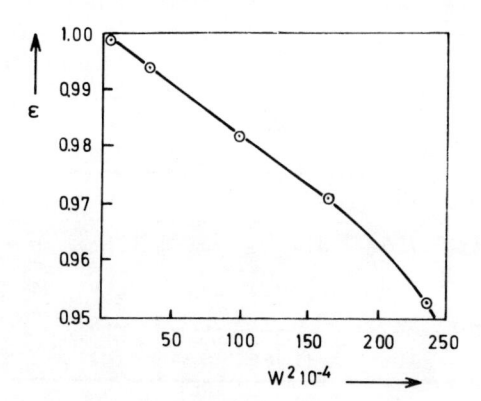

Figure 3.10 Expansion coefficient ϵ plotted as a function of the flow rate squared, w^2 (kg/h)^2 (points from Table 3.3).

Table 3.4

w kg/h	300	400	500	600	700	800	900	Notes
$(w/100)^2$	9	16	25	36	49	64	81	
ϵ	0.998	0.998	0.996	0.994	0.991	0.988	0.985	from Figure 3.10
h mm	26.4	46.9	73.6	106.5	145.8	191.6	244	$h = (w/58.5\epsilon)^2$

w kg/h	1000	1100	1200	1300	1400	1500	Notes
$(w/100)^2$	100	121	142	169	196	225	
ϵ	0.982	0.979	0.975	0.970	0.965	0.956	from Figure 3.10
h mm	303	369	443	525	615	719	$h = (w/58.7\epsilon)^2$

EXAMPLE 3.3. VELOCITY MEASUREMENTS WITH A PITOT TUBE IN A RECTANGULAR DUCT

Air at a pressure of 780 mm mercury and temperature 80°C flows through a duct of square cross section with area $(0.3)(0.3) = 0.09$ m². Measurements with a pitot tube at points 1 to 4 (Figure 3.11) gave the results tabulated:

Point number	1	2	3	4
Pressure difference, mm water gauge	7.5	11.5	15.5	13.5

Find the average air velocity in the duct.

Solution

Air density,

$$\rho = \frac{Mp}{RT} = \frac{(29)(0.78)(13{,}600)(9.82)}{(8314)(273+80)} = 1.03 \text{ kg/m}^3.$$

Insertion into equation (3.10) with the correction factor $\alpha = 0.99$ gives

$$V = 0.99\sqrt{2\Delta p/1.03}$$
$$= 1.38\sqrt{\Delta p}. \qquad (a)$$

Average air velocity,

$$V_{av} = \frac{1}{A}\int_0^A V\,dA. \qquad (b)$$

Point number	1	2	3	4	Notes
Pressure difference Δp N/m²	73.7	113	152	133	Δp = (mm water gauge) × 9.82
Velocity V m/s	11.8	14.7	17.0	15.9	$V = 1.38\sqrt{\Delta p}$
Distance from centre x m	0.14	0.12	0.02	0.08	
Vx	1.65	1.76	0.34	1.27	

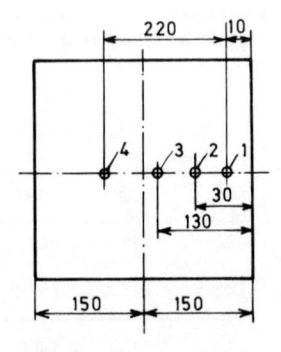

Figure 3.11 Rectangular duct: pitot tube measurements at the points 1, 2, 3, and 4. The dimensions are in mm.

A is the cross sectional area of the duct, $A = L^2 = 0.3^2 = 0.09$ m^2. The velocity profile is assumed to be symmetrical. That is, the velocity is assumed to be the same at all points equidistant from the closest wall. Hence, dA in equation (b) can be substituted by $4(2x) dx$ where x is the distance from the centre as shown in Figure 3.12. Equation (b) can be integrated,

$$V_{av} = \frac{8}{L^2} \int_{x=0}^{x=L/2} Vx \, dx. \tag{c}$$

V and Vx are calculated in the table on page 66 and plotted in Figure 3.12.

Equation (c) is solved graphically as $\int_{x=0}^{x=L/2} Vx \, dx$ is the area below the curve Vx in Figure 3.12. With the scales used in Figure 3.12 this area is 18 cm^2.

1 cm on the abscissa corresponds to $x = 0.03$ m,
1 cm on the ordinate corresponds to $Vx = 0.3$ m^2/s,

i.e. 1 cm^2 in the diagram corresponds to $(0.03)(0.3) = 0.009$ m^3/s, and 18 cm^2 corresponds to $(0.009)(18) = 0.162$ m^3/s. The latter value inserted into equation (c) with $L = 0.3$ m gives

$$V_{av} = \frac{8}{0.3^2} 0.162 = 14.4 \text{ m/s}.$$

Problems

3.1. The rate of water flow ($\rho = 1000$ kg/m^3, $\mu = 0.0009$ N s/m^2) through a pipe is to be measured by means of a standard orifice meter. The inner diameter of the pipe $D = 50$ mm. The rate of flow can be varied between 1.5 m^3/h and 12 m^3/h.

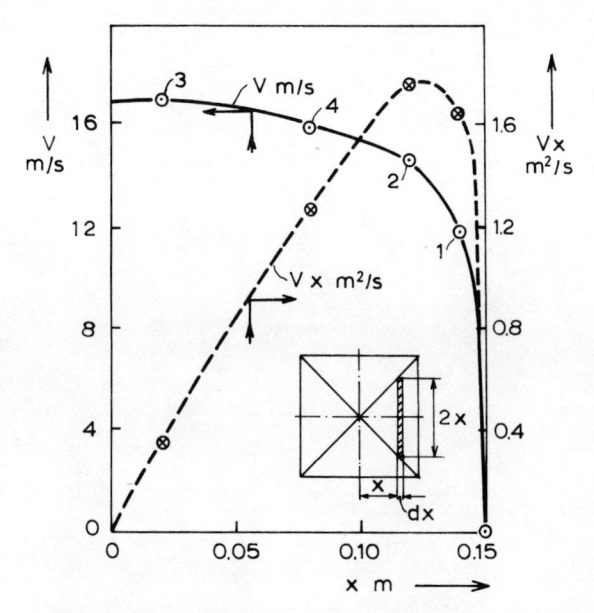

Figure 3.12 Velocity V and product Vx plotted as a function of the distance x.

The pressure difference across the orifice is to be measured by means of a differential gauge, and the difference between the liquid levels in the gauge should be a minimum of 20 mm and maximum 200 mm.

(a) Why is it readily evident that more than one orifice is needed?

(b) Determine the diameter of the largest orifice when it is to give a reading of 200 mm in a regular mercury differential gauge at maximum rate of flow.

(c) Suggest an arrangement of differential gauge(s) that makes it possible to use only one orifice over the entire flow range.

3.2 A rotameter tube is 300 mm long with an internal diameter of 20 mm at the bottom and 25 mm at the top. The float weighs 4.0 grams (density 8 g/cm^3) and has a diameter of 19.5 mm. At a flow rate of 140 litres of water per hour, the float is 180 mm above the bottom of the rotameter tube.

(a) Calculate the discharge coefficient for the flow past the float.

(b) Assuming the same discharge coefficient, calculate the flow rate for water, when the float is at its top position.

References

1 *DIN 1952, Durchflussmessung mit genormten Düsen, Blenden und Venturidüsen*, Beuth-Vertrieb GmbH, Berlin W15, 1969.
2 Dean jr., R. C.: *Aerodynamic Measurement*, Gas Turbine Laboratory, MIT, Boston, Massachusetts, 1956.
3 Herning, F.: *Grundlagen und Praxis der Durchflussmessung*, VDI-Verlag, Düsseldorf, 3rd Edn., 1967.
4 Ower, E.: Measurement of the flow of liquids and gases, *Trans. Instn. Chem. Eng.*, **18**, 47 (1940).

CHAPTER 4
Pumping, Compression, and Expansion

Pumping

An energy balance per unit mass for the process indicated within the broken lines of Figure 4.1 is

$$U_1 + p_1 v_1 + \frac{V_1^2}{2} + z_1 g + q = U_2 + p_2 v_2 + \frac{V_2^2}{2} + z_2 g + w + \Delta E \tag{4.1}$$

where

U = internal energy, J/kg = N m/kg
p = pressure, Pa = N/m^2
v = specific volume per unit mass, m^3/kg
V = velocity, m/s
g = acceleration due to gravity, m/s^2
z = height above horizontal reference level, m
q = heat added, J/kg = N m/kg
w = mechanical work, N m/kg = J/kg
ΔE = accumulated energy, in the following assumed to be zero.

The term pv represents the flow work involved when the fluid is displaced through a volumetric space v against the restraint of pressure p, $V_1^2/2$ or $mV_1^2/2$ (where m is the mass = 1 kg) is the kinetic energy, and (zg) or (mzg) is the potential energy.

Pumping of an incompressible fluid can be calculated conveniently by introduction of the density $\rho = 1/v$ kg/m^3. Furthermore if the term q in equation (4.1) is zero, the change in internal energy, $U_2 - U_1$, can be replaced by $\Delta p_f / \rho$ in equation (4.1). Equation (4.1) then becomes

$$p_1 + \rho \frac{V_1^2}{2} + z_1 \rho g = p_2 + \rho \frac{V_2^2}{2} + z_2 \rho g + \rho w + \Delta p_f \tag{4.2}$$

or

$$(-w) = \frac{\Delta p_f + p_2 - p_1}{\rho} + \frac{V_2^2 - V_1^2}{2} + (z_2 - z_1)g. \tag{4.3}$$

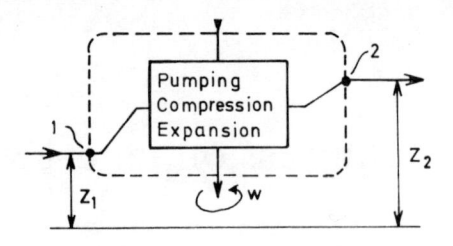

Figure 4.1 Energy balance for flow from point 1 to point 2. q is the heat added and w the mechanical work done.

In some cases the kinetic energy term can also be neglected. Equation (4.3) then becomes

$$(-w) = \frac{\Delta p}{\rho} + (z_2 - z_1)g \tag{4.4}$$

where $\Delta p/\rho = (\Delta p_f + p_2 - p_1)/\rho$ and w are expressed in N m/kg, J/kg or W s/kg.

In pumping $\dot{Q}$ m^3/s or $\dot{Q}\rho$ kg/s, the theoretical power consumption according to equation (4.4), is

$$-P_{theor} = \dot{Q}\rho(-w) = \dot{Q}[\Delta p + (z_2 - z_1)\rho g] \text{ N m/s} \tag{4.5}$$

where the figures are to be divided by 1000 to give kW, by 735.5 to give metric horsepower, or by 745.7 to give British horsepower. The pumping efficiency η is the ratio of theoretical to real power consumption. The efficiency η varies from 0.3 for the smallest, up to 0.89 for the largest centrifugal pumps, and from 0.70 to 0.94 for piston, plunger, screw, gear wheel, or membrane pumps (Figures 4.2 to 4.7). It is good engineering practice to install a motor having power 15 to 25% greater than the calculated requirement. The motor driving the centrifugal pump should in general be chosen large enough to operate the pump at a back pressure corresponding to its maximum power consumption.

Centrifugal pumps

Because of their simplicity, low initial cost, uniform flow, quiet operation and low maintenance costs, centrifugal pumps are the most widely used in the chemical industry. They are available in different sizes and designs, with capacities ranging from 0.5 cubic metre per hour to 6 cubic metres per second.

The rotating vanes of the impeller force the liquid through the pump. Centrifugal force together with velocity head at the outlet of the impeller are converted to pressure head in the volute. The impeller vanes can be open or shrouded with a cover plate as shown in Figure 4.2. Figure 4.3a shows the characteristic performance

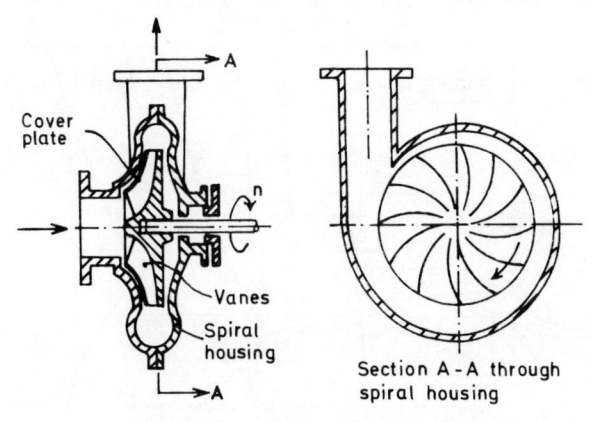

Figure 4.2   A centrifugal pump.

of a centrifugal pump with a shrouded impeller, and Figure 4.3b shows efficiencies attained at various 'specific speeds' n_s by single-stage, single-suction, impellers for different capacities Q. Multi-stage pumps are used to obtain higher pressure heads. For limited ranges, the liquid head is roughly proportional to the square of the peripheral speed of the impeller.

Centrifugal pumps are constructed of cast iron, brass, stainless steel, hastelloy, porcelain, etc. For pumping of sea water a stainless steel shaft is recommended, with an impeller in sea water-resistant brass or with rubber lining, and cast iron volute.

The shaft seal is the most vulnerable part of a centrifugal pump. This is particularly the case for pumps running at high speeds, or for those transporting liquids with solid particles in suspension. The shaft seal may be omitted in submerged pumps and in canned-motor pumps. The latter have the impeller and motor rotor in the same housing. The bearings in a canned-motor type are immersed in the liquid transported by the pump, or in vapour of this liquid. This type of centrifugal pump is widely used for pumping organic solvents, organic heat-transfer liquids, light oils, refrigerants and other particle-free liquids where it is important that leakage be avoided.

Positive displacement pumps

Positive displacement pumps contain various devices for propelling fluids, namely reciprocating pistons, plungers, diaphragms, rotating gear wheels, screws or special rotating pistons. These pumps are used to provide liquid heads higher than those obtainable with centrifugal pumps operating at moderate speeds, for metering or for proportioning of liquids, and for transport of highly viscous fluids.

Reciprocating pumps as shown in Figure 4.4 have their main application in the pressure range from 2 to 10 bar. They are also used as metering pumps.

The *plunger pump* has a reciprocating rod with clearance at the cylinder wall as shown in Figure 4.5. It is used as a metering pump and as a high pressure pump

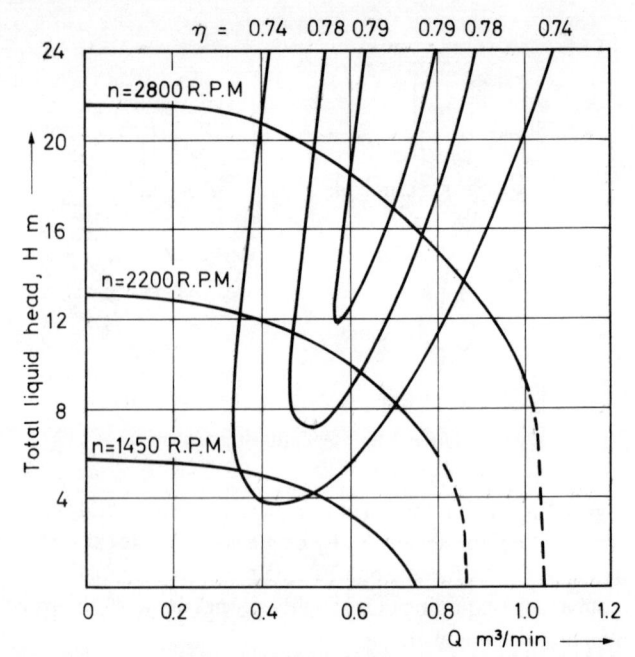

Figure 4.3a Characteristic performance curves (total liquid head H against volumetric rate of flow Q) for a particular centrifugal pump with 125 mm impeller rotating at n = 1450, 2200, and 2800 revolutions per minute. η = efficiency.

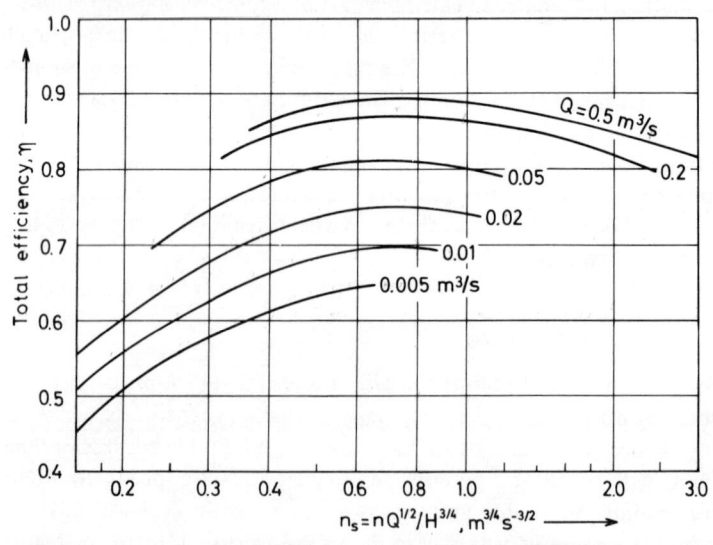

Figure 4.3b Total efficiencies of centrifugal pumps as a function of 'specific speed', n_s m$^{3/4}$/s$^{3/2}$. (From A. J. Stepanoff, *Centrifugal and Axial Flow Pumps*, John Wiley & Sons, Inc., 1967).

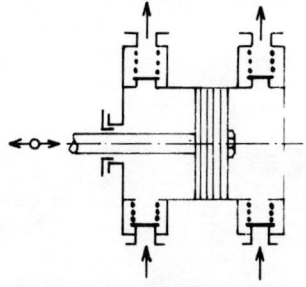

Figure 4.4 Double acting reciprocating pump with spring-loaded valve plates.

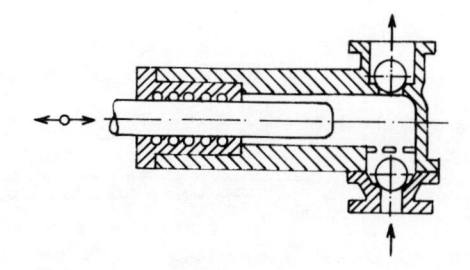

Figure 4.5 Plunger pump with ball valves.

for discharge pressures up to 1000 bar. It is also well suited as a proportioning pump, and it may have several plungers driven from one shaft.

Both piston and plunger pumps may be equipped with an air or gas chamber on the pressure side to give a more even rate of flow in the pressure line.

Screw pumps with two parallel screws in contact with each other as well as *gear pumps* as shown in Figure 4.6 are well suited for pumping lubricants. Gear pumps with one metal and one plastic gear wheel are used for pumping non-lubricating fluids such as refrigerants.

The *membrane pump* (Figure 4.7) is used mainly as a metering pump for liquids

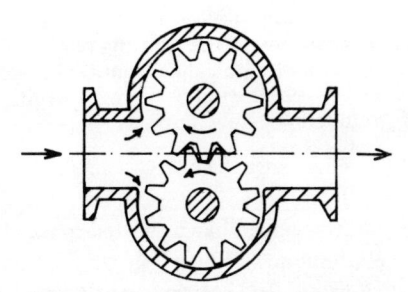

Figure 4.6 A gear wheel pump.

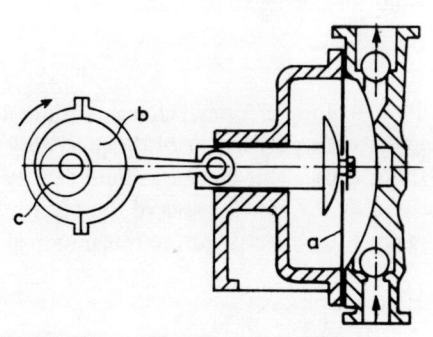

Figure 4.7 Membrane pump with adjustable stroke and ball valves: a membrane; b and c, eccentrics that can be adjusted relative to each other.

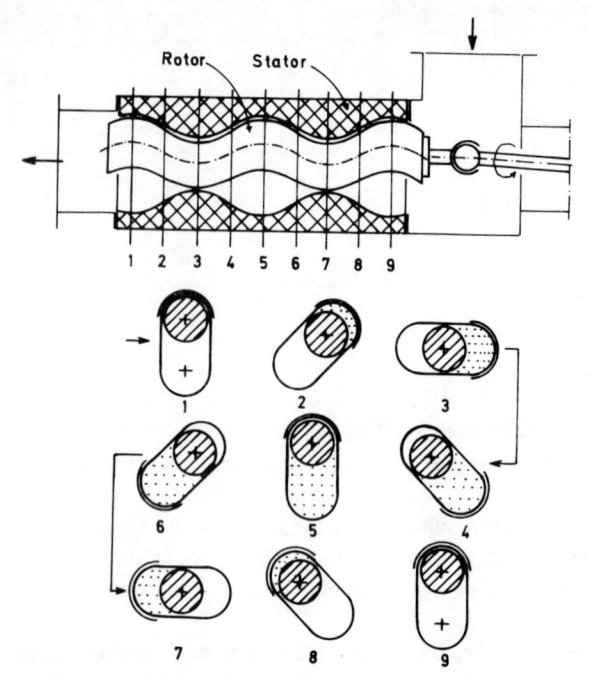

Figure 4.8 A mono pump: the lower drawing shows nine positions of the rotating part of the pump and the elastic housing (from G. Leuschner[1]). Reproduced by permission of Verlag Chemie, GmbH, Weinheim – New York.

where it is important to avoid leakage. The largest size can deliver up to 20 cubic metres per hour.

The *mono pump* has a rotor shaped like a twisted snake (Figure 4.8). It rotates without clearance in an elastomeric stationary sleeve. This pump is used for clear liquids, slurries, and dry powders. The even rate of flow makes this pump especially well suited as a feed pump for slurries to filter presses.

Compression

Compression of a fluid resulting in a small change of fluid density can be calculated with sufficient accuracy by means of the preceding equations for incompressible liquids. For appreciable change in fluid density the work of compression is represented by the hatched area of the pv-diagram shown in Figure 4.9. The corresponding equations resulting from integration along the v- and p-axis are,

$$(-w) = \int_{v_1}^{v_0} (p - p_0) dv + (p_1 - p_0) v_1 \tag{4.6}$$

$$(-w) = \int_{p_0}^{p_1} v \, dp. \tag{4.7}$$

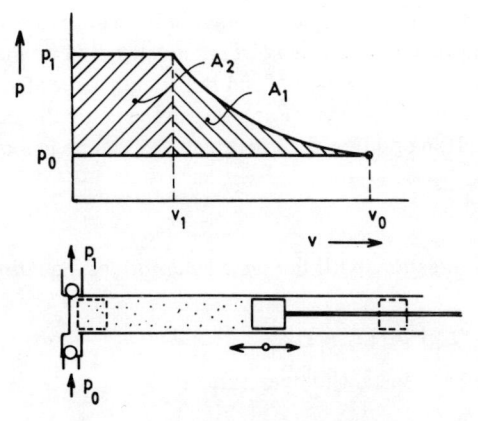

Figure 4.9 Compression work represented by the area $A_1 + A_2$ by continuous compression. A_1 is the compression work by compression from v_0 to v_1 and A_2 is the work associated with the displacement.

In the *isothermal* compression of an ideal gas $v = v_0 p_0/p$, or from equation (4.7),

$$(-w)_T = p_0 v_0 \ln \frac{p_1}{p_0}. \tag{4.8}$$

Ideal adiabatic compression is also called *isentropic* compression to indicate that the entropy remains constant.

For ideal adiabatic compression of an ideal gas $v = v_0(p_0/p)^{1/\kappa}$. Insertion of this into equation (4.7) and integration gives

$$(-w)_s = \frac{-\kappa}{\kappa - 1} p_0 v_0 \left[1 - \left(\frac{p_1}{p_0}\right)^{(\kappa-1)/\kappa}\right] \tag{4.9}$$

where $\kappa = c_p/c_v$. For air $\kappa = 1.4$ and for superheated water vapour $\kappa = 1.3$.

The change in temperature during adiabatic compression of an ideal gas is given by combination of the equations

$$p_1 v_1 = RT_1$$
$$p_0 v_0 = RT_0$$

and

$$p_1 v_1^\kappa = p_0 v_0^\kappa, \text{ i.e.}$$
$$\frac{T_1}{T_0} = \left(\frac{p_1}{p_0}\right)^{(\kappa-1)/\kappa} \tag{4.10}$$

where T is the absolute temperature.

In *polytropic* compression the heat removed is less than in isothermal compression. Polytropic compression is calculated in the same way as adiabatic compression, but with the polytropic exponent n instead of κ, where $1 < n < \kappa$.

The work of compression of a *non-ideal* gas is calculated by means of equation (4.1). The sum of the internal energy U and the energy associated with diplacement pv, is the enthalpy, $h = U + pv$. For gases and vapours the potential energy zg is usually negligible. If in addition the kinetic energy can be ignored, equation (4.1) reduces to

$$h_1 + q = h_2 + w. \tag{4.11}$$

In *isothermal* compression, with the same assumptions, equation (4.11) reduces to

$$(-w)_T = (h_2 - h_1) - T(s_2 - s_1) \tag{4.12a}$$

where the heat added at constant temperature

$$q = -T(s_2 - s_1) \tag{4.12b}$$

The enthalpies h_1 and h_2 and the entropies s_1 and s_2 are taken from thermodynamic diagrams, usually of the types shown in Figures 4.10 and 4.11,[2,3] or they may be estimated.[4]

For *isentropic* compression (*reversible, adiabatic* compression) equation (4.11) may be written as

$$(-w)_s = (h_2 - h_1)_s. \tag{4.13}$$

This gives the theoretical work of compression, $[(h_2)_s - h_1]$ in Figure 4.10.

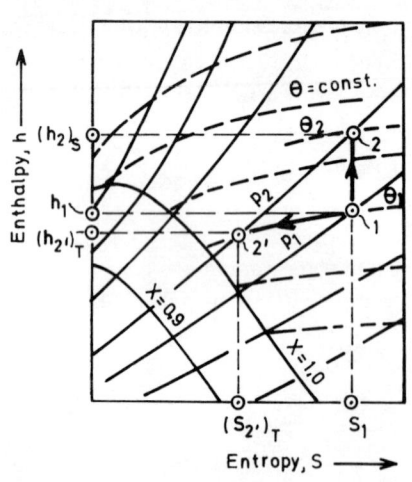

Figure 4.10 Isentropic compression from pressure p_1 and temperature θ_1 (point 1) to pressure p_2 (point 2), and isothermal compression from the same point to the same pressure (point 2') in an enthalpy against entropy diagram. The temperature after isentropic compression is θ_2.

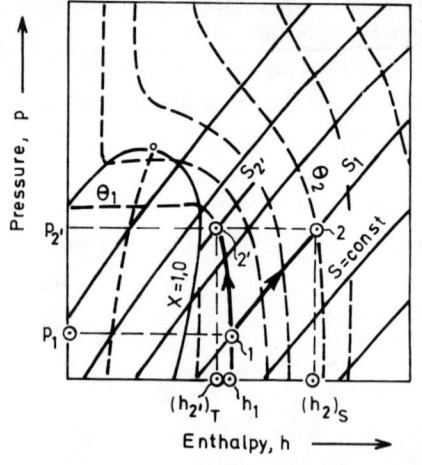

Figure 4.11 The compressions in Figure 4.10, shown in a pressure against enthalpy diagram.

The real work of compression, $(-w)_{real}$, is the ratio of the ideal isothermal compression work and the isothermal efficiency η_{is},

$$(-w)_{real} = (-w)_T/\eta_{is} \tag{4.14}$$

or the ratio of the isentropic compression work and the adiabatic efficiency, η_{ad},

$$(-w)_{real} = (-w)_s/\eta_{ad}. \tag{4.15}$$

Values of isothermal and adiabatic efficiency for some of the common types of fans and compressors (Figures 4.12 to 4.17) are given in Table 4.1 together with data for capacities and pressure differences.

Single-stage *propeller fans* (Figure 4.12) are suitable in installations with low pressure drops. They are used in ventilation, cooling towers, flue gas ducts, air-cooled heat exchangers, and for circulation of air through heating and cooling units. Multi-stage propeller fans with stationary guide vanes are available for pressure differences up to 500 mm water (0.05 bar) when operating with air at atmospheric pressure.

Centrifugal fans are similar to centrifugal pumps. Centrifugal force and velocity head converted to pressure head in the volute give the increase in static head. Their characteristic curves are similar to those for centrifugal pumps (Figure 4.3a). The curves supplied by manufacturers are usually based on air with density $\rho_0 = 1.2$ kg/m^3. The pressure difference read from such curves must be multiplied by $\rho/1.2$ for air or gas with density ρ different from 1.2 kg/m^3.

Fans with curved vanes give pressure heads up to 1200 mm water (0.12 bar) when used for gas with density $\rho_0 = 1.2$ kg/m^3. Centrifugal fans with straight vanes as shown in Figure 4.13, give less pressure heads. They are used for pneumatic transport when particles pass through the fan.

The *two-lobe blower* (Figure 4.14) is a positive displacement compressor. Gear wheels on the outside give both lobes the same speed. The capacity ranges from 250 to 25,000 m^3/h and the discharge pressure from 0.1 to 0.8 bar. The volume can only be varied by changing speed or by by-passing. Axial-flow lobe blowers with

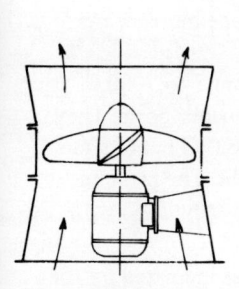

Figure 4.12
Propeller fan with motor connected directly.

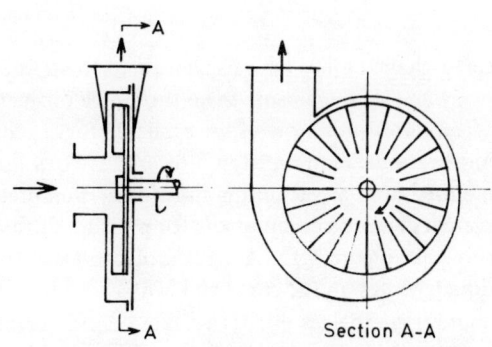

Figure 4.13 Centrifugal fan with straight vanes. (These fans are also made with curved vanes.)

Table 4.1 Data for some fans and compressors*

Type	Figure	Normal pressure difference, mm water or bar, or pressure ratio, p_2/p_1	Suction capacity, m^3/s	Efficiency, η_{is} or η_{ad}
Propeller fan,				
one stage	4.12	0–30 mm	0.13–30	η_{is} = 0.5–0.7
multi-stage		up to 500 mm		
Centrifugal fan,	4.13	15–1200 mm	0.1–30	η_{is} = 0.5–0.8
two in series		800–1800 mm	0.17–3	η_{is} = 0.6–0.7
Rotary blower	4.14	1000–8000 mm	0.07–7	η_{is} = 0.5–0.75
Liquid-piston				
rotary blower	4.15	up to 5 bar	0.001–2	η_{ad} = 0.6–0.7
Helical screw compressor, dry type				
one stage	4.16	p_2/p_1 = 1.5–4	0.2–11	η_{ad} = 0.75–0.85
two stages		p_2/p_1 = 3–15		
oil injected				
one stage	4.16	p_2/p_1 = 2.5–12	0.04–1.7	η_{ad} = 0.55–0.8
two stages		p_2/p_1 = 10–30		
Centrifugal compressors, multi-stage, water cooled	4.17	pressure ratio per stage, p_2/p_1 = 1.2–1.4 (up to 500 bar)	0.5–30	η_{ad} = 0.66–0.8
Reciprocating compressors,				
one stage		p_2/p_1 = 2–8	up to 0.3	η_{ad} = 0.76–0.84
multi-stage		6–4000 bar		

*Based on data from suppliers. The obtainable pressure differences for centrifugal fans and centrifugal compressors are proportional to the gas density and the rotational speed squared, n^2. Their characteristics are as for centrigual pumps, Figure 4.3a.

screw-shaped lobes have capacities up to 40,000 m^3/h. Some lobe blowers are noisy and workers using them should be protected from noise.

Liquid-piston blowers are available for capacities ranging from 5 to 7000 m^3/h. Moderate size liquid-piston blowers have low initial and maintenance costs. They are used both as vacuum pumps for suction pressures down to 0.033 bar (25 mm mercury) and as compressors for pressure differences up to 5 bar. The principle is shown in Figure 4.15. A rotor is placed excentric in a housing partly filled with liquid. Vanes on the rotor and centrifugal force causes the liquid to form a rotating liquid ring. The gas inlet is a sickle shaped opening in the back as indicated by the dotted area in Figure 4.15.

Gas trapped between the rotor, its vanes, the sides of the housing, and the liquid ring decreases in volume and its pressure increases as the 'pocket' approaches the outlet on the right side (Figure 4.15). Too high back pressure will force the liquid ring out from the top of the rotor, causing gas to be transported back from the pressure

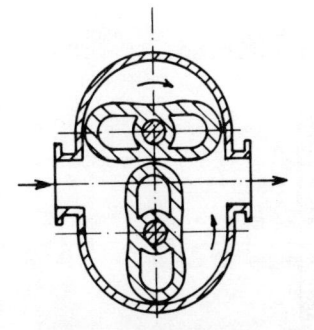

Figure 4.14 Two-impeller-type of positive rotary blower. Gear wheels outside the housing give the same rotational speed for both wheels.

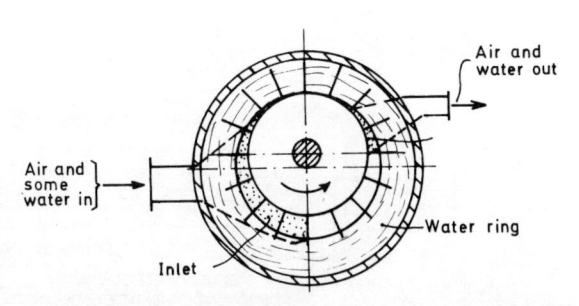

Figure 4.15 Liquid-piston type of rotary blower. Inlet to, and outlet from, the openings between the liquid ring and the rotor and the blades are located in the back of the housing.

to the suction side. Water is the common sealing liquid, but sulphuric acid is used in the compression of chlorine. It is important that some cold liquid is admitted to the suction side continuously together with the gas, to keep the blower cool.

Helical screw compressors are rotary positive displacement machines with a built-in compression ratio. They bridge the gap between reciprocating machines and turbo compressors, and they are also used as *expansion machines* with gases, steam, and other vapours as propellants. Figure 4.16(a) shows the main components and Figure 4.16(b) the two rotors. When used as a compressor, power is supplied to the male rotor. When used as an expander, power is delivered by the male rotor shaft as the machine's output. In the dry type compressor the female rotor is driven by the timing gears. In the type with oil injection the female rotor is driven by the male rotor.

The gas to be compressed is trapped in the gullies between the rotor lobes. Each rotor is sealed by close fit with the housing, both at the fore and aft ends and at the circumference. Similarly, the fit between the rotors seals the gullies at the mesh points. After one sixth rotation of the female rotor the gas in position 3 in Figure 4.16(b) is moved forwards to position 4. The volume has decreased and the pressure increased. The next one sixth rotation of the female rotor moves the gas to the position of pocket 5 where it is discharged through the outlet port shown on the right hand side in Figure 4.16(b).

A radial *turbocompressor* is in principle the same as a centrifugal blower, but more carefully machined. It has several stages on the same shaft, and stationary guide vanes between the impellers (Figure 4.17). Turbocompressors are used to handle large volumes, usually with discharge pressures higher than obtainable by the compressor types described earlier. Typical applications are vapour recompression for heat pumps, compression of refrigerant vapours in large refrigeration plants, and compression and recirculation of ammonia synthesis gas.

The pressure ratio over one stage is proportional to the density of the gas. It has a maximum of 1.4 for compression of air with density 1.2 kg/m^3.

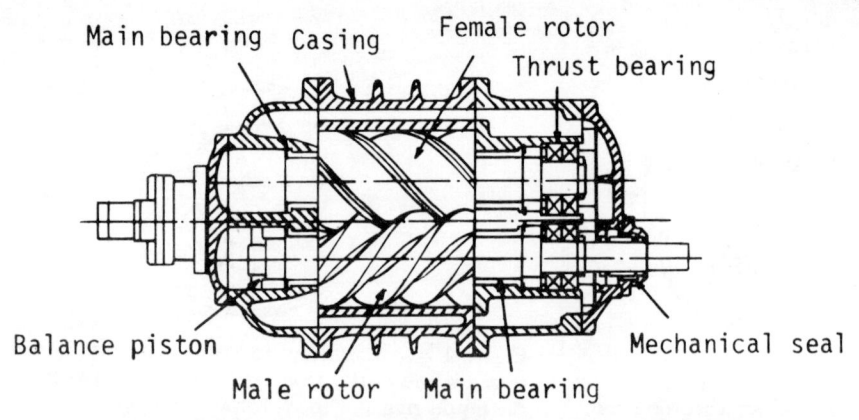

Figure 4.16a Section through a helical screw compressor. Reproduced by permission of MYCOM Europe S.A. for Mayekawa Mfg. Co., Japan.

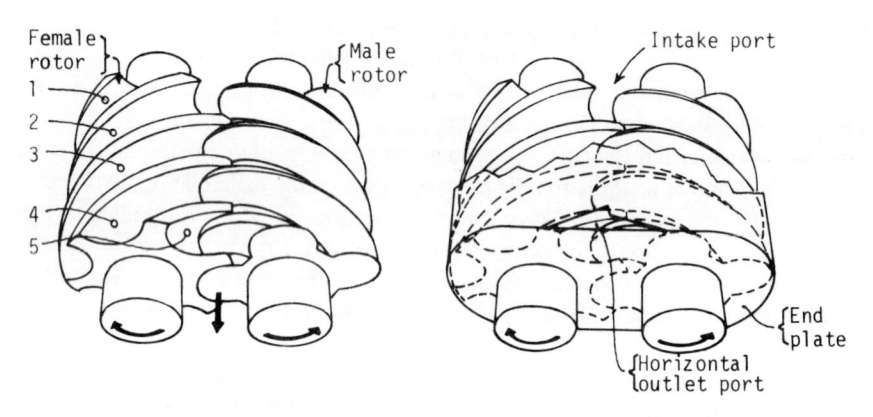

Figure 4.16b Rotors of a helical screw compressor.

Reciprocating compressors operate in the same way as reciprocating pumps (Figure 4.4). They are widely used in the chemical industry for volumes smaller than turbocompressors can handle. Multi-stage reciprocating compressors are built with discharge pressures up to 4000 bar and with plungers (Figure 4.5) instead of pistons in the last two stages. Multi-stage machines are provided with coolers between the stages to remove the heat of compression developed in the gas coming from the previous stage. Compressors with piston rings of steel require oil lubrication in the cylinder. Special reciprocating machines are available for compression of noble gases, chlorine, oxygen, and other gases which must be kept free from oil. They either have Teflon-covered piston rings or have clearance between the cylinder wall and the piston with grooves to provide a 'labyrinth' seal.

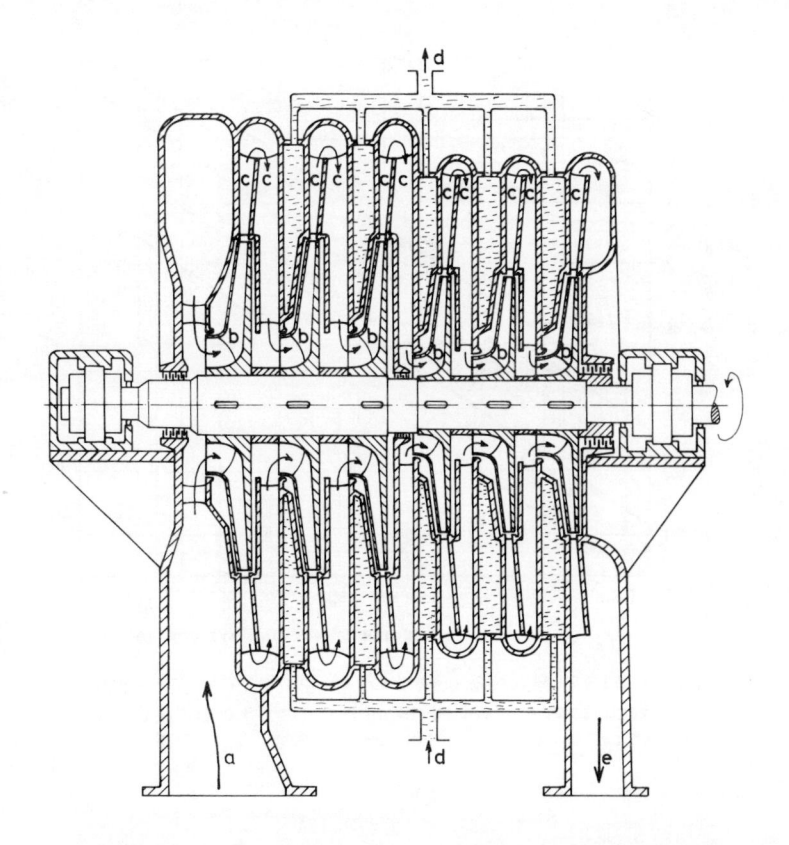

Figure 4.17 Six-stage, water-cooled turbocompressor (radial compressor): a, gas inlet; b, vanes on the impeller; c, stationary vanes; d, cooling water; e, gas outlet.

Vacuum pumps

Vacuum pumps are used to maintain lower than atmospheric pressures in process equipment by removing gases generated in the process, as well as by removing initial air present and air leaking into the equipment. For a vacuum system to be considered commercially tight, the air leakage should not exceed values shown in Figure 4.18. In addition 2 kg of air per hour should be allowed for leakage through seals about shafts rotating under vacuum.[5] Wiegand[6] recommends that from 0.2 to 0.4 kg of air per hour leakage should be allowed for each normal flange connection and 0.05 to 0.1 kg per hour for especially close fitting flange connections. For volumes of equipment between 0.2 and 500 m^3 the leakage in kg per hour may be estimated from the following:

$$w_0 = 0.45 v^{0.65} \text{ kg/h (minimum)} \tag{4.16a}$$

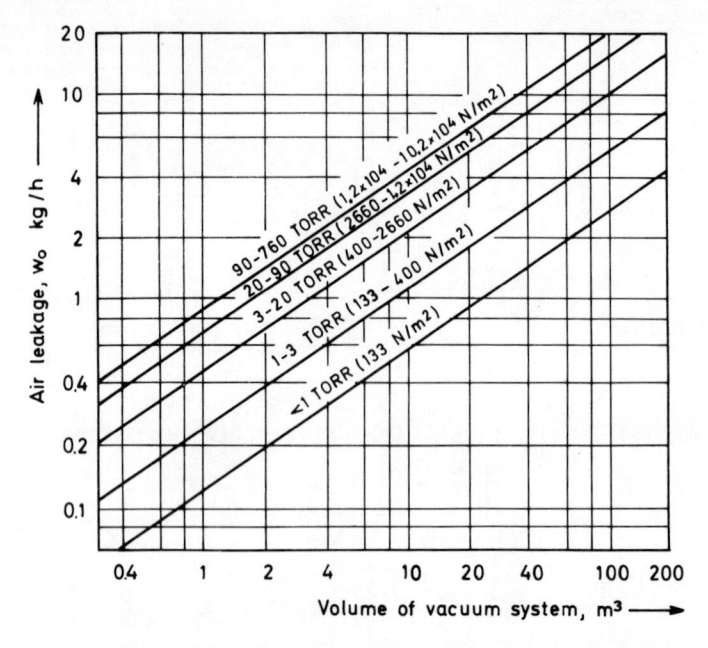

Figure 4.18 Maximum air leakage values for commercially tight vacuum systems without an agitator.[5] Reproduced by permission of the Heat Exchange Institute.

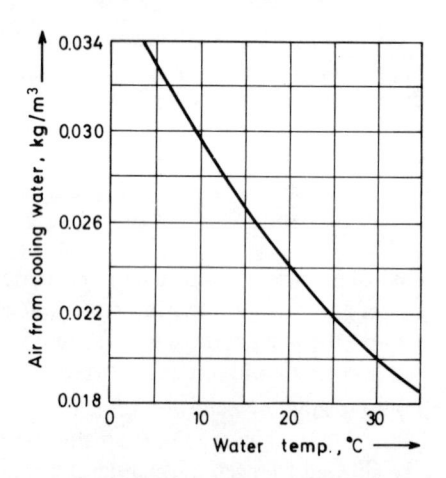

Figure 4.19 Allowance to be made for air in condensing water, kg air per m^3 water.[7] Reproduced by permission of the Heat Exchange Institute.

or

$$w_0 = 1.0v^{0.65} \text{ kg/h (maximum)} \qquad (4.16b)$$

where v = volume, m^3.

In *barometric condensers* air is liberated from the cooling water. Figure 4.19 gives the amounts of air evolved from cooling water.

Table 4.2 gives data for some common types of vacuum pumps. Sliding-vane compressors, liquid-piston rotary blowers, and two-impeller positive rotary blowers have adiabatic efficiencies in the range $\eta_{ad} = 0.5-0.75$ when operated at moderate vacuum. The efficiency decreases further at extremely low pressures.

Table 4.2 Common types of vacuum pumps. Minimum suction pressure is obtained when the suction line is closed. Except for the diffusion pumps, the values are given for a back pressure of one atmosphere. The capacities given were obtained from suppliers

Compressor type	Figure	Minimum suction pressure, mmHg	Maximum suction volume for air at the highest of the given suction pressures, m^3/s
Single stage steam ejector	4.27	80–150	0.11
Liquid piston rotary blower	4.15	25–50	2.2
Single stage sliding vane rotary blower	4.20	3–5.5	1.7
Steam ejector with liquid piston rotary blower for pre-vacuum		2–7	0.7
Five stage steam ejector		0.5–1.0	1.7
Kinney pump (gas balast)	4.21	0.1–1.0	0.25
Oil diffusion pump with 0.3 mmHg pre-vacuum	4.22	0.0001–0.00045	47
Mercury diffusion pump with pre-vacuum		10^{-9}	100

Water jet ejectors are similar to steam ejectors (Figure 4.27). They have low efficiencies and are mainly used as laboratory pumps, to create vacuum in centrifugal pumps before start-up, and for occasional evacuation of equipment. Estimated capacities and water consumptions are given in Figures 4.23 and 4.24. Additional details are given in the literature.[8,9]

Steam ejectors are often used as vacuum pumps in the chemical process industries because of their simple construction and lack of moving parts. They can handle wet, dry, or corrosive vapour mixtures and have low installation and maintenance costs. For calculations refer to page 87. For details about vacuum valves and flanges, cold traps, baffles, and pressure meters consult books on vacuum technology.[10,11]

84

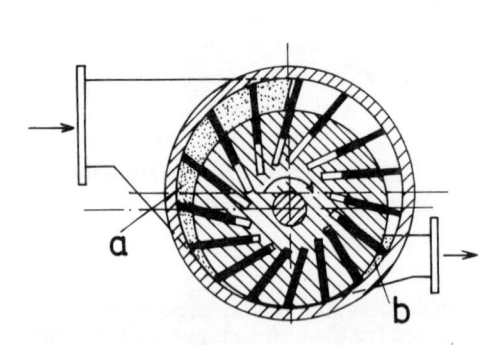

Figure 4.20 Sliding vane rotary blower: inlet opening a and outlet opening b are marked with dots.

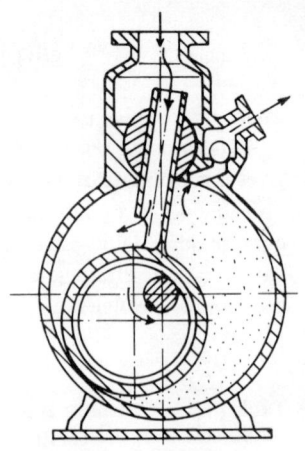

Figure 4.21 Kinney pump: gas under compression is marked with dots. Air may be introduced during the compression stage to avoid condensation (gas ballast pump).

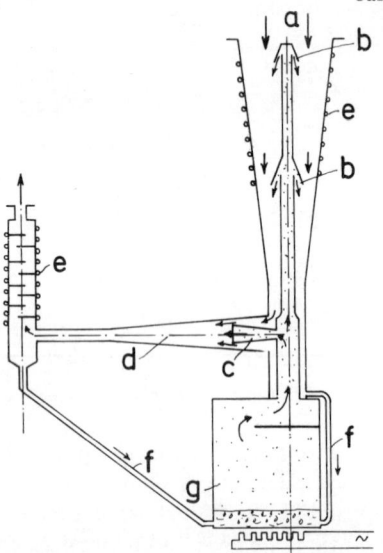

Figure 4.22 Oil diffusion pump: gas from the suction inlet a is mixed with the oil vapour from the jets b and c followed by a pressure increase in d before passing to the condenser e where the oil vapour is condensed and returned to the boiler g via tube f.

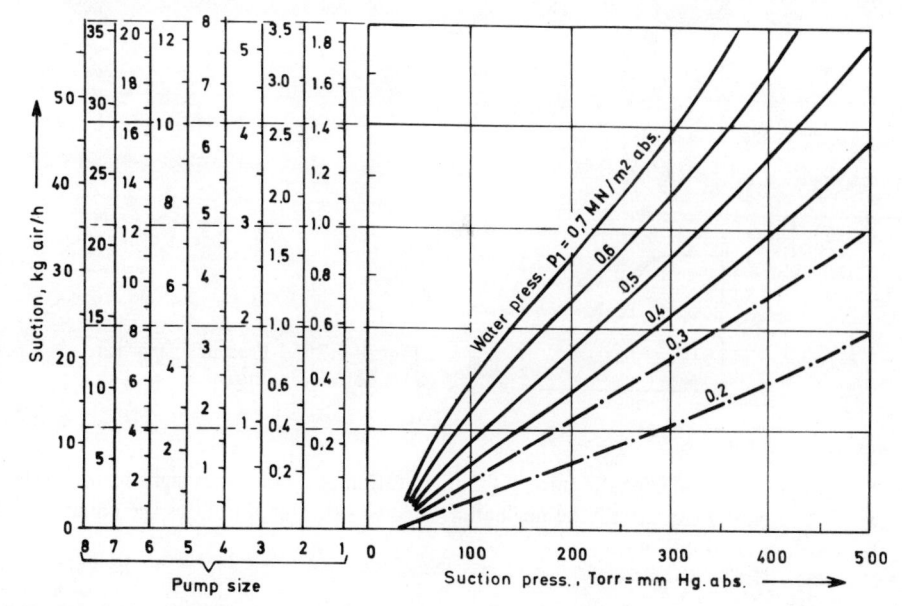

Figure 4.23 Suction volume for water jet ejectors with discharge pressure 0.1 MN/m² (1 bar) plotted against the suction pressure and the water pressure for 8 different pumps.[6] The lower curves are for low-pressure nozzles. Reproduced by permission of Wiegand Karlsruhe, GmbH, West Germany.

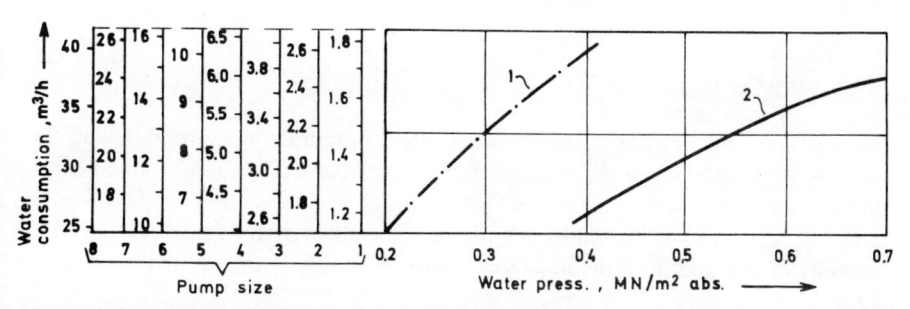

Figure 4.24 Water consumption for the water jet pumps in Figure 4.23. Curve 1 is for pumps with low-pressure nozzles and 2 for those with high-pressure nozzles.[6] Reproduced by permission of Wiegand Karlsruhe, GmbH, West Germany.

Expansion

The theoretical work of expansion of an expansion engine is calculated by means of the same equations as work of compression, but with the opposite sign. The real or useful work of expansion is found by multiplying the theoretical work of expansion by the adiabatic efficiency of the machine. For turbines this efficiency is from 2 to 6% higher than for the corresponding turbocompressor.

For expansion in a throttle valve with negligible kinetic energies before and after the expansion and with negligible heat exchange with the surroundings, the energy balance [equation (4.11)] simplifies to

$$h_1 = h_2 \tag{4.17}$$

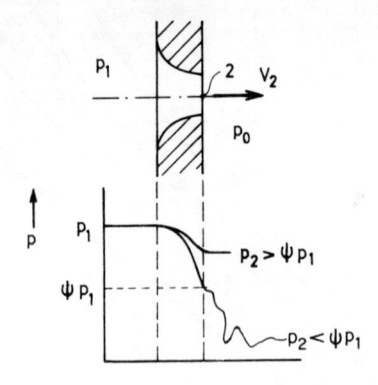

Figure 4.25 Pressure variation in a converging nozzle.

For expansion in nozzles the term $V_2^2/2$ in equation (4.1) becomes significant, while $V_1^2/2, z, g$, and q can usually be neglected. Also $w = 0$. This simplifies the equation for ideal, adiabatic flow to

$$V_2 = \sqrt{2(h_1 - h_2)_s}. \tag{4.18}$$

The pressure at the exit of a converging nozzle (point 2 Figure 4.25) will be p_0 if

$$p_0 \geqslant \psi p_1 \tag{4.19}$$

where ψ is the *critical pressure ratio*,

$$\psi = \left(\frac{2}{\kappa + 1}\right)^{\kappa/(\kappa-1)}. \tag{4.20}$$

If $p_0 < \psi p_1$, the pressure at point 2 is equal to ψp_1 and the velocity V_2 equals the velocity of sound for the gas in question. For most gases ψ lies between 0.53 and 0.58.

$\kappa = c_p/c_v$ can be estimated from a thermodynamic diagram. Temperature and pressure at two points on the same isentrope (s = constant) are inserted into the equation

$$\frac{T_1}{T_2} = \left(\frac{p_1}{p_2}\right)^{(\kappa-1)/\kappa}. \tag{4.21}$$

Use of this equation is shown in Example 4.11.

Converging–diverging nozzles (Laval nozzles, Figure 4.26) must be used in cases where $p_2 < \psi p_1$, to make use of the total available enthalpy difference. The pressure

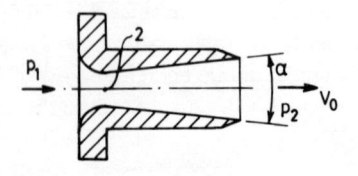

Figure 4.26 Converging–diverging nozzle (Laval nozzle).

in the throat adjusts itself automatically to

$$p_2 = \psi p_1. \tag{4.22}$$

This corresponds to the velocity of sound in the throat. The velocity in the diverging part of the nozzle is supersonic.

The angle α in Figure 4.26 should be kept sufficiently small to ensure that the jet does not depart from the walls of the nozzle. On the other hand, the angle should not be too small, lest it result in excessive friction loss. With $\alpha = 12°$, about 96–98% of the available enthalpy difference is converted into kinetic energy, i.e.

$$V_0 \approx \sqrt{(2)(0.97)(h_1 - h_0)_s} \tag{4.23}$$

where 0.97 is the fraction of the available enthalpy difference that is converted into kinetic energy. About 3% loss occurs in the divergent part of the nozzle.

Nozzles for steam ejectors should be designed for expansion to a pressure slightly below the suction pressure of the ejector.

Steam Ejectors

Steam ejectors are compressors without moving parts. The energy source for these compressors comes from vapour under pressure. They are built as shown in Figure 4.27, where the lower diagram gives the pressure change through the ejector.

Figure 4.28 shows the corresponding theoretical change of state for an ejector where the primary fluid and the secondary fluid are the same. With the nomenclature of Figure 4.28 the available energy for the primary fluid between point 1 at pressure p_1 and point 3 at pressure p_2 is

$h_1 - h_3$ kJ/kg primary fluid.

The theoretical compression work is

$h_4 - h_0$ kJ/kg secondary fluid

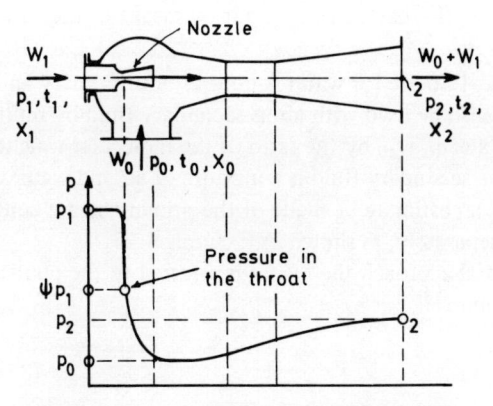

Figure 4.27 A steam ejector. The pressure diagram is shown below.

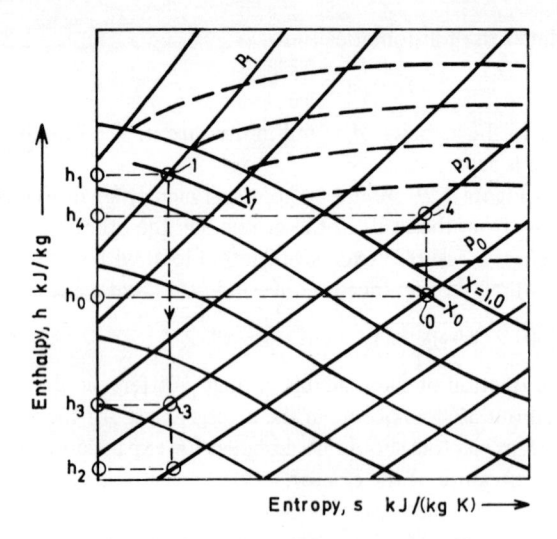

Figure 4.28 Enthalpy against entropy diagram corresponding to Figure 4.27.

where point 0 (Figure 4.28) indicates the state for the secondary fluid in front of the ejector. With w_1 kg/h primary fluid and w_0 kg/h secondary fluid the ejector efficiency is

$$\eta_e = \frac{w_0(h_4 - h_0)}{w_1(h_1 - h_2)}. \qquad (4.24)$$

This efficiency is, however, unsuited for ejector calculations. Most of the losses occur in the mixing of primary and secondary gas or vapour, and in the preceding compression of the total amount of fluid. Estimates of the amount of primary vapour needed, w_1 kg/h, may be carried out by means of the data represented in Figures 4.29 and 4.30.[12]

Figures 4.29 and 4.30 are for water vapour as both primary and secondary fluid. The diagrams may also be used with air as secondary fluid by multiplying the value found for primary steam, w_1, by the ratio of the molecular weights of air and water, 29/18 = 1.61. If the secondary fluid is a mixture of air and water vapour, it is recommended that an estimate be made of the primary steam consumption for each of the two media separately, as shown in Example 4.13.

For a fundamental approach the reader is referred to the publications by Flügel,[13] Wiegand,[14] and Bauer.[15]

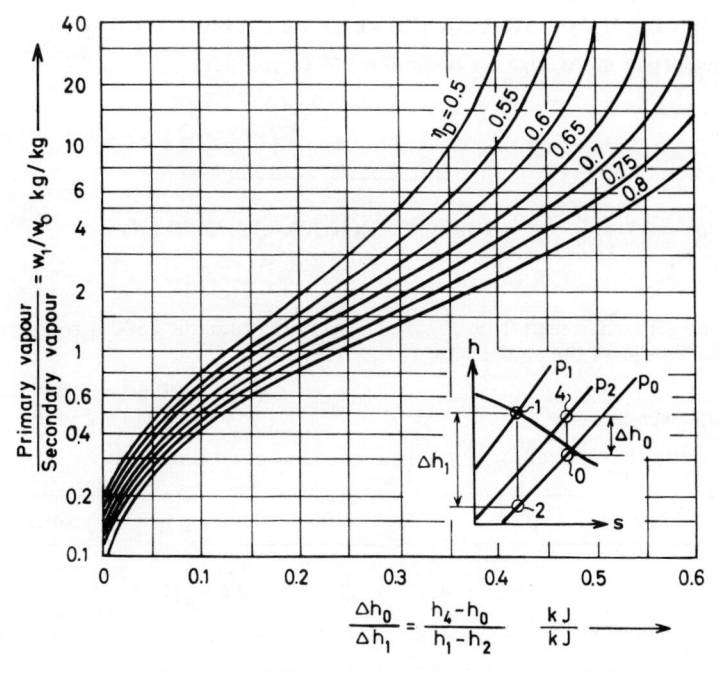

Figure 4.29 The ratio between the amount of primary vapour, w_1, and secondary vapour, w_0, for steam ejectors, plotted against the enthalpy ratio $\Delta h_0/\Delta h_1$ with the diffuser efficiency η_D given in Figure 4.30.[12]

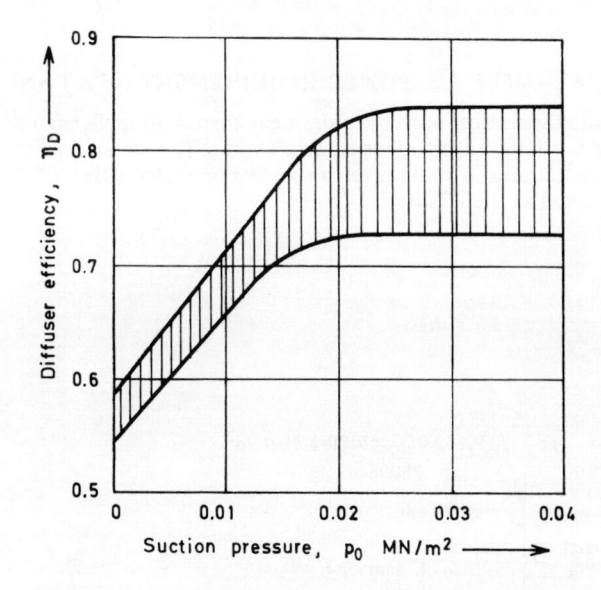

Figure 4.30 Empirical diffuser efficiency η_D plotted as a function of the suction pressure p_0 MN/m², according to Petzold.[12]

EXAMPLE 4.1. POWER REQUIREMENT OF A CENTRIFUGAL PUMP

Find the correct motor size for the pump in Example 1.1.

Solution

Equation (4.5) gives the theoretical power consumption

$$\dot{Q}\rho(-w) = \frac{2500}{(1000)(60)}(30.2)(9.82)(1030) = 12{,}730 \text{ N m/s}$$

or $\qquad 12{,}730/1000 = 12.7 \text{ kW}.$

The pump efficiency read from Figure 1.20 at a flow rate of 2500 litres per minute $\eta = 0.60$. This gives the actual power requirement

$$12.7/0.60 = 21 \text{ kW}$$

With 20% over-capacity, $\qquad P = (1.2)(21) = 25 \text{ kW}$

We can verify that the motor is not over-loaded at other flow rates.

Flow rate, $\dot{Q}$ litre/min	1000	1500	2000	2500	3000	Notes
Efficiency, η	0.50	0.59	0.625	0.60	0.35	from Figure 1.20
Head, H m water	49	45.5	39.5	30.2	13	from Figure 1.20
Power consumption, kW	16.5	19.5	21.3	19.2	18.8	$\frac{\dot{Q}H}{60\eta}g\rho 10^{-6}$

The table shows that the maximum power requirement is only slightly above the calculated value.

EXAMPLE 4.2. POWER REQUIREMENT OF A FAN

Calculate the theoretical power requirement for the propeller fan of Figure 4.31, both with and without a diffuser, at an air flow rate $\dot{Q} = 18 \text{ m}^3/\text{s}$ with density $\rho_1 = 1.15 \text{ kg/m}^3$, against a static pressure of 23 mm water gauge. 60% of the kinetic energy is recovered in the diffuser.

Solution

$$V_2 = \frac{18}{(\pi/4)1.2^2} = 15.9 \text{ m/s}.$$

$P_2 = P_1 + 0.023 \times 1000 \times 9.81 \text{ N/m}^2$
or 760 mm Hg

$\rho_1 = 1.15 \text{ kg/m}^3$

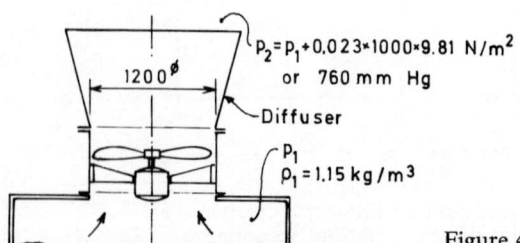

Figure 4.31 Propeller fan with diffuser

The air density is almost constant, and equation (4.3) multiplied by the mass flow rate, $\dot{Q}\rho$ kg/s, gives the power requirement without diffuser,

$$P = \dot{Q}\rho(-w) = (18)(1.15)\frac{(23)(9.82)}{1.15} + \frac{15.9^2}{2} + 0 = 6680 \text{ N m/s} = 6.7 \text{ kW}.$$

Here Δp_f, $V_1^{2/2}$, and $(z_2 - z_1)$ are neglected. With diffuser, only 40% of the kinetic energy is lost, and

$$P = (18)(1.15)\frac{(23)(9.82)}{1.15} + 0.4\frac{15.9^2}{2} = 5110 \text{ N m/s} \quad \text{or} \quad 5.1 \text{ kW}.$$

EXAMPLE 4.3. COMPRESSION OF AN IDEAL GAS

In a two-stage compressor 2000 m³/h air, temperature 20°C and pressure $p_1 = 1$ bar = 0.1 MN/m², is compressed to $p_3 = 1.6$ MN/m² with interstage pressure $p_2 = 0.4$ MN/m². At the interstage pressure the air is cooled to 20°C before it is recompressed.

(a) Calculate the adiabatic efficiency of the compressor when the power consumption measured at the compressor shaft is 236 kW.

(b) Calculate the terminal air temperature after reversible, adiabatic compression.

Solution

(a) Equation (4.9) with $v_0 = 2000/3600$ m³/s gives

$$(-w)_s = \frac{-1.4}{1.4 - 1}(0.1)(10^6)\frac{2000}{3600}(1 - 4^{(1.4-1)/1.4}) = 94,500 \text{ N m/s}$$

$$= 94.5 \text{ kW}.$$

For an ideal gas with the same pressure ratio and suction temperature in each stage, the power consumption in each stage is also the same. The corresponding power consumption for the two stages is (2)(94.5) = 189.0 kW, and the adiabatic efficiency,

$$\eta_{ad} = 189/236 = 0.80.$$

(b) Equation (4.10) gives

$$\frac{273 + 20}{T_2} = \left(\frac{1}{4}\right)^{0.4/1.4}, \quad T_2 = 435 \text{ K} \quad \text{or} \quad \theta = 435 - 273 = 162°\text{C}$$

EXAMPLE 4.4. COMPRESSION OF A NON-IDEAL GAS, APPLICATION OF h AGAINST s DIAGRAM.

12,000 kg water vapour per hour at pressure 0.1 MN/m² with water content 2.5% ($x = 0.975$) is to be compressed to 0.2 MN/m² in a turbo-compressor with an adiabatic efficiency $\eta_{ad} = 0.74$.

(a) Calculate the power consumption of the compressor.
(b) Find the vapour temperature after compression.

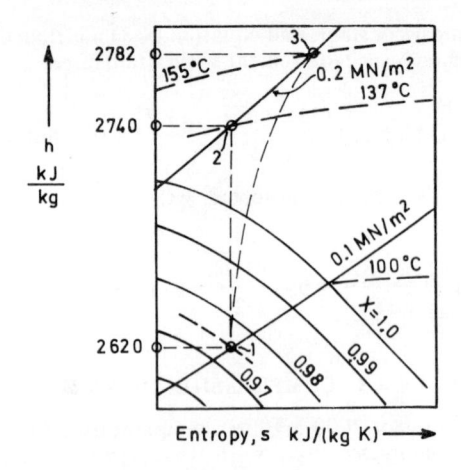

Figure 4.32   Enthalpy against entropy diagram for water vapour.

Solution

(a)   The enthalpies in Figure 4.32 are:

Enthalpy before compression, $h_1 = 2620$ kJ/kg.
Enthalpy after isentropic compression, $h_2 = 2740$ kJ/kg.
Theoretical work of compression, $h_2 - h_1 = 120$ kJ/kg.
Real enthalpy difference, $120/0.74 = 162$ kJ/kg.
Work of compression, $\dfrac{12{,}000}{3600} \cdot 162 = 540$ kW.

(b)   The temperature after compression can be read from Figure 4.32 at pressure $p_3 = p_2 = 0.2$ MN/m² and enthalpy

$h_3 = 2620 + 162 = 2782$ MN/m²,

$\theta_3 = 155°$C.

EXAMPLE 4.5. COMPRESSION OF A NON-IDEAL GAS. APPLICATION OF p AGAINST h DIAGRAM

$W = 350$ kg ammonia vapour per hour containing 2% liquid ($x = 0.98$) is to be compressed from pressure 0.16 MN/m² and saturation temperature $-23.7°$C to pressure 1.0 MN/m².

(a)   Calculate the theoretical power consumption in kW in ideal, adiabatic compression of ammonia, assuming ideal gas behaviour ($\kappa = 1.31$).

(b)   Calculate by means of the p against h diagram the theoretical power consumption during ideal adiabatic compression and the actual power consumption when the adiabatic efficiency of the compressor is $\eta_{ad} = 0.83$.

(c)   Find the gas temperature after compression.

(d)   Calculate the heat recovered for room heating when the compressed vapour is cooled to 60°C in a radiator before it passes on to the condenser.

Solution

(a) The volumetric flow rate of gas to the compressor is

$$v_0 = \frac{WRT}{Mp} = \frac{(350)(8314)(273.2 - 23.7)}{(17)(0.16)(10^6)} = 267 \text{ m}^3/\text{h}.$$

Insertion in equation (4.9) gives

$$(-w)_s = \frac{-1.31}{1.31 - 1}(0.16)(10^6)(267)\left[1 - \left(\frac{1}{0.16}\right)^{(1.31-1)/1.31}\right]$$

$$= (97.9)(10^6) \text{ N m/h}$$

or

$$P = (97.9)(10^6)/(1000)(3600)$$
$$= 27 \text{ kW}.$$

(b) With the symbols in Figure 4.33, from a p against h diagram:

Enthalpy after adiabatic compression, $h_2 = 1563$ kJ/kg.
Enthalpy before compression, $h_1 = 1306$ kJ/kg.
Theoretical work of compression, $h_2 - h_1 = 257$ kJ/kg.

$$\text{or } P = \frac{350}{3600}(257) = 25 \text{ kW}.$$

The real enthalpy–difference

$$h_3 - h_1 = \frac{h_2 - h_1}{\eta_{ad}} = \frac{257}{0.83} = 310 \text{ kJ/kg}$$

and power consumption

$$P = \frac{350}{3600} 310 = 30 \text{ kW}.$$

(c) Enthalpy after compression,

$$h_3 = 1306 + 310 = 1616 \text{ kJ/kg}.$$

From the enthalpy diagram, at the intersection of the pressure line at $p = 1.0$ MN/m² with the enthalpy line at $h_3 = 1306 + 310 = 1616$ kJ/kg, the temperature is

$$\theta_3 = 113°C.$$

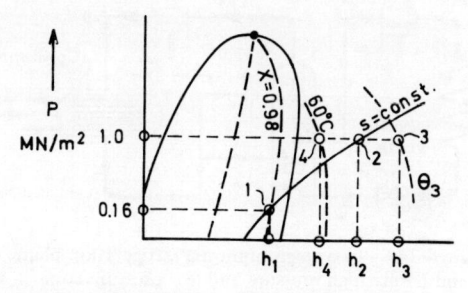

Figure 4.33 Pressure against enthalpy diagram for ammonia.

(d) Heat removed per second by the radiator,

$$\frac{350}{3600}(h_3 - h_4) = \frac{350}{3600}(1616 - 1482)$$

$$= 13 \text{ kJ/s}$$

$$\therefore P = 13 \text{ kW}.$$

EXAMPLE 4.6. TWO-STAGE REFRIGERATION PLANT

Figure 4.34 shows a two-stage ammonia refrigeration plant with a capacity

$Q_0 = 400,000$ kJ/h (95,500 kcal/h).

The pressure at the low pressure side is 0.08 MN/m² (points 5 and 6), in IP 0.28 MN/m², and at the high pressure side (points 9, 1, 2, 3, and 4) 1.0 MN/m². The liquid from the condenser (points 1, 2 and 3) is 6°C colder than the saturation temperature in the condenser, and the liquid at point 4 is 3°C warmer than the liquid in the interstage receiver. The adiabatic efficiency of each compressor $\eta_{ad} = 0.83$. Heat exchange with the ambient air and pressure drop in the pipes and heat exchangers are negligible. The vapour at points 6 and 8 is assumed to be saturated.

(a) Find the power required (in kW) for the low pressure compressor and the temperature of the ammonia at point 7.

(b) Find similar information for the high pressure compressor.

(c) Calculate the amount of heat $\dot{Q}_1$ removed from the condenser.

See the pressure against enthalpy diagram for ammonia. (Figure 4.35).

Solution

(a) The enthalpies read from Figure 4.35 are:

Enthalpy of ammonia after the evaporator,	$h_6 = 1310$ kJ/kg.
Enthalpy of ammonia in front of the evaporator,	$h_4 = h_5 = \underline{60 \text{ kJ/kg}}$
Enthalpy change in the evaporator,	$h_6 - h_5 = 1250$ kJ/kg.

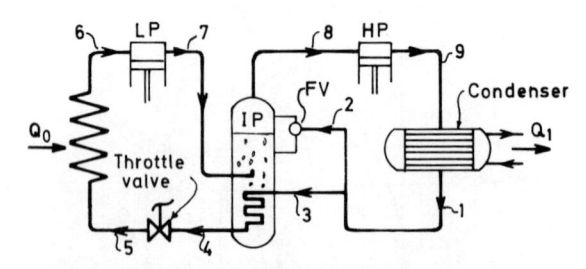

Figure 4.34 Two stage ammonia refrigeration plant. HP and LP are high pressure and low pressure compressors and IP an interstage receiver where the liquid is kept at a steady level by means of the float valve FV.

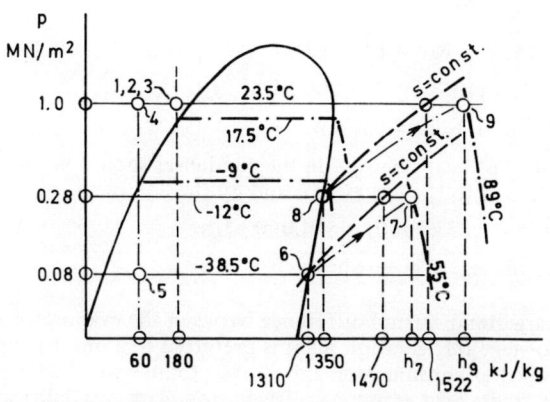

Figure 4.35 Pressure against enthalpy diagram with points corresponding to those marked on Figure 4.34.

The amount of ammonia circulating on the low pressure side,

$$\frac{400{,}000}{1250} = 320 \text{ kg/h}.$$

Enthalpy after adiabatic compression (Figure 4.35),   1470 kJ/kg,

Enthalpy before compression, $h_6 = $ 1310 kJ/kg.

Work of adiabatic compression,   160 kJ/kg.
Actual work of compression, 160/0.83   = 193 kJ/kg.
    or (320)(193)/3600 = 17 kW.

Enthalpy after compression

$h_7 = h_6 + 193 = 1310 + 193 = 1503$ kJ/kg.

Temperature at point 7 (p against h diagram),

$\theta_7 = 55°$C.

(b)   With y kg ammonia flowing through pipe 8 for every kg ammonia in the low pressure system, the enthalpy balance around the interstage receiver is

$h_7 + h_3 + (y - 1)h_2 = h_4 + yh_8$

$1503 + 180 + (y - 1)180 = 60 + y1350$,

$y = 1.233$ kg/kg.

The rate of compression in HP, (1.233)(320) = 395 kg/h.
   For the high pressure compressor:
   Enthalpy before compression, $h_8 = $ 1350 kJ/kg.
   Enthalpy after ideal adiabatic compression,
                      (Figure 4.35),   1522 kJ/kg.

Theoretical work of compression,   172 kJ/kg.
Actual work of compression,   172/0.83 = 207 kJ/kg
    or (395)(207)/3600 = 23 kW

Enthalpy after compression, $h_9 = h_8 + 207$ = 1557 kJ/kg.
Temperature at point 9 (Figure 4.35), θ_9 = 89°C.

(c) Heat removed in the condenser,

$$h_9 - h_1 = 1557 - 180 = 1377 \text{ kJ/kg}$$

or $(395)(1377) = 544{,}000 \text{ kJ/h}$

or $544{,}000/3600 = 151 \text{ kW}.$

Check The amount of heat removed in the condenser should be equal to the amount of heat added to the evaporator and by the compressors,

$$400{,}000 + (17 + 23)(3600) = 544{,}000 \text{ kJ/h}$$

which is equal to the figure given above.

Note With the large temperature difference between the evaporator and the condenser, a two-stage refrigeration plant is preferred to a one-stage plant, because it gives both less power consumption and a lower temperature after compression. The latter is an advantage as ammonia refrigeration plants use lubricating oils with low viscosity. Lubrication may be poor when the temperature exceeds $120-130°C$.

EXAMPLE 4.7. EVACUATION WITH A WATER JET EJECTOR

A water jet ejector is used to prime the centrifugal pump in Example 1.1 by means of evacuation. The average volume to be evacuated (pipes, pump, and heat exchanger) is about 2.2 m^3.

(a) Calculate the suction pressure for the water jet ejector when the centrifugal pump is filled with sea-water.

(b) Estimate the time it will take to reach this pressure under the following conditions. Atmospheric pressure is $750 \text{ mmHg} = (133.3)(750) = 10^5 \text{ N/m}^2$. The pump is evacuated at low tide, and is located 5.5 metres above low tide level. Air is assumed to be an ideal gas with molecular weight $M = 29$ and constant temperature $17°C$ is maintained. The density of sea water is $\rho = 1030 \text{ kg/m}^3$. The water jet ejector uses city water with pressure 0.4 MN/m^2 and its water consumption is approximately $13 \text{ m}^3/\text{h}$. The gas constant $R = 8{,}314 \text{ J/(kmol K)}$. See also Figures 4.23 and 4.24.

Solution

(a) Atmospheric pressure, p_1 $\qquad = 100{,}000 \text{ N/m}^2.$
Suction pressure, $p_1 - p_2 = (5.5)(1030)(9.82) = 55{,}600 \text{ N/m}^2.$

With primed pump, $p_2 = p_1 - (p_1 - p_2)$ $\qquad = 44{,}400 \text{ N/m}^2.$

(b) Isothermal change of state for an ideal gas with constant volume v_0 yields

$$v_0 \, dp = -RT \, dn = -RT \frac{dw}{M} = -\frac{RT}{M}\left(\frac{dw}{dt}\right) dt$$

where n kmol or w kg is positive for air removed. By rearrangement

$$\int_{p_2}^{p_1}\left(\frac{dt}{dw}\right) dp = \frac{RT}{Mv_0} \int_0^t dt = \frac{(8314)(290)}{(29)(2.2)} t = 37{,}800 t. \qquad (a)$$

Experimental values of dw/dt as a function of pressure p, taken from Figure 4.23, are given in the following table. The reciprocal values dt/dw are plotted in Figure 4.36 as a function of the suction pressure p.

Suction pressure p_a mmHg	300	400	500	600	700	Notes
Suction pressure, p N/m²	39,990	53,320	66,650	79,980	93,310	$p = 133.3 p_a$
$\dfrac{dw}{dt}$ kg/h	8.5	12	15.7	20	25	from Figure 4.23
$\dfrac{dt}{dw}$ h/kg	0.118	0.083	0.064	0.050	0.040	

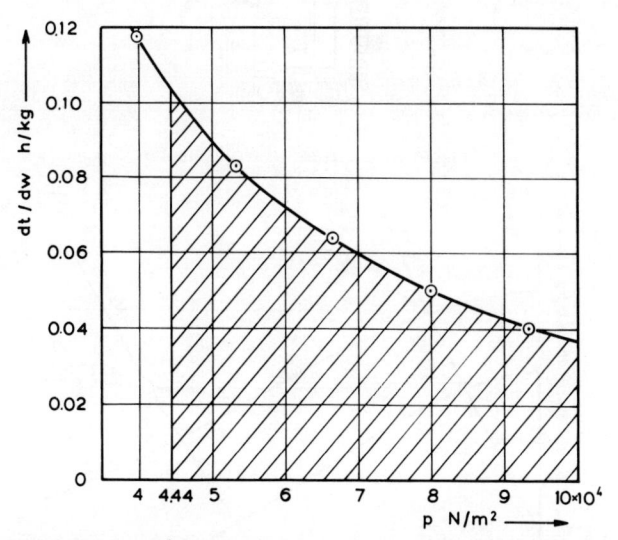

Figure 4.36 dt/dw in hours per kg air removed, plotted as a function of the pressure p in N/m². The hatched area corresponds to $\int_{p_2}^{p_1} \left(\dfrac{dt}{dw}\right) dp$ in equation (a).

The hatched area in Figure 4.36 is equal to 17.1 squares. A square corresponds to
$$(10{,}000 \text{ N/m}^2)(0.02 \text{ h/kg}) = 200 \text{ (N/m}^2\text{)(h/kg)}.$$
Hence,
$$\int_{p_2}^{p_1} \left(\dfrac{dt}{dw}\right) dp = (17.1)(200) = 3420 \text{ (N/m}^2\text{)(h/kg)}.$$
Insertion of this value in equation (a) gives
$$t = 3420/37{,}800 = 0.09 \text{ h}$$
$$\text{or } 5.4 \text{ min.}$$

EXAMPLE 4.8. JOULE–THOMSON EXPANSION

To determine the water content of wet steam, a small side-stream of wet steam is throttled down to atmospheric pressure in a throttle calorimeter as shown in Figure 4.37.

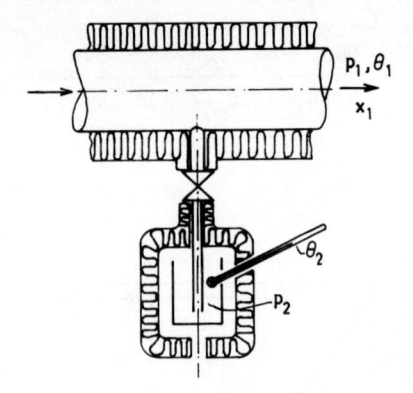

Figure 4.37 A throttle calorimeter.

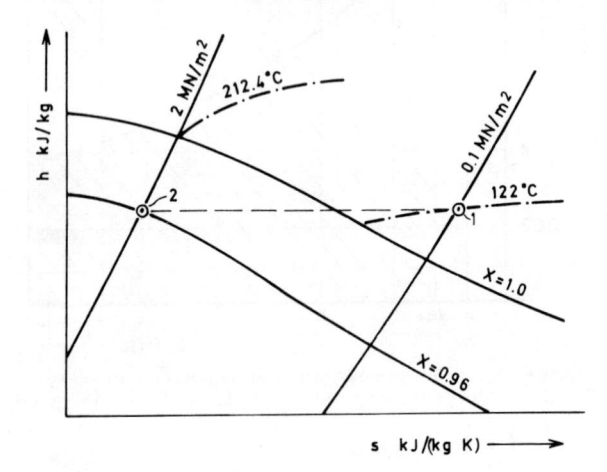

Figure 4.38 h against s diagram for water vapour.

Find the water content of the wet steam at pressure $p_1 = 2.0$ MN/m² when the temperature in the throttle calorimeter θ_2 is $122°$C.

Refer to the h against s diagram for water vapour (Figure 4.38).

Solution

Equation (4.17) gives $h_2 = h_1$ corresponding to point 2 in Figure 4.38, i.e. steam quality $x = 0.96$ or water content $1 - 0.96 = 0.04$ kg water/kg wet steam.

EXAMPLE 4.9. WORK OF COMPRESSION AND EXPANSION IN THE LIQUEFACTION OF METHANE BY THE CLAUDE PROCESS

Figure 4.39 shows the Claude process combined with Joule–Thomson expansion (Linde process) for liquefaction of methane at atmospheric pressure. Methane gas (pressure 14 MN/m² and temperature $15°$C at point 1) enters the expansion turbine E where it expands to atmospheric pressure and then passes through the heat exchanger I, where the gas is heated to $-25°$C before recompression by compressor C to 14 MN/m². The compressed gas is cooled in the heat exchanger I to $-158°$C

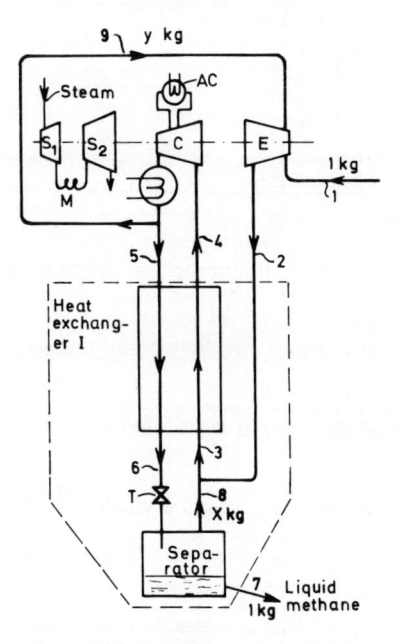

Figure 4.39 Claude process for liquefaction of methane.

at point 6 (6°C above the normal boiling point of methane) and throttled down to atmospheric pressure by throttle T. The liquid formed in the expansion is withdrawn from the separator at 7. The uncondensed vapour from the separator is mixed with the gas from the turbine E.

The compressor C has three interstage coolers (in Figure 4.39 only one is shown). Each after-cooler, AC, brings the temperature down to 15°C. The adiabatic efficiency of the compressor and the turbine is 0.75 and 0.80, respectively.

Part of the gas is recirculated (stream 9 Figure 4.39) to remove heat sufficient to reduce the temperature of stream 6 to −158°C before expansion at T. The pressure drop in the heat exchangers may be neglected.

(a) Calculate x in stream 8 and y in stream 9 in kg per kg liquid methane.

(b) Calculate the mechanical energy produced by the expansion turbine E and the energy required for the compressor C in kWh per kg liquid methane.

(c) The compressor C and the expansion turbine E are connected through a common shaft to a two-stage steam turbine with a high pressure stage S_1 and a low pressure stage S_2. An inter-stage superheater M is placed between stages S_1 and S_2 in order to reduce the liquid content of the vapour in the low pressure stage S_2. Calculate the steam consumption for the steam turbines in kg/kg liquid methane when the pressure in front of the high pressure stage, S_1, is $p_1 = 12$ MN/m², and the temperature θ_1 is 520°C. The adiabatic efficiency of the high pressure stage is $\eta_{ad} = 0.77$. The corresponding values for the low pressure stage, S_2, are $p_2 = 0.3$ MN/m², $\theta_2 = 390°$C, and $\eta_{ad} = 0.82$. The pressure after the low pressure stage is $p_3 = 0.0025$ MN/m².

Refer to the pressure against enthalpy diagram for methane and enthalpy against entropy diagram for water vapour.

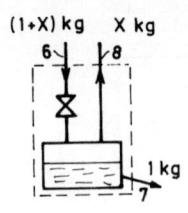

Figure 4.40
The liquid receiver of Figure 4.39.

Solution

(a) The enthalpy balance for the liquid receiver as shown in Figure 4.40 is

$$(1 + x)h_6 = h_7 + xh_8 \tag{a}$$

where h_6 is the enthalpy at pressure 14 MN/m^2 and temperature $-164 + 6 = -158°$C,

$$h_6 = -250 \text{ kJ/kg},$$

h_7 is the enthalpy of liquid methane at atmospheric pressure (0.1 MN/m^2),

$$h_7 = -282 \text{ kJ/kg},$$

and h_8 is the enthalpy of gas at $-164°$C and atmospheric pressure,

$$h_8 = 228 \text{ kJ/kg}.$$

These values are inserted in equation (a),

$$(1 + x)(-250) = -282 + 228x, \quad x = 0.067 \text{ kg/kg liquid methane.}$$

The corresponding enthalpy balance around the part of the apparatus framed by the broken line in Figure 4.39 gives

$$(1 + y)h_2 + (1 + x)h_5 = h_7 + (1 + x + y)h_4 \tag{b}$$

where h_2 is the enthalpy after the expansion turbine (Figure 4.41)

$h_2 = h_1 - 0.80(h_1 - h_{20}) = 453 - 0.80(453 - 112) = 180$ kJ/kg
$h_5 = h_1 =$ enthalpy before expansion, $h_5 = 453$ kJ/kg
$h_7 =$ enthalpy of the liquid methane, $h_7 = -282$ kJ/kg
$h_4 =$ enthalpy at 0.1 MN/m^2 and $-25°$C, $h_4 = 515$ kJ/kg.

These values and $x = 0.067$ are inserted into equation (b),

$$(1 + y)180 + (1 + 0.067)(453) = -282 + (1 + 0.067 + y)515$$
$$\therefore \quad y = 1.182 \text{ kg/kg liquid methane.}$$

(b) Using values obtained from Figure 4.41, the following are obtained.

Enthalpy in front of the expansion turbine, $h_1 = 453$ kJ/kg.
Enthalpy after ideal adiabatic expansion, $h_{20} = 112$ kJ/kg.
Theoretically available, $h_1 - h_{20} = 341$ kJ/kg.
Actually used, $\eta_{ad}(h_1 - h_{20}) = (0.80)(341)$ $= 273$ kJ/kg.

With $1 + y = 2.182$ kg gas through the expansion turbine per kg liquid methane the energy consumption is

$$(2.182)(273) = 596 \text{ kJ/kg liquid methane}$$

or

$$596/3600 = 0.166 \text{ kWh/kg liquid methane.}$$

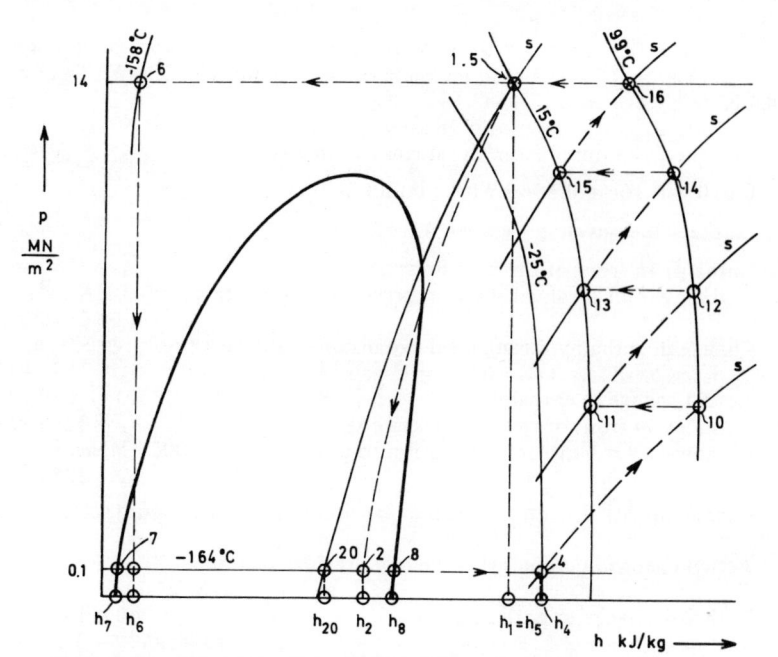

Figure 4.41 Pressure against enthalpy diagram for methane: curves s are for constant entropy.

The corresponding amount of methane gas passing through the compressor is

$1 + x + y = 1 + 0.067 + 1.182 = 2.249$ kg/kg liquid methane.

The compression path in ideal adiabatic compression in a four stage compressor with the same exit temperature after each cooler, is shown by the broken zig-zag path 4–10–11–12–13–14–15–16 in Figure 4.41. The temperature after compression, 99°C, is found by trial and error on the diagram. The corresponding theoretical work of compression per kg liquid methane is

$(1 + x + y)[(h_{10} - h_4) + (h_{12} - h_{11}) + (h_{14} - h_{13}) + (h_{16} - h_{15})]$
$2.249[(795 - 515) + (784 - 600) + (763 - 589) + (718 - 552)]$
$= 2.249(280 + 184 + 174 + 166) = 1808$ kJ/kg liquid methane.

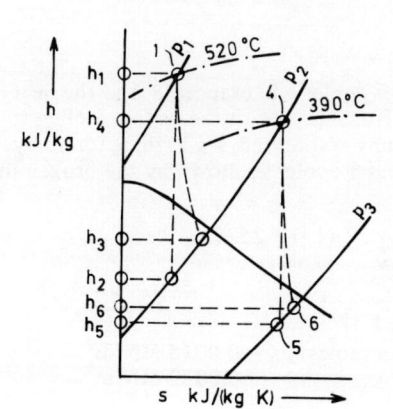

Figure 4.42 Enthalpy against entropy diagram for water vapour.

With adiabatic efficiency $\eta_{ad} = 0.75$ the corresponding actual work of compression is $1808/0.75 = 2411$ kJ/kg liquid methane or $2411/3600 = 0.670$ kWh/kg liquid methane.

(c) The energy required from the steam turbines is

$0.670 - 0.166 = 0.504$ kWh/kg liquid methane.

From the values given in Figure 4.42:

Enthalpy in front of the high pressure section, $\qquad h_1 = 3350$ kJ/kg.
Enthalpy after ideal, adiabatic expression to pressure $p_2 = 0.3$ MN/m^2,
$\qquad\qquad h_2 = 2520$ kJ/kg.
Change in enthalpy during ideal expansion in the high pressure section,
$\qquad\qquad h_1 - h_2 = 830$ kJ/kg.
Actual change in enthalphy, $\qquad h_1 - h_3 = (0.77)(830) = 639$ kJ/kg.
Enthalpy in front of the low pressure section, $\qquad h_4 = 3255$ kJ/kg.
Enthalpy after isentropic expansion to pressure $p_3 = 0.0025$ MN/m^2,
$\qquad\qquad h_5 = 2352$ kJ/kg.

Change in enthalpy during ideal expansion in the low pressure section,
$\qquad\qquad h_4 - h_5 = 903$ kJ/kg.
Actual change in enthalpy in the low pressure section,
$\qquad h_4 - h_6 = (0.82)(903) = 740$ kJ/kg.
Enthalpy change in both sections, $\qquad 639 + 740 = 1379$ kJ/kg.
or $\qquad\qquad 1379/3600 = 0.383$ kWh/kg steam.

The corresponding steam consumption is

$0.504/0.383 = 1.32$ kg steam/kg liquid methane.

EXAMPLE 4.10. REFRIGERATION PLANT WITH STEAM EJECTOR

A factory needs 24 m^3/h cooling water at 13°C, but in especially hot weather the city water temperature can increase to 18°C. The factory has, however, an excess of steam at pressure 0.4 MN/m^2 and water content 2% ($x = 0.98$), and an existing vacuum pump which can maintain a suction pressure of 25 mmHg.

A steam ejector refrigeration plant as shown in Figure 4.43 is planned in order to get cold cooling water in hot weather. Estimate the steam consumption for the ejector.

Refer to steam tables and the enthalpy diagram for water vapour and the Figures 4.29 and 4.30.

Solution

Some of the water injected into the water cooler will evaporate, and the heat required for vaporization will be extracted from the rest of the water, cooling it down to the desired temperature. The amount evaporated, w_0 kg/h, is found by means of an enthalpy balance around the water cooler as shown by the broken line in Figure 4.44.

$(24,000 + w_0)(4.19)(18) = (24,000)(4.19)(13) + 2527w_0$

as h_0 is 2527 kJ/kg (saturated vapour at 13°C),

$w_0 = 205$ kg/h.

The saturation pressure at 13°C (from steam tables) $p_0 = 0.0015$ MN/m^2.
25 mmHg corresponds to $(133.3)(25)(10^{-6})\qquad = 0.0033$ MN/m^2.
With the symbols in Figure 4.29 and 4.30:

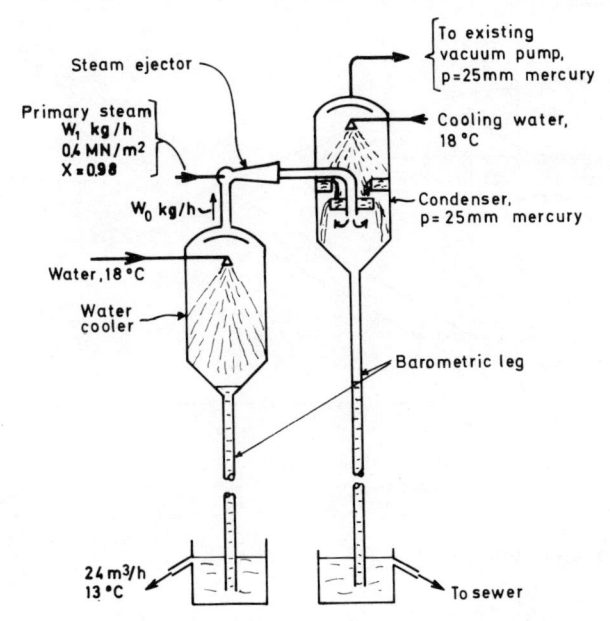

Figure 4.43 Refrigeration plant to supply 24 m³/h water at a temperature of 13 °C.

Enthalpy of primary steam $\qquad h_1 = 2695$ kJ/kg.
Enthalpy of primary steam after isentropic expansion to
0.0015 MN/m², $\qquad h_2 = 1945$ kJ/kg.

Enthalpy difference, $\qquad \Delta h_1 = h_1 - h_2 = 750$ kJ/kg.
Enthalpy of the secondary vapour, $\qquad h_0 = 2527$ kJ/kg.
Enthalpy after ideal, adiabatic compression, $\qquad h_4 = 2635$ kJ/kg.

Enthalpy difference, $\qquad \Delta h_0 = h_4 - h_0 = 108$ kJ/kg,
which gives $\qquad \Delta h_0/\Delta h_1 = 108/750 = 0.144$.

According to Figure 4.30 at a suction pressure of 0.0015 MN/m² the diffuser efficiency is $\eta_D = 0.56 - 0.60$. With $\Delta h_0/h_1 = 0.144$ and $\eta_D = 0.56 - 0.60$ the mass ratio according to Figure 4.29 is $w_1/w_0 = 0.89 - 0.95$ kg/kg. The highest diffuser efficiency gives the steam consumption

$$w_1 = (205)(0.89) = 182 \text{ kg/h}.$$

The lowest diffuser efficiency gives the steam consumption

$$w_1 = (205)(0.95) = 195 \text{ kg/h}.$$

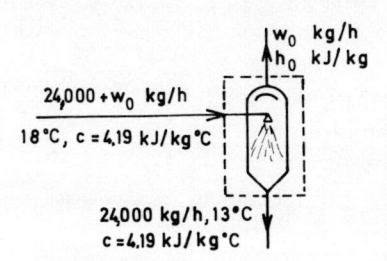

Figure 4.44 Enthalpy balance around the water cooler.

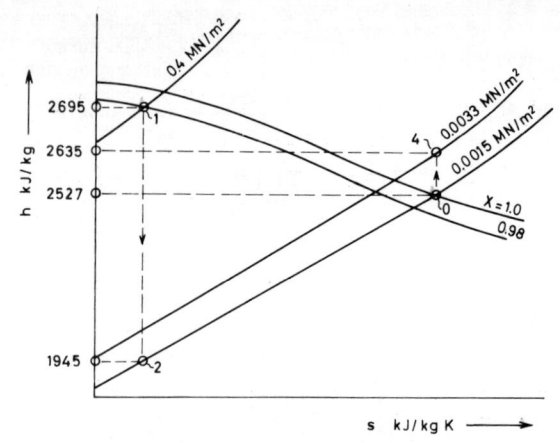

Figure 4.45 Enthalpy against entropy diagram for water vapour.

Primary steam consumption,

$$182 < w_1 < 195 \text{ kg/h}.$$

Note It is interesting to check the efficiency as defined by equation (4.24),

$$\eta_e = \frac{(205)(2635 - 2527)}{(188.5)(2695 - 2026)} = 0.176$$

where w_1 is the average steam consumption, $(182 + 195)/2 = 188.5$ kg/h. This shows that only 17.6% of the theoretically available energy in the primary steam is used for compression of the secondary vapour. However, this type of compressor is cheap and requires little maintenance. Therefore, a steam ejector may be a good solution despite its low efficiency.

EXAMPLE 4.11. NOZZLE CALCULATION

Calculate the throat diameter D_2 and the outlet diameter D_3 of the nozzle in Figure 4.46, which is the nozzle for primary steam in Example 4.10. The nozzle efficiency is 0.97 and the steam consumption 190 kg/h.

Solution

As a first trial the pressure in the throat is estimated to be $p_2 = (0.58)(0.4) = 0.23$ MN/m². Steam tables give the saturation temperatures

at $p_1 = 0.4$ MN/m², $\theta_1 = 143.6°C$
at $p_2 = 0.23$ MN/m², $\theta_2 = 124.7°C$.

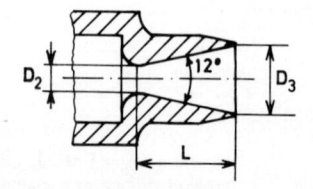

Figure 4.46 A converging–diverging nozzle.

Equation (4.21),

$$\frac{273.2 + 143.6}{273.2 + 124.7} = \left(\frac{0.4}{0.23}\right)^{(\kappa-1)/\kappa},$$

$$\kappa = 1.09.$$

This value inserted into equation (4.20) gives

$$\psi = \left(\frac{2}{1.09 + 1}\right)^{1.09/(1.09-1)} = 0.587$$

$p_2 = (0.587)(0.4) = 0.235 \text{ MN/m}^2$.

From Figure 4.47:

Enthalpy in front of the nozzle,	$h_1 = 2695$ kJ/kg.
Enthalpy after expansion to pressure $p_c = 0.235$ MN/m^2,	$h_2 = 2600$ kJ/kg.
Enthalpy difference,	$h_1 - h_2 = 95$ kJ/kg.

Equation (4.18) gives the velocity in the throat,

$$V_2 = \sqrt{(2)(95)(10^3)} = 436 \text{ m/s}.$$

Steam tables give the volume of saturated steam at pressure $p_2 = 0.235$ MN/m^2, $v_2'' = 0.762$ m^3/kg. With quality $x_2 = 0.948$ (read from a steam diagram) the specific volume in the throat is,

$$v_2 \approx (0.948)(0.762) = 0.722 \text{ m}^3/\text{kg}.$$

Volume per second through the throat

$$436 \frac{\pi}{4} D_2^2 = \frac{190}{3600} 0.722,$$

$$D_2 = 0.0105 \text{ m}$$
$$= 10.5 \text{ mm}.$$

For isentropic expansion the available enthalpy difference (Figure 4.47),

$$\Delta h_1 = 2695 - 1945 = 750 \text{ kJ/kg}.$$

With efficiency $\eta = 0.97$ enthalpy converted into mechanical energy $(0.97)(750) = 728$ kJ/kg.

$$V_3 = \sqrt{(2)(728)(10^3)} = 1207 \text{ m/s}.$$

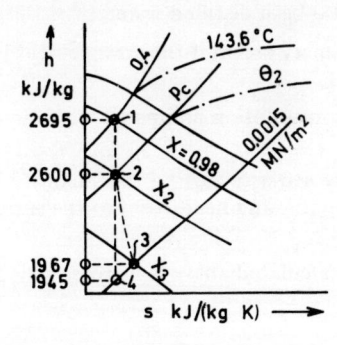

Figure 4.47 Enthalpy against entropy diagram for steam.

Enthalpy after expansion, $h_3 = 2695 - 728 = 1967$ kJ/kg.
Steam quality (Figure 4.47), $x_3 = 0.774$
Volume of saturated steam, $v_3'' = 88.4$ m^3/kg.

Volume at the outlet,

$$v_3 \approx (0.774)(88.4) = 68.4 \text{ m}^3/\text{kg}.$$

$$1207 \frac{\pi}{4} D_3^2 = \frac{190}{3600} 68.4$$

$$\therefore D_3 = 0.0617 \text{ m}$$

$$= 61.7 \text{ mm}.$$

Length of divergent section,

$$L = \frac{61.7 - 10.5}{2 \text{ tg } 6°}$$

$$= 244 \text{ mm}.$$

Note The nozzle is unusually large. A better mixing with the secondary vapour and a better efficiency is obtained if smaller nozzles are used as indicated in Figure 4.48. If 6 nozzles are used in parallel, their dimensions will be

$$D_2 = 10.5/\sqrt{6} = 4.3 \text{ mm}$$
$$D_3 = 61.7/\sqrt{6} = 25.2 \text{ mm}$$
$$L = 244/\sqrt{6} = 99.6 \text{ mm}.$$

The smallest diameter, 4.3 mm, is still well above the $\sim$2 mm generally accepted as the lower limit to avoid clogging by particles.

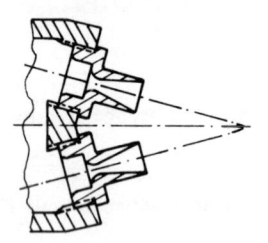

Figure 4.48   Head with several nozzles.

EXAMPLE 4.12. HEAT PUMP FOR DISTILLED WATER

A factory for plastic products needs 700 kg/h distilled water.

(*a*)   Estimate the consumption of primary steam if the arrangement in Figure 4.49 is used.

(*b*)   What is the reduction in steam consumption compared with direct use of condensate from primary steam?

(*c*)   Estimate the consumption of city water, y kg/h, if this quantity is adjusted to condense all vapour from the steam ejector, and heat losses to the surroundings are negligible.

(*d*)   What happens if the quantity y calculated above is increased by 50%?

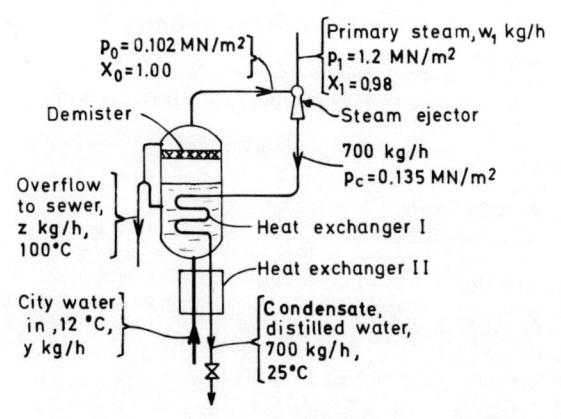

107

[Figure 4.49: A heat pump for production of distilled water — diagram showing Demister, Steam ejector, Heat exchanger I, Heat exchanger II, with labels: $p_0 = 0.102\ MN/m^2$, $X_0 = 1.00$; Primary steam, w_1 kg/h, $p_1 = 1.2\ MN/m^2$, $X_1 = 0.98$; 700 kg/h, $p_c = 0.135\ MN/m^2$; Overflow to sewer, z kg/h, 100°C; City water in, 12°C, y kg/h; Condensate, distilled water, 700 kg/h, 25°C.]

Figure 4.49   A heat pump for production of distilled water.

Solution

(a) The enthalpies are (Figure 4.50):

Primary steam before expansion,	$h_1 =$	2745 kJ/kg.
After isentropic expansion,	$h_2 =$	2373 kJ/kg.
Theoretically available difference,	$\Delta h_1 = h_1 - h_2 =$	372 kJ/kg.
Secondary vapour,	$h_0 =$	2676 kJ/kg.
Secondary vapour after isentropic compression,	$h_4 =$	2730 kJ/kg.
Theoretical work of compression,	$\Delta h_0 = h_4 - h_0 =$	54 kJ/kg,

$$\therefore \Delta h_0 / \Delta h_1 = 54/372 = 0.145$$

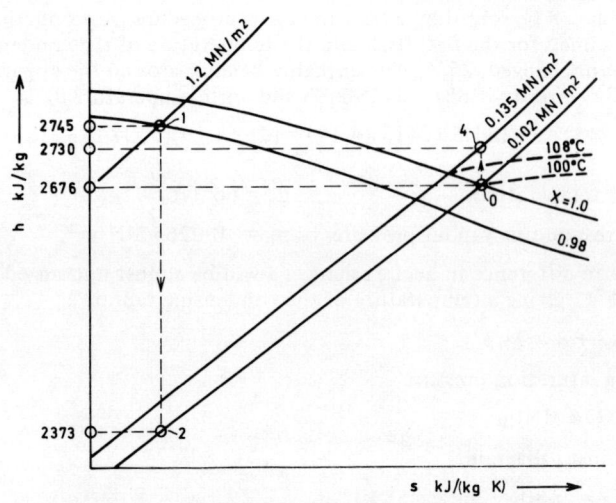

[Figure 4.50: Enthalpy–entropy (h vs s) diagram with curves at 1.2 MN/m², 0.135 MN/m², 0.102 MN/m², showing points at $h = 2745$ (point 1), 2730 (point 4), 2676 (point 0), 2373 (point 2); temperatures 108°C and 100°C; $X = 1.0$ and 0.98 lines.]

Figure 4.50   Enthalpy against entropy diagram for the process in Figure 4.49.

Suction pressure 0.102 MN/m² corresponds to a diffuser efficiency (Figure 4.30),

$\eta_D = 0.73$ to 0.84.

These values inserted into Figure 4.29 give $w_1/w_0 = 0.54$–0.70 kg primary steam/kg secondary vapour, corresponding to

$$w_1 + w_1/0.54 = 700,$$
$$w_1 = 245 \text{ kg/h}$$

and

$$w_1 + w_1/0.70 = 700,$$
$$w_1 = 288 \text{ kg/h}.$$

(b) The saving in primary steam based on the lowest efficiency is

$$\frac{700 - 288}{700} 100 = 59\%.$$

(c) Calculations based on the lowest efficiency give the material balance around the apparatus,

$$288 + y = 700 + z \tag{a}$$

and the enthalpy balance

$$(288)(2745) + (12)(4.19)y = (25)(4.19)(700) + (100)(4.19)z. \tag{b}$$

Equations (a) and (b) give

$$y = 2413 \text{ kg/h}.$$

(d) The temperature and pressure in the apparatus and, as a result, the amount of evaporated water will all decrease. With unchanged nozzle the amount of primary steam is enchanged, i.e. the amount of distilled water will decrease.

The problem can be solved by a trial and error procedure. A production of 650 kg/h is assumed for the first trial, and the temperature of the condensate is assumed to be unchanged, 25°C. The enthalpy balance around the apparatus with $z = (1.5)(2413) - (650 - 288) = 3258$ kg/h and with temperature θ_z is

$$(288)(2745) + (12)(1.5)(2413)(4.19) = (25)(4.19)(650) + (3258)(4.19)\theta_z \tag{c}$$

or water temperature, $\qquad\qquad \theta_z = 66.3°C.$

The corresponding vapour pressure $\quad p_0 = 0.0265 \text{ MN/m}^2.$

The temperature difference in heat exchanger I will be almost unchanged, $108 - 100 = 8°C$, giving a temperature of the condensing vapour

$$\theta_c = 66.3 + 8 = 74.3°C.$$

Corresponding saturation pressure,

$$p_c = 0.0374 \text{ MN/m}^2.$$

From an h against s diagram,

$$\Delta h_0 = 2680 - 2622 = 58 \text{ kJ/kg}$$
$$\Delta h_0/\Delta h_1 = 58/372 = 0.156.$$

Diffuser efficiency at $p_0 = 0.0265$ MN/m² (Figure 4.30),

$\eta_D = 0.73 - 0.84$.

The lowest of these efficiencies gives (Figure 4.29)

$w_1/w_0 = 0.77$

and

$w_1 + w_0 = 288 + 288/0.77 = 662$ kg/h.

This value inserted into equation (c) in place of 650 gives $\theta_z = 66.2°C$, i.e. practically the same value as before. Hence, 50% increase in water flow to the apparatus decrease the amount of distilled water by

$(100)(700 - 662)/700 = 5.4\%$.

Note This shows that an increase of the flow rate to the apparatus has little influence on the production rate of distilled water. However, it does create a vacuum, i.e. the condensate has to be withdrawn by means of a pump or a barometric leg.

EXAMPLE 4.13. STEAM EJECTOR FOR EVACUATION

Figure 4.51 shows a distillation column for fractional distillation of fatty acids. The volume of column and condenser is 18 m³. Pressure and temperature in front of the first steam ejector are 200 N/m² (1.5 mmHg) and 60°C, respectively. The pressure after the last ejector is 3330 N/m³ (25 mmHg). The primary steam for both ejectors is saturated at pressure 0.7 MN/m² (7 bar, quality $x = 1.0$).

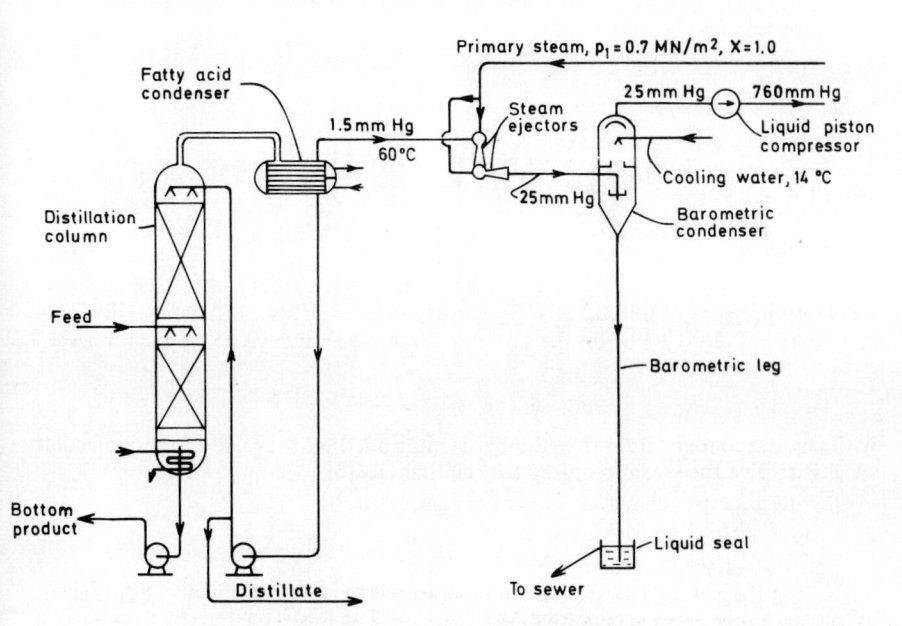

Figure 4.51 Vacuum distillation column with a two-stage steam ejector discharging into a barometric condenser evacuated by a liquid piston compressor.

(a) Estimate the volumetric air flow to the first ejector.

(b) Estimate the amount of primary steam necessary for the two-stage ejector when the first ejector has a compression ratio of 7.5.

(c) Estimate the flow of cooling water to the barometric condenser and the suction capacity of the liquid piston compressor if the cooling water is heated from 14°C to 17°C.

(d) Select a suitable motor size for the liquid piston compressor having an efficiency $\eta_{ad} = 0.50$.

Solution

(a) With a volume of 18 m³ and pressure 200 N/m², Figure 4.18 gives an air leakage of 1.6 kg/h. Additional leakage in the pump for bottoms is 2.0 kg/h. Total leakage,

$$w_0 = 3.6 \text{ kg/h}.$$

(Leakage from the condensate pump is not included. The long, liquid filled suction line may give atmospheric pressure at the shaft seal.)

Equation (4.16b) gives the leakage $w = 18^{0.65} = 6.5$ kg/h. To be on the safe side, this value is chosen as the design value, giving an air volume in front of the first ejector,

$$v_a = n \frac{RT}{p} = \frac{6.5}{29} \frac{(8314)(273+60)}{200}$$

$$= 3100 \text{ m}^3/\text{h}.$$

(b) The first ejector compresses from pressure 200 N/m² to $(7.5)(200) = 1500$ N/m². The enthalpy difference for isentropic expansion of the primary steam is (Figure 4.52)

$$\Delta h_1 = 2763 - 1732 = 1031 \text{ kJ/kg}.$$

Isentropic compression of the air requires [equation (4.9)]

$$\frac{-1.4}{1.4-1}(200)(3100)\left[1 - \left(\frac{1500}{200}\right)^{(1.4-1)/1.4}\right] 10^{-3} = 1690 \text{ kJ/h}$$

or $\Delta h_{oa} = 1690/6.5 = 260$ kJ/kg

$\therefore \Delta h_{oa}/\Delta h_1 = 260/1031 = 0.252$.

A suction pressure of 0.0002 MN/m² corresponds to diffuser efficiency (Figure 4.30) $\eta_D = 0.55 - 0.585$. To be on the safe side the lowest value (0.55) is used. Figure 4.29 gives

$$w_1/w_0 = 2.3.$$

With air as secondary fluid this number is multiplied by the ratio of the molecular weights to give the steam required for the first ejector,

$$w_1 = \frac{29}{18}(2.3)(6.5) = 24 \text{ kg/h}.$$

Neglecting the kinetic energies, the temperature after the first ejector, θ_3, is given by the enthalpy balance (Figure 4.53),

$$(24)(2763) + (6.5)(1.0)(60) = 24 h_3 + (6.5)(1.0)\theta_3$$

$$h_3 = 2779 - 0.271 \theta_3, \tag{a}$$

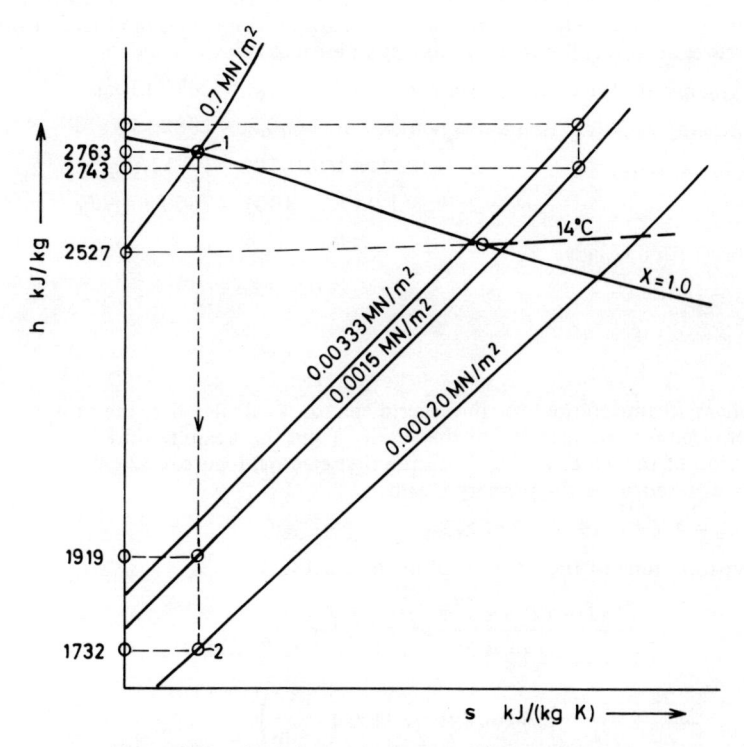

Figure 4.52 Enthalpy against entropy diagram for steam.

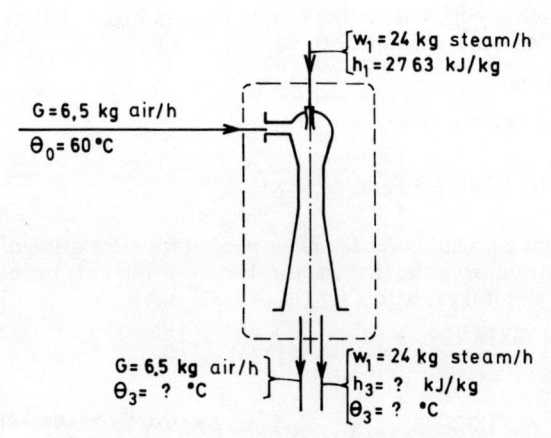

Figure 4.53 Enthalpy balance around the first ejector.

where a temperature of 0°C is used as reference state. The second 'equation' is the h against s diagram for steam, where h_3 can be found as a function of θ_3. At this low pressure, however, the steam is almost an ideal gas:

Enthalpy at 100°C (from diagram), $\quad h_{100} = 2690$ kJ/kg.

Enthalpy at 200°C (from diagram), $\quad h_{200} = 2880$ kJ/kg.

Heat capacity, $\quad c = (2880 - 2690)/100 = 1.9$ kJ/(kg °C).

$$h_3 = 2690 + 1.9\,(\theta_3 - 100) = 2500 + 1.9\theta_3. \quad (b)$$

Equations (a) and (b) give

$$\theta_3 = 128.5°C$$

and

$$h_3 = 2744 \text{ kJ/kg}.$$

The primary steam required for the second ejector is calculated as the sum of steam required for compression of the air alone and the steam required for compression of the water vapour from the first ejector. Figure 4.52 gives the enthalpy difference for the primary steam,

$$\Delta h_1 = 2763 - 1919 = 844 \text{ kJ/kg}.$$

The specific volume of the air in front of the ejector is,

$$v_a = \frac{(8314)(273 + 128.5)}{(29)(1500)} = 76.7 \text{ m}^3/\text{kg}$$

$$\Delta h_{oa} = \frac{-1.4}{1.4 - 1}(1500)(76.7)\left[1 - \left(\frac{3330}{1500}\right)^{(1.4-1)/1.4}\right] 10^{-3}$$

$$= 103 \text{ kJ/kg}$$

$\Delta h_{oa}/\Delta h_1 = 103/844 = 0.122$.

With diffuser efficiency $\eta_D = 0.56$ (the lowest curve in Figure 4.30 for suction pressure 0.0015 MN/m²) (Figure 4.29),

$$w_1/w_0 = 0.80.$$

Primary steam for compression of the air,

$$w_a \frac{29}{18}(0.80)(6.5) = 8.4 \text{ kg/h}.$$

The steam diagram does not cover the range needed for calculation of the compression of steam from the first ejector. However, this compression can be calculated as for an ideal gas with $\kappa = c_p/c_v = 1.32$.

$$v_s = \frac{(8314)(273 + 128.5)}{(18)(1500)} = 123.6 \text{ m}^3/\text{kg}$$

$$\Delta h_o = \frac{-1.32}{1.32 - 1}(1500)(123.6)\left[1 - \left(\frac{3330}{1500}\right)^{(1.32-1)/1.32}\right] 10^{-3}$$

$$= 163 \text{ kJ/kg}$$

$\Delta h_o/\Delta h_1 = 163/844 = 0.193$.

Figures 4.29 and 4.30 give $w_1/w_0 = 1.5$, $w_1 = (1.5)(24) = 36$ kg/h.

Primary steam for the second ejector, $8.4 + 36 = 44.4$ kg/h.

Primary steam required for both ejectors, $24 + 44.4 = 68.4$ kg/h.

Note The first ejector has a higher pressure ratio than the second one to reduce the primary steam required. The same pressure ratio in both ejectors (interstage pressure 813 N/m^3) gives primary steam consumption 13 kg/h in the first and 74 kg/h in the second stage. The steam temperature at the outlet of the nozzle in the first ejector will be $-15°$C. Heat conduction through the metal of the nozzle will probably keep the surface temperature above $0°$C, but there is a danger of ice forming on the walls of the mixing chamber after the nozzle. Hence, this part of the ejector should be heated, by a steam jacket, say.

(c)   The consumption of cooling water for the barometric condenser is determined by material and an enthalpy balance (Figure 4.54). Here w_0 is water vapour in the air from the condenser. The air leaving the condenser is assumed to have the same temperature as the cold water, $14°$C. The corresponding partial pressure of water vapour ($14°$C) is 1500 N/m^2. With partial pressure of air $3330 - 1500 = 1830$ N/m^2 and both air and water vapour as ideal gases is

$$\frac{6.5/29}{w_0/18} = \frac{1830}{1500},$$

$w_0 = 3.3$ kg/h.

Material balance (Figure 4.54) gives

$$44.4 + 6.5 + 24 + w_k = 6.5 + 3.3 + w_u$$

$$w_u = 65.1 + w_k \tag{c}$$

and the enthalpy balance

$$(44.4)(2763) + (6.5)(1.0)(60) + (24)(2763) + (4.19)(14)w_k$$
$$= (6.5)(1.0)(14) + (3.3)(2527) + (4.19)(17)w_u$$

$$w_u = 2540 + 0.8235 w_k. \tag{d}$$

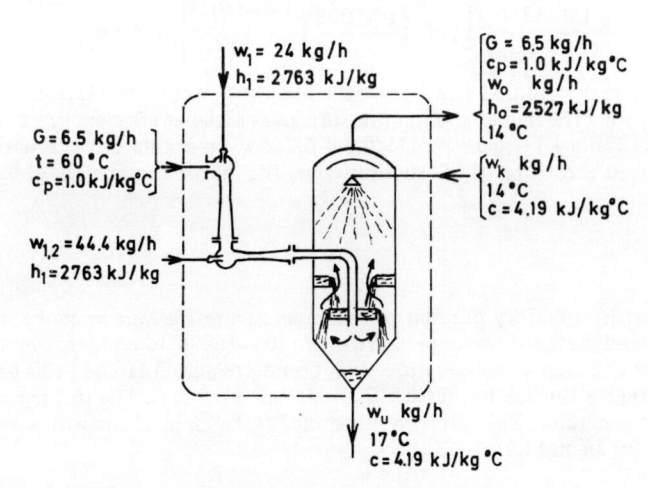

Figure 4.54   Enthalpy balance around ejector and condenser.

From equations (c) and (d),

$$w_k = 14{,}000 \text{ kg/h}$$
$$= 14 \text{ m}^3/\text{h}.$$

Figure 4.19 gives the air liberated from the cooling water at temperature $14°C$, 0.027 kg/m^3 water, or in this case $(14)(0.027) = 0.4 \text{ kg/h}$.
(This means that the calculation should be repeated with 6.9 instead of 6.5 kg air/h, but the error in w_k is insignificant.)

Air from the condenser, $\quad\quad\quad 6.5 + 0.4 = 6.9$ kg/h.
Water vapour from the condenser, $(3.3)(6.9)/6.5 = 3.5$ kg/h.

Total volume,

$$v = \left(\frac{6.9}{29} + \frac{3.5}{18}\right) \frac{8134(273 + 14)}{3330} = 310 \text{ m}^3/\text{h}.$$

(d) The adiabatic exponent for the mixture is

$$\kappa = \frac{1830}{3330} 1.4 + \frac{1500}{3330} 1.32$$
$$= 1.364.$$

Power consumption in isentropic compression [equation (4.9)],

$$P = \frac{-1.364}{1.364 - 1} 3330 \frac{310}{3600} \left[1 - \left(\frac{100{,}000}{3330}\right)^{(1.364-1)/1.364}\right] 10^{-3}$$
$$= 1.6 \text{ kW}$$

or real power consumption,

$$1.6/0.50 = 3.2 \text{ kW}.$$

The liquid piston compressor will also be used to evacuate the equipment, i.e. the motor chosen also takes the maximum load during evacuation. The power consumption with air and suction pressure p is

$$P = \frac{1}{\eta_{ad}} \frac{-1.4}{1.4 - 1} \frac{310}{3600} \left[1 - \left(\frac{100{,}000}{p}\right)^{(1.4-1)/1.4}\right] 10^{-3} \text{ kW}. \quad\quad (e)$$

Suction pressure $p \approx 30{,}000 \text{ N/m}^2$ gives the highest power consumption $P = 3.71/\eta_{ad}$ kW. The higher suction pressure gives a higher efficiency, assumed to be $\eta_{ad} = 0.6$ (Table 4.1), and $P = 3.71/0.6 = 6.2$ kW. To obtain fast evacuation, i.e. start-up without throttling of the suction valve, the motor chosen should be of at least 6.2 kW.

Problems

4.1. In a fertilizer factory pure nitrogen is used as a refrigerant to cool a mixture of hydrogen and methane under pressure down to $-160°C$ to condense most of the methane. The nitrogen at temperature $-90°C$ and pressure 3 MN/m^2 (30 bar) expands through a turbine to pressure 0.15 MN/m^2 (1.5 bar). The turbine is coupled to an electric generator. The nitrogen is assumed to be an ideal gas with $\kappa = c_p/c_v = 1.4$ and $c_p = 29$ J/(mol K).

(a) Calculate the temperature of the nitrogen after isentropic expansion.

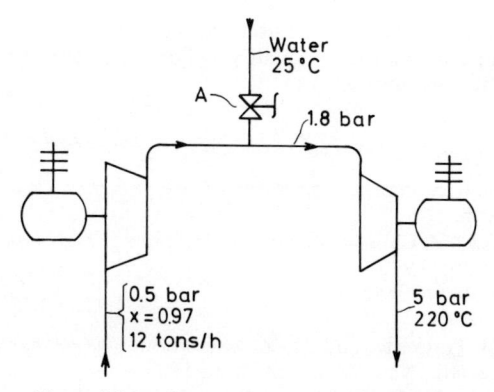

Figure 4.55 Two-stage compression of steam.

(b) Calculate the work of isentropic expansion in kJ/kmol and in kWh/kmol.

(c) The adiabatic efficiency of the turbine is $\eta_{ad} = 0.72$. What is the temperature of the nitrogen after the expansion?

4.2 Figure 4.55 shows a system for two-stage compression of 12 ton/h of steam with water injection between the two stages to keep the discharge temperature from the high pressure compressor at 220°C. The adiabatic efficiency of both compressors is $\eta_{ad} = 0.73$.

(a) Calculate the power input of the low pressure compressor (from 0.5 to 1.8 bar) measured at the shaft, and the temperature before water injection.

(b) Calculate the power input of the high pressure compressor (from 1.8 to 5 bar).

4.3. The electric motor of an ammonia compressor uses 65 kW, and the efficiency of motor and transmission is estimated to be $\eta = 0.85$. The mass flow of ammonia is 600 kg/h. Temperature and pressure before compression are $\theta_1 = 10°C$ and $p_1 = 0.1$ MN/m^2 (1.0 bar), respectively, and the pressure after compression $p_2 = 0.5$ MN/m^2 (5.0 bar).

4.4 What is the adiabatic efficiency η_{ad} and the temperature after compression?

A compressor (Figure 4.56) compresses 350 kg/h ammonia gas from 1.6 bar and $-20°C$ to 10 bar. The adiabatic efficiency is $\eta_{ad} = 0.82$.

(a) Find the gas temperature after compression with valve A closed.

(b) Liquid ammonia (10 bar and 24°C) is throttled through valve A to pressure 1.6 bar and injected into the suction line of the compressor. How much ammonia (in kg/h) has to pass through valve A to give a gas temperature $\theta_2 = 95°C$ after compression?

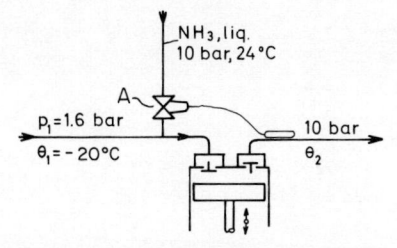

Figure 4.56 An ammonia compressor.

4.5. Calculate the outlet diameter and the throat diameter of a nozzle for a steam ejector with pressure 6 bar and steam quality $x = 0.98$ in front of the nozzle and pressure 0.12 bar after the nozzle.

References

1. Leuschner, G: *Kleines Pumpenhandbuch für Chemie und Technik*, Verlag Chemie, Weinheim, 1967.
2. Perry, R. H. and C. H. Chilton: *Chemical Engineers' Handbook*, Mc-Graw-Hill, New York, 5th Edn., 1973.
3. Natural Gas Suppliers Association: *Engineering Data Book*, Tulsa, Oklahoma, 9th Edn., 1972.
4. Reid, R. C., J. M. Prausnitz, and Th. K. Sherwood: *The Properties of Gases and Liquids*, McGraw-Hill, New York, 3rd Edn., 1977.
5. The Heat Exchange Institute: *Standards for Steam Jet Ejectors*, 3rd Edn., 1967.
6. Wiegand: *Einstufige und mehrstufige Strahlpumpen*, Wiegand Karlsruhe, 1970.
7. The Heat Exchange Institute: *Standards for direct contact barometric and low level condensers*, 5th Edn., 1970.
8. Blenke, H., K. Bohners, and E. Vollmerhaus: Untersuchungen zur Berechnung des Betriebsverhaltens von Treibstrahlförderern, *Chem.-Ing.-Techn.*, 35, 201–208 (1963).
9. Witte, J.: Efficiency and Design of Liquid-gas Ejectors, *Brit. Chem. Eng.*, 10, 602–607 (1965).
10. Diels, K. and R. Jaeckel: *Leybold Vakuum-Taschenbuch*, Springer-Verlag, Berlin, 2nd Edn., 1962.
11. Brunner, W. F. and Th. H. Batzer: *Practical Vacuum Techniques*, Reinhold, New York, 1965.
12. Petzold, M.: Graphische Methode zur Bestimmung des Dampfverbrauches von Brüdenverdichtern, *Chem.-Ing.-Techn.*, 22, 147–150 (1950).
13. Flügel, G.: *Berechnung von Strahlapparaten*, VDI-Forschungsheft 395, VDI-Verlag, Berlin, 1939.
14. Wiegand, J.: *Bemessung von Dampfstrahl-Verdichtern und ihr Verhalten bei wechselnden Betriebsbedingungen*, VDI-Forschungsheft 401, VDI-Verlag, Berlin, 1940.
15. Bauer, B.: *Theoretische und experimentelle Untersuchungen an Strahlapparaten für kompressible Strömungsmittel*, VDI-Forschungsheft 514, VDI-Verlag, Düsseldorf, 1966.

CHAPTER 5

Agitation

Liquids are agitated so that a suspension of particles can be obtained, liquids can be blended, bubbles can be dispersed, homogeneous mixtures can be obtained, or so that heat and/or mass transfer can be promoted.

Agitators

The best type of *impeller* to be used depends on the liquid viscosity. *Propellers*, Figure 5.1a, and *turbine impellers*, Figure 5.1b, are efficient in liquids of low viscosity ($\mu < 10$ cP). In highly viscous liquids ($\mu > 150$ P = 15 N s/m^2) the impeller should pass through a large fraction of the volume. Suitable ones are *gate-type paddles* (Figure 5.1c), *anchor impellers* (Figure 5.1d), and *finger agitators* (Figure 5.1e). For Non-Newtonian liquids the *helical ribbon impeller* is employed.[1]

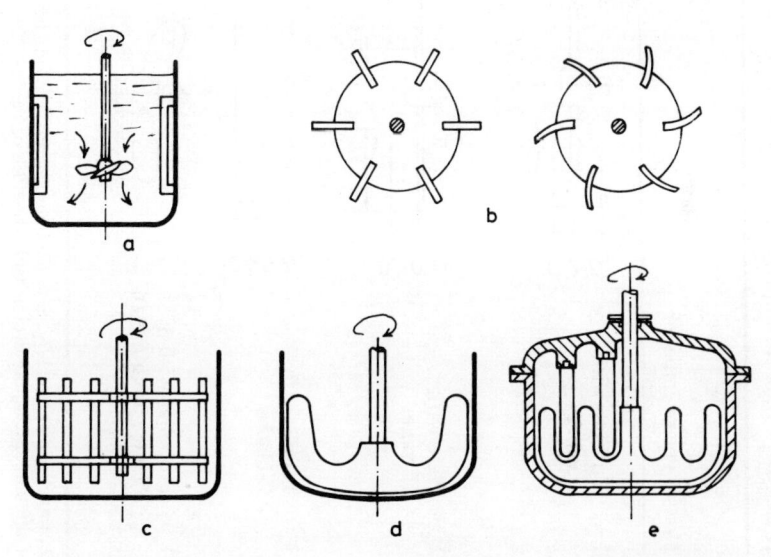

Figure 5.1 Agitators: a, tank with propeller and baffles to prevent rotation of the liquid. Clearance between the baffles and the tank prevents formation of stagnant pockets of liquid. b, turbine impellers seen from above; c, gate-type paddle; d, anchor impeller; e, finger agitator

Table 5.1 Data for impellers

No.	Type of impeller		$\dfrac{D_t}{D_i}$	$\dfrac{z_e}{z_i}$	$\dfrac{z_i}{D_i}$	Baffles Number	B/D_i
1	Turbine with 6 vanes	0.2 D_i, 0.25 D_i	3	2.7 –3.9	0.75 –1.3	4	0.10
2	Plate with 16 vanes	0.10 D_i, 0.35 D_i	2.5	2.5	0.75	4	0.25
3	Paddle with 2 blades	0.25 D_i	3	2.7 –3.9	0.75 –1.3	4	0.10
4	Propeller with 3 blades, pitch $h = D_i$		3	2.7 –3.9	0.75 –1.3	4	0.10

D_i, impeller diameter; D_t, tank diameter; h, propeller pitch; z_i, height of impeller above bottom of the tank; z_e, liquid level in tank; B, width of baffles.

Power consumption

The power consumption P in agitation for a certain type of impeller with given geometrical relations to the tank is determined by

D a characteristic length
n number of revolutions per unit time
ρ density of liquid
μ dynamic viscosity

The dimensional analysis in Example 2.2 gave

$$\frac{P}{\rho D^5 n^3} = f\left(\frac{\rho n D^2}{\mu}\right). \tag{5.1}$$

The function $\rho n D^2/\mu$ in equation (5.1) can be determined experimentally for several types of impellers. Results for some are given in Table 5.1 and Figure 5.2.

Equation (5.1) is derived assuming that gravitational forces have a negligible influence. In general, this assumption is satisfied if baffles are installed.

Two straight-blade turbines on the same shaft one impeller diameter apart, draw almost twice the power of one turbine.

Effects of agitation

The effect of the agitation increases with the *agitation intensity*, R, defined as power consumption per unit liquid volume in the tank:

	Agitation intensity R kW/m^3	Recommended tip speed for turbine agitators,[3] m/s
Very low agitation	0.02	
Low agitation	0.06	2.5 to 3.3
Medium agitation	0.15	3.3 to 4.1
High agitation	0.4	4.1 to 5.6
Very high agitation	0.9	

Measurements show that stage efficiencies in liquid–liquid extraction,[4,5] and dissolution rates for particles kept in suspension[6] are functions of the agitation intensity R.

Figure 5.2 shows that the power number $P/\rho D_i^5 n^3$ in geometrically similar tanks with baffles is almost constant for $\rho n D_i^2/\mu > 2000$, i.e.

$$P \approx C_1 \rho D_i^5 n^3 \tag{5.2}$$

where C_1 is a constant. For $\rho n D_i^2/\mu > 2000$ this gives

$$R = C_2 \frac{\rho D_i^5 n^3}{D_i^3} = C_2 \rho D_i^2 n^3. \tag{5.3}$$

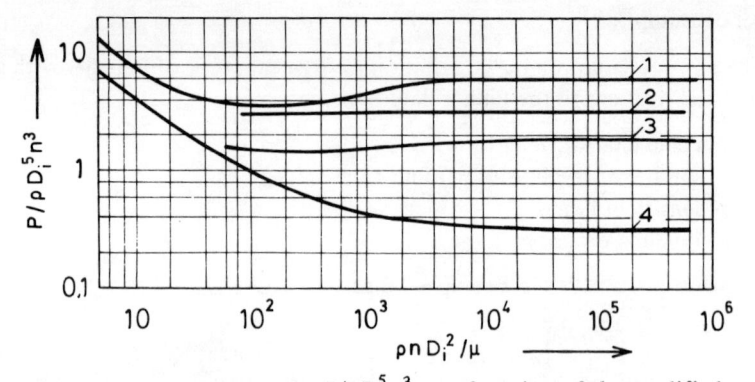

Figure 5.2 Power number $P/\rho D_i^5 n^3$ as a function of the modified Reynolds number $\rho n D_i^2/\mu$ for different impellers. The numbers on the curves refer to the numbers in Table 5.1.[2]

With the same agitation intensity in model and prototype, equation (5.3) gives

$$\frac{n_p}{n_m} = \left(\frac{L_m}{L_p}\right)^{2/3}. \tag{5.4}$$

For scale-up of mixing tanks an unchanged ratio between the pumping capacity of the impeller and the rate of liquid flow through the tank is recommended.[7]

Dispersed gas reduces the power consumption of agitators. The power consumed by a turbine impeller dispersing a gas can be approximated by the equations[7]

$$P_a = \left(1 - 2.1 \frac{\dot{Q}}{nD_i^3}\right) P \tag{5.5}$$

for

$$\frac{\dot{Q}}{nD_i^3} < 0.21$$

and

$$P_a = \left(0.62 - 3.08 \frac{\dot{Q}}{nD_i^3}\right) P \tag{5.6}$$

for

$$0.21 < \frac{\dot{Q}}{nD_i^3} < 0.66$$

where P is the power consumed by ungassed liquid and $\dot{Q}$ is the volumetric gas flow rate, m³/s. The equations are valid for Froude numbers, $Fr = n^2 D_i/g$, from 0.1 to 2.0. Equation (5.6) corresponds to flooding.

The fluid *jet mixer* (Figure 5.3) is an alternative to an impeller mixer. The primary fluid can be liquid, vapour, or gas. With liquid as primary fluid, the minimum pressure in front of the nozzle is given in Table 5.2.

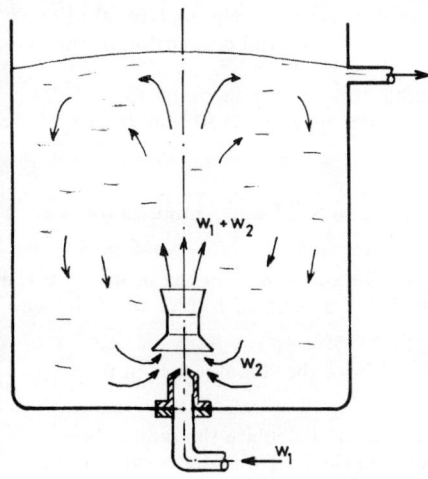

Figure 5.3 A fluid jet mixer.

Table 5.2 Recommended minimum liquid pressure in front of the nozzle.[9]

Liquid viscosity, cP	Minimum liquid pressure, MN/m^2
<50	0.07–0.14
200–800	0.3 –0.45
2000–6000	0.6 –0.7

The quantity of circulating liquid is estimated by the momentum equation,

$$w_1 V_1 + w_2 V_2 = (w_1 + w_2) V_3 \tag{5.7}$$

where

w_1 = mass flow of primary fluid, kg/s
w_2 = mass of entrained liquid, kg/s
V_1 = primary fluid velocity at the exit of the nozzle, m/s
V_2 = inlet velocity of entrained liquid, m/s
V_3 = liquid velocity through the throat of the diffuser, m/s.

Fluid jet mixers are usually installed in pipe lines.[9]
For mixing of powders the reader is referred to the article by Fan et al.[10]

EXAMPLE 5.1. TANK FOR DISSOLUTION OF CRYSTALS

200 kg/h of crystals are to be dissolved at a temperature of 75°C in a tank with propeller and steam jacket. Solvent preheated to 75°C is pumped into the tank continuously and the solution removed as overflow (Figure 5.4).

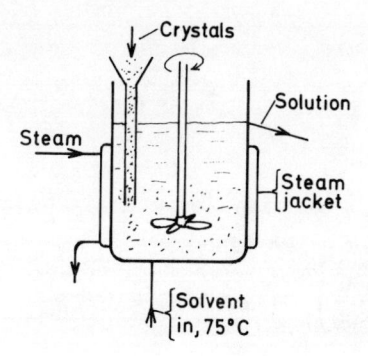

Figure 5.4 Dissolution tank of Example 5.1.

The specific densities of both solvent and solution are approximately 900 kg/m³ and their specific heat capacities 2.7 kJ/(kg °C). Heating and dissolution of the crystals requires 460 kJ/kg.

The following results were obtained in batch tests with a geometrically similar model containing 30 litres of liquid, with propeller speed $n = 1500$ rev./min $= 25$ rev./s and power consumption 0.015 kW.

1. At 75°C the dissolution rate was 4 kg/h.
2. The time required to heat 30 litres from 70°C to 80°C was 5 minutes for the solvent and 6 minutes for the solution.

(a) Estimate the volume of the prototype and the value of n for the propeller if the agitation intensity R kW/m³ is unchanged and this is assumed to give an unchanged dissolution rate.

(b) What quantity of heat will be transferred from the steam jacket in the prototype if the liquid is heated from 70 to 80°C?

(c) The 200 kg/h crystals are added continuously. Calculate the tank volume needed if the agitation intensity is unchanged and the heat transfer is sufficient to keep the solution at a constant temperature of 75°C.

Agitation will be in the fully turbulent region, i.e. the power number $P/\rho n^3 D^5$ is a constant. The calorific value of the tanks is negligible. The heat transferred is proportional to $Ah\Delta\theta$ where A is the steam-heated surface area and h the heat transfer coefficient of the agitated tank,

$$h = \text{const.} \left(\frac{k}{D_i}\right) \left(\frac{\rho n D_i^2}{\mu}\right)^{0.6} \left(\frac{c\mu}{k}\right)^{0.3}$$

or with constant density ρ, heat capacity c, viscosity μ, and thermal conductivity k,

$$h = \text{const.} \, n^{0.6} D_i^{0.2}.$$

The temperature difference $\Delta\theta$ is unchanged.

Solution

(a)   Volume needed,

$$v = 30 \, \frac{200}{4} = 1500 \text{ litres.}$$

Corresponding length ratio,

$$\lambda = L_p/L_m = \sqrt[3]{1500/30} = 3.68.$$

Equation (5.4) gives

$$n_p = n_m \left(\frac{L_m}{L_p}\right)^{2/3} = 1500 \left(\frac{1}{3.68}\right)^{2/3}$$

$$= 630 \text{ rpm.}$$

(b)   Heat transferred in the model,

$$\dot{Q}_m = (0.03)(900)(2.7)(80-70)(60)/6 = 7290 \text{ kJ/h}$$

$$h_p = h_m \left(\frac{n_p}{n_m}\right)^{0.6} \left(\frac{D_p}{D_m}\right)^{0.2} = \left(\frac{630}{1500}\right)^{0.6} (3.68)^{0.2} h_m = 0.77 \, h_m$$

$$\frac{\dot{Q}_p}{\dot{Q}_m} = \frac{A_p h_p}{A_m h_m} = \left(\frac{L_p}{L_m}\right)^2 \frac{h_p}{h_m} = (3.68^2)(0.77) = 10.43$$

$\dot{Q}_p = (10.43)(7290) = 76{,}000$ kJ/h.

(c) Heat required, $Q_p = (460)(200) = 92{,}000$ kJ/h. The new length ratio λ is sought:

$$\frac{\dot{Q}_p}{\dot{Q}_m} = \frac{A_p h_p}{A_m h_m}$$

where

$A_p/A_m = \lambda^2$ and $h_p/h_m = (n_p/n_m)^{0.6}\lambda^{0.2} = \lambda^{0.2}/(\lambda^{2/3})^{0.6} = \lambda^{-0.2}$.

$\dot{Q}_p = \dot{Q}_m \lambda^2 \lambda^{-0.2} = \lambda^{1.8}\, \dot{Q}_m$

$92{,}000 = \lambda^{1.8}\, 7290$,

$\lambda = 4.09$

or volume $v_p = \lambda^3 v_m = (4.09^3)(30) = 2050$ litre.

Note In scale-up of tanks with heating or cooling jackets the heating surface area must be checked. The volume increases in proportion to L^3 and the surface only to L^2.

EXAMPLE 5.2. EMULSIFYING TANK

An emulsion of heavy tar oil in water is to be produced in batches in a steam jacketed pot with impeller. The water is introduced first and heated to 60°C. During continued agitation the preheated tar oil and emulsifier are added. Tests with a model, with diameter $D_m = 0.25$ m, liquid volume 12.5 litre, and $n = 1425$ rev./min for the impeller, gave the following results:

average power consumption during the emulsifying process, $P_m = 0.006$ kW	
heating time	8.5 min
emulsifying time	15 min

To obtain the same emulsifying time the prototype must have the same agitation intensity, R kW/m³.

Calculate for the prototype with diameter $D_p = 1.5$ m:

(a) the average power consumption during the emulsifying process;

(b) n_p rev./min;

(c) the heating time in minutes;

(d) the maximum production in an 8 hour day. 20 minutes are required to empty the pot and fill it again with water.

The equation for heat transfer is the same as in Example 5.1.

Solution

(a) $P_p = P_m (D_p/D_m)^3 = 0.006(1.5/0.25)^3$
      $= 1.3$ kW.

Note The power consumption can vary considerably during a batch process, and the motor must be sized for the maximum and not the average power consumption.

(b) Equation (5.4) gives,
$$n_p = n_m(D_m/D_p)^{2/3} = 1425\,(0.25/1.5)^{2/3} = 432 \text{ rev./min.} \quad (a)$$

(c) The heat required is proportional to the volume,
$$\frac{\dot{Q}_p}{\dot{Q}_m} = \left(\frac{D_p}{D_m}\right)^3 = \frac{A_p h_p \Delta\theta\, t_p}{A_m h_m \Delta\theta\, t_m} \quad (b)$$

where t_p and t_m are the heating time in prototype and model, respectively.
In equation (b)
$$A_p/A_m = (L_p/L_m)^2. \quad (c)$$

As in Example 5.1
$$h_p/h_m = (n_p/n_m)^{0.6}(D_p/D_m)^{0.2}. \quad (d)$$

Equations (a), (b), (c), and (d) give
$$t_p = (D_p/D_m)^{1.2} t_m = (1.5/0.25)^{1.2}(8.5)$$
$$= 73 \text{ min.}$$

(d) Total time per batch, 73 + 15 + 20 = 108 min. Number of batches per 8 hours, (8)(60)/108 = 4.4, i.e. 4 batches or $(4)(12.5)(1.5/0.25)^3$ = 10,800 litre.

Note The time for heating is (100)(73)/108 = 68% of the total time per batch. An increased production can easily be obtained either by preheating the water or by direct steam injection into the water.

EXAMPLE 5.3. GAS ABSORPTION

An oxygen-consuming fermentation was investigated in the tank of Figure 5.5. The liquid volume in the tank is 100 litres. The absorption of oxygen turned out to be the rate controlling step in the fermentation. The absorption rate is given by the equation
$$w = k_1 a(c_i - c_1) \text{ kg/(h m}^3)$$
where
  c = oxygen concentration in the bulk of the liquid, kg/m^3
  c_i = oxygen concentration in liquid in equilibrium with the gas in the bubbles, kg/(h m^3)
  k_1 = liquid film coefficient, m/h
  a = specific interface area of the bubbles, m^2/m^3.

Results from other investigations[11] with liquid film controlled absorption and moderate gas velocities indicate the relationship
$$k_l a \approx f(V_g) R$$

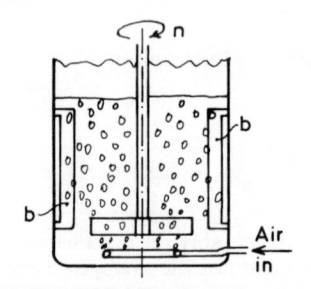

Figure 5.5 Tank with gas injection, paddle and baffles b.

where V_g is the gas velocity referred to the total cross section, $V_g < 0.05$ m/s and R is the agitation intensity without gas injection, kW/m^3.

Tests carried out with the tank of Figure 5.5 gave the following results.

Measurement	3	4	7	8	9	10
Impeller speed, n_m rev./min	92	92	145	145	145	145
Air velocity, V_g m/min.	0.03	0.12	0.024	0.06	0.18	0.216
$k_1 a$ h^{-1}	5.5	10	13	26	56	56

(a) What values of $k_1 a$ can be expected in a similar tank with liquid volume 1500 litres, as a function of the superficial air velocity, V_g, and the revolutions per minute of the impeller, n_p? The agitation is in the region where the power number $P/\rho D_i^5 n^3$ is constant.

(b) What are the minimum limits of error to be expected?

Solution

(a) Equation (5.3) with unchanged ρ, $R = $ (const.)$D_i^2 n^3$ or $k_1 a \approx$ (const.)$f(V_g)D_i^2 n^3 = F(V_g)D_i^2 n^3$

$$k_1 a / D_i^2 n^3 \approx F(V_g). \qquad (a)$$

Values of the left hand side of this equation are shown in the following table where k_1 is in m/s, n in rev./s and the third root of the volume, $L = 0.1^{1/3} = 0.464$ m, is inserted instead of D_i.

Measurement	3	4	7	8	9	10
Velocity, V_g m/min	0.03	0.12	0.024	0.06	0.18	0.216
$F(V_g) = k_1 a/L^2 n^3$ s^2/m^2	0.00197	0.00358	0.00119	0.00238	0.00512	0.00512

The values from this table are plotted in Figure 5.6, and a curve is drawn through them and the origin.

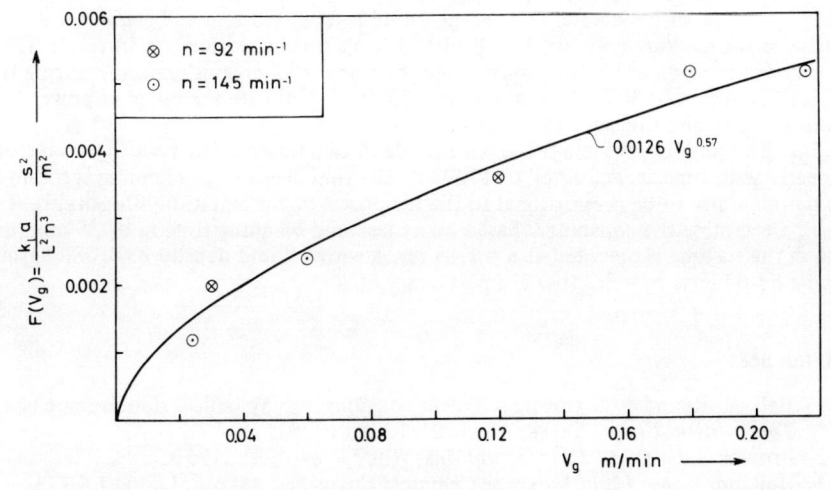

Figure 5.6 The left-hand side of equation (a), $k_1 a/L^2 n^3$ s^2/m^2, plotted as a function of the gas velocity V_g m/min in the model.

The curve in Figure 5.6 corresponds to

$$F(V_g) = 0.0126 \, V_g^{0.57} \, \text{s}^2/\text{m}^2 \tag{b}$$

which gives

$$(k_1 a)_p = F(V_g) L_p^2 n_p^3 = 0.0126 \, V_g^{0.57} 1.5^{2/3} n_p^3 = 0.0165 V_g^{0.57} n_p^3 \quad \text{s}^{-1}$$

or

$$(k_1 a)_p = 59 V_g^{0.57} n_p^3 \quad \text{h}^{-1} \tag{c}$$

Note This answer is limited to the same geometry, the same superficial gas velocity V_g and the same agitation intensity as in the model, i.e.

$$0.03 < V_g < 0.22 \, \text{m/min} \text{ and } 50 < n_p < 80.$$

The limits for n_p are given by equation (5.4),

$$n_p = \left(\frac{L_m}{L_p}\right)^{2/3} n_m.$$

(b) The standard deviation in $F(V_g)$ for the 6 measurements is

$$\sigma = \sqrt{\sum_1^6 [F(V_g)_{\text{exp}} - F(V_g)_{\text{calc}}]^2/6} = \pm 0.00025 \, \text{s}^2/\text{m}^2$$

where $F(V_g)_{\text{exp}}$ is from the table under (a) and $F(V_g)_{\text{calc}}$ is from equation (b). This gives a 95% confidence interval for $F(V_g)$ of $2\sigma = \pm 0.0005 \, \text{s}^2/\text{m}^2$. The corresponding confidence limits for $(k_1 a)_p$ are

$$(0.0005)(1.5^{2/3})(3600) n_p^3 = 2.4 n_p^3 \quad \text{h}^{-1}$$

or

$$(k_1 a)_p = (59 V_g^{0.57} \pm 2.4) n_p^3 \, \text{h}^{-1}.$$

Problems

5.1. A tank with diameter $D_t = 30$ m, liquid level $z_e = 3.5$ m and geometrical ratios as for measurement no. 1 in Table 5.1 has a turbine with $n = 1.14$ rev./s. The density of the liquid $\rho = 1080$ kg/m^3 and the viscosity increases gradually during the operation from $\mu = 300$ cP initially to $\mu = 4000$ cP. Estimate the range of power consumption and the agitation intensity.

5.2. The tank in problem 5.1 is used to blend two liquids. The viscosity increases linearly with time from 300 cP to 4000 cP. The time needed for blending is found experimentally to be proportional to the reciprocal of the agitation intensity, $t_b = C/R$. Determine the constant C based on a measured blending time t_b of 25 minutes when the turbine is operated at $n = 1.14$ rev./s with a liquid density $\rho = 1080$ kg/m^3 and a liquid level $z_e = 3.2$ (use a time-average of R).

References

1 Hall, K. R. and J. C. Godfrey: Power consumption by helical ribbon impellers, *Trans. Instn. Chem. Engrs.*, **48**, T201-T208 (1970).
2 Brown, G. G. et al.: *Unit Operations*, Wiley, New York, 1950.
3 Holland, F. A.: *Fluid Flow for Chemical Engineers*, Arnold, London, 1973.
4 Flynn, A. W. and R. E. Treybal: Liquid–liquid extraction in continuous-flow agitated extractors, *A.I.Ch.E. J.*, **1**, 324–328 (1955).

5 Valentin, F. H. H.: Mass transfer in agitated tanks, *Brit. Chem. Eng.*, **12**, 1213–1218 (1967).
6 Büche, W.: Leistungsbedarf von Rührwerken, *Zeitschr. VDI*, **81**, 1065–1069 (1937).
7 Berresford, H. I. *et al.*: Continuous blending of low viscosity fluids: an assessment of scale-up criteria, *Trans. Instn. Chem. Engrs.*, **48**, T21–T27 (1970).
8 Calderbank, P. H., in V. W. Uhl and J. B. Gray: *Mixing: Theory and Practice*, vol. II, Academic Press, New York, 1967.
9 Harris, L. S.: Jet eductor mixers, *Chem. Eng.*, **73**, No. 21, 216–222 (1966).
10 Fan, L. T., S. J. Chen, and C. A. Watson: Solids mixing, *Ind. Eng. Chem.*, **62**, No. 7, 53–69 (1970).
11 Friedman, A. M. and E. N. Lightfoot jr.: Oxygen absorption in agitated tanks, *Ind. Eng. Chem.*, **49**. 1227–1230 (1957).

CHAPTER 6
Particle and Drop Mechanics

The movement of particles, drops, and bubbles in fluids is of vital importance in many processes. It is also a field in which we have to rely upon empirical data. Theory, however, may help to plan experiments and to check the experimental results.

The particle sizes and properties of liquid and solid dispersoids depend on the source and nature of the operation generating the particles. Figure 6.1 gives a list of some particle sizes, methods for their analysis, and types of gas cleaning equipment.[1] For details of gas cleaning equipment refer to the literature.[2,3,4]

The frictional force or drag on a particle or drop that moves with velocity V relative to surrounding fluid is

$$F = f_D A_p \rho \frac{V^2}{2} \qquad (6.1)$$

where

A_p = area of particle projected on a plane normal to direction of motion, m^2
f_D = drag factor given in Figure 6.2 as a function of Reynolds number and shape factor Ψ

$$\Psi = \frac{\text{surface of a sphere with the same volume as the particle}}{\text{surface of the particle}} \qquad (6.2)$$

D_s in Reynolds number in Figure 6.2, the diameter of a sphere with the same volume as the particle.

For calculation purposes it is convenient to substitute the curve for spheres ($\Psi = 1.0$) in Figure 6.2 by a series of straight lines,

for $500 < \text{Re}_D < 200{,}000$, $f_D = 0.44$ \qquad (6.3)

for $2 < \text{Re}_D < 500$, $f_D = 18.5 \, \text{Re}_D^{-0.5}$ \qquad (6.4)

for $0.00001 < \text{Re}_D < 2$, $f_D = 24 \, \text{Re}_D^{-1}$. \qquad (6.5)

Terminal velocity

The terminal velocity, V_t, of a falling drop or particle is the velocity when the gravitational force equals the drag force; for spheres

$$\frac{\pi}{6} D^3 (\rho_p - \rho) g = f_D \frac{\pi}{4} D^2 \rho \frac{V_t^2}{2}$$

$$V_t = 2 \sqrt{\frac{D(\rho_p - \rho)g}{3 f_D \rho}} \tag{6.6}$$

where ρ_p is the density of a particle and ρ that of the surrounding fluid, kg/m^3. The equations (6.3), (6.4), (6.5), and (6.6) give

for $\quad 500 < \text{Re}_D < 200{,}000, \quad V_t = 1.74 \sqrt{\dfrac{D(\rho_p - \rho)g}{\rho}} \tag{6.7}$

for $\quad 2 < \text{Re}_D < 500, \quad V_t = \left[0.072 D^{1.6} \dfrac{\rho_p - \rho}{\mu^{0.6} \rho^{0.4}} g \right]^{1/1.4} \tag{6.8}$

for $0.00001 < \text{Re}_D < 2, \quad V_t = 0.0556 D^2 \dfrac{\rho_p - \rho}{\mu} g. \tag{6.9}$

Equation (6.9) or Stoke's law is not valid for $\text{Re} < 10^{-5}$ where 'slip' between the gas molecules gives an increased sedimentation rate. This is taken into account by the *Cunningham correction factor*,[6]

$$V_c = \left[1 + \frac{\lambda_m}{D} (1.644 + 0.522 e^{-0.656 D/\lambda_m}) \right] V_t \tag{6.10}$$

where

V_c = sedimentation velocity, m/s

V_t = sedimentation velocity according to equation (6.9), m/s

λ_m = mean free path of the molecules, m.

Based on the kinetic theory of gases and the theory of viscosity of gases of low density, the mean free path is[7]

$$\lambda_m = 3 \frac{\mu}{\rho} \sqrt{\frac{\pi M}{8 R T}} \tag{6.11}$$

where M is the molecular weight of the gas.

The movement of particles of diameter less than 0.1 μm is dominated by *Brownian motion*, i.e. irregular movement due to impact with the fluid molecules.

Equations (6.6) to (6.10) also apply to equipment with centrifugal forces when the acceleration of gravity, g, is replaced by the centrifugal acceleration, V^2/r or $\omega^2 r$.

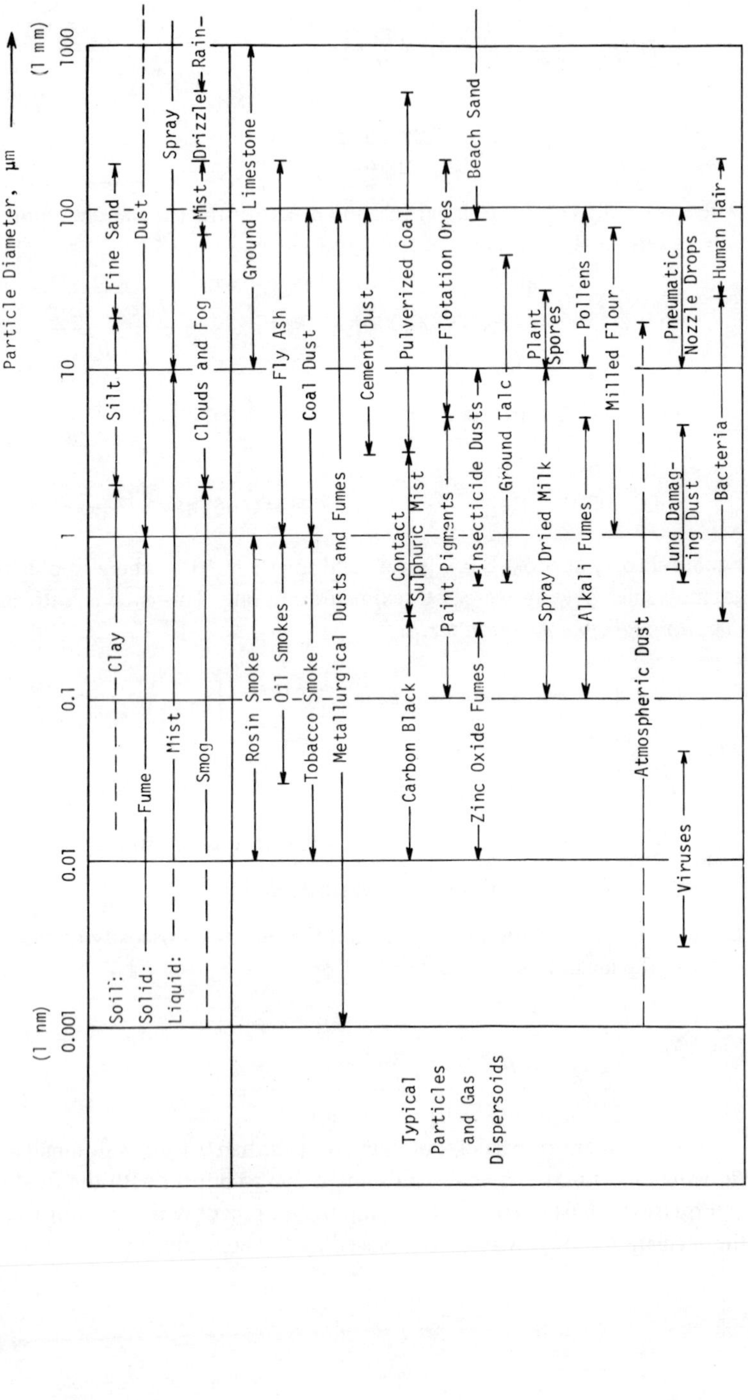
Particle Diameter, μm
0.001
(1 nm)
0.01
0.1
1
10
100
1000
(1 mm)
Soil:
Solid:
Liquid:
Clay
Silt
Fine Sand
Fume
Dust
Mist
Spray
Smog
Clouds and Fog
Mist
Drizzle
Rain
Typical
Particles
and Gas
Dispersoids
Rosin Smoke
Ground Limestone
Oil Smokes
Fly Ash
Tobacco Smoke
Coal Dust
Metallurgical Dusts and Fumes
Cement Dust
Carbon Black
Contact
Sulphuric Mist
Pulverized Coal
Paint Pigments
Flotation Ores
Zinc Oxide Fumes
Insecticide Dusts
Beach Sand
Ground Talc
Spray Dried Milk
Plant
Spores
Alkali Fumes
Pollens
Milled Flour
Atmospheric Dust
Viruses
Lung Damag-
ing Dust
Pneumatic
Nozzle Drops
Bacteria
Human Hair

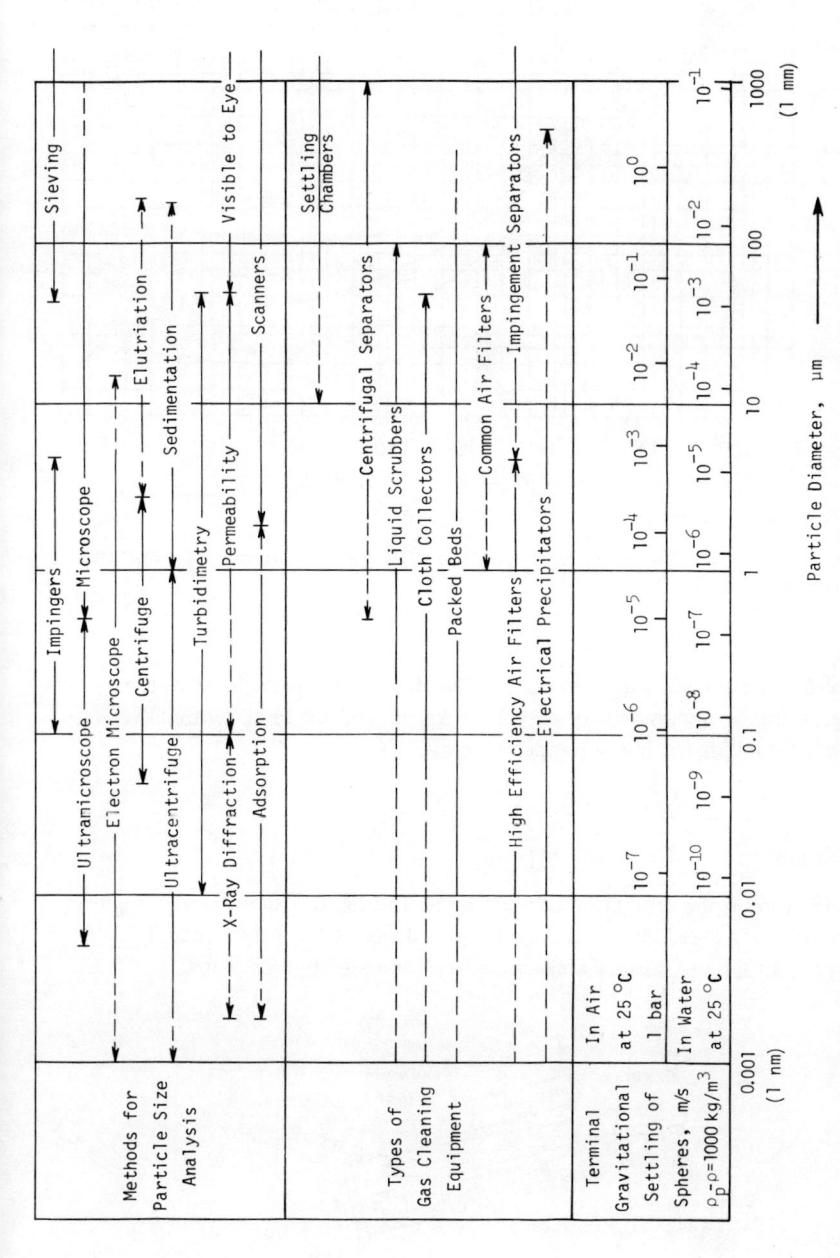

Figure 6.1 Characteristics of particles and particle dispersoids.[1] Reproduced from Lapple C. E., *Stanford Res. Inst. J.*, 5, 94 (1961) by permission of the Stanford Research Institute.

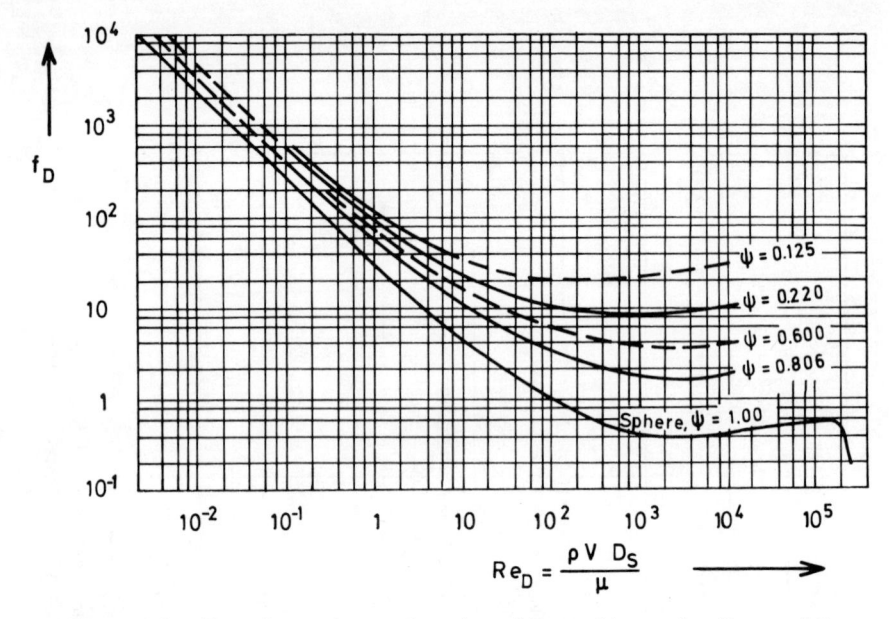

Figure 6.2   Drag factor f_D as a function of Reynolds number Re_D and the shape factor Ψ.[5]

Fall velocity equivalent diameter is used to characterize particles. It is the diameter of a sphere with density $\rho_p = 1000$ kg/m^3 and the same terminal velocity as the particle falling in air at a temperature of 20°C.

Classification

Classification is the separation of particles according to size or density. Figure 6.3 is an example of a classifier based on gravity, and Figure 6.4 a thickener. Table 6.1 give empirical values of surface areas necessary for a selection of pulps.

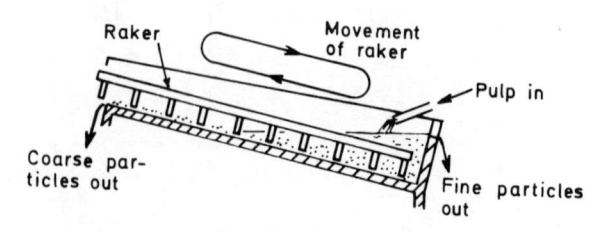

Figure 6.3   Rake classifier: the rake is lifted up before it returns to the right.

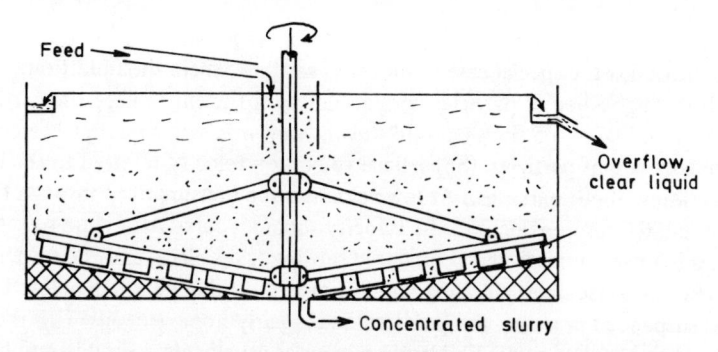

Figure 6.4 Thickener: the rake is lifted up when the thickener stops to prevent it sticking in the sediment.

Table 6.1 Empirical data for thickener surfaces.[8]

Type of pulp	% solids feed	underflow	Area, m^2/ton/h
Concentrate from copper fraction	14–50	40–75	7–45
Sludge from sugar beet	3–8	17–22	10–45
Cement sludge	16–20	60–70	35–55
Clay in water	1–4	15–45	100–500
Magnesium hydroxide from brine	8–10	25–50	135–220
Magnesium hydroxide from sea water	0,3–0,6	11–22	450–1100
Titanium hydroxide	10–20	40–45	135–250
Bauxite from sulphuric acid treatment	5–9	20–30	170–340

Hindered settling

Particle concentrations above 0.1% by volume give reduced settling velocities due to increase both in the apparent viscosity and in the density. The hindered settling velocity of particles in non-flocculated suspension is estimated by the equation

$$V_h = \epsilon^n V_t \tag{6.12}$$

where

V_t = terminal velocity in dilute suspensions, m/s
ϵ = (volume of liquid)/(volume of suspension)
n = exponent, according to Scholl,[9] $n = 3.65$ for $\epsilon > 0.6$, and according to Maude and Whitmore,[10]

$n = 4.4\, Re^{-0.08}$ for $1 < Re < 2500$
$n = 2.35$ for $Re > 2500$.

Fluidization

Fluidization is the special case of hindered settling, where the fluid flows upwards through a bed of particles with a velocity sufficient to keep the particles in suspension. Figure 6.5 shows a fluid flowing upwards with superficial velocity V_s through a bed of particles. (Superficial velocity = velocity referred to the total cross-section without particles.) At low velocities the pressure drop through the bed is approximately proportional to the velocity squared, curve AB Figure 6.6. At velocities between point C and D hindered settling takes place, and the suspended mass sometimes resembles a boiling liquid. For Froude numbers well below 1.0, as for sand suspended in water, the fluidized bed is fairly homogeneous. This is the *particulate fluidization* state. Fluidization in gases usually gives Froude numbers above 1.0. Gas 'bubbles' rises through the bed and give the impression of a boiling liquid with jets of particles spouting upwards as the 'bubbles' break. This is the

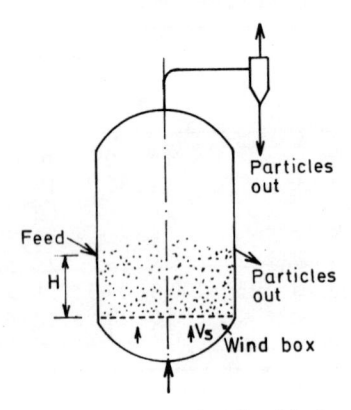

Figure 6.5 A fluidized bed.

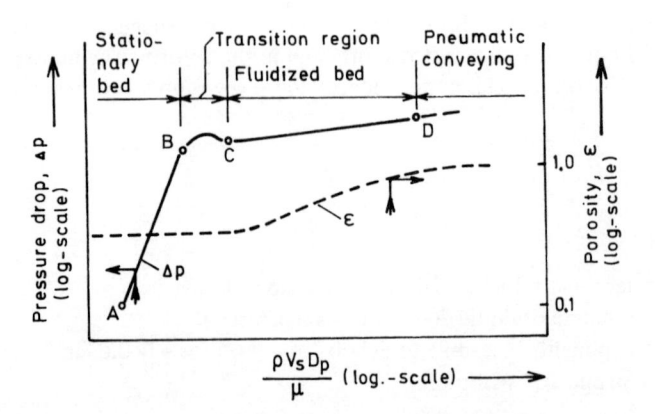

Figure 6.6 Pressure drop Δp and void fraction ϵ in a bed of particles as a function of the superficial Reynolds number $\mathrm{Re}_s = \rho V_s D_p/\mu$.

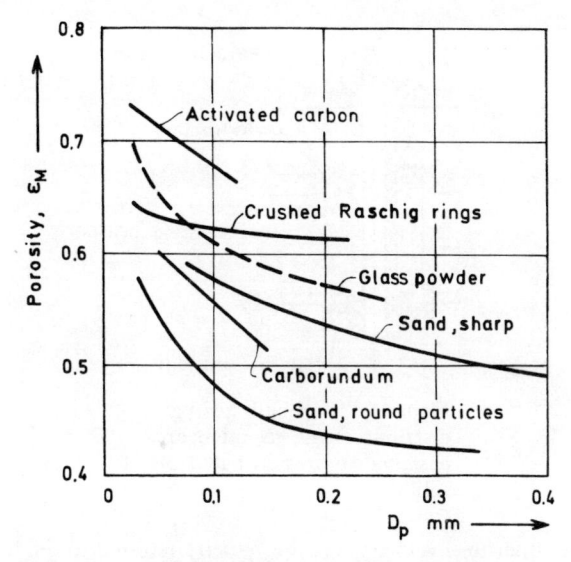

Figure 6.7 Minimum porosity during fluidization, ϵ_M, plotted as a function of particle diameter, D_p. Reproduced by permission of McGraw-Hill Book Co., from Leva, M., Fluidization, © 1959 by McGraw-Hill Inc.

aggregative fluidization state. An excellent description of fluidization is given by Flood and Lee.[11]

Fluidized beds were introduced for catalytic cracking of mineral oils. Today they are also used for other purposes, such as, mineral roasting, drying, heat transfer, crushing, and coating of particles.

Porosity is the ratio of the fluid volume to the total volume. The porosity when fluidization starts, ϵ_M, is given in Figure 6.7 as a function of particle diameter, D_p.[12]

The pressure drop in a fluidized bed corresponds to the weight of the particles, and is almost constant over the fluidization range. The pressure drop when fluidization starts is

$$\Delta p = H_M(1 - \epsilon_M)(\rho_p - \rho)g \tag{6.13}$$

where H_M = bed height when fluidization starts. In addition there can be pressure drop through the fluid distributor, which may consist of perforated plates, grids, or specially designed nozzles. Figure 6.8 shows one design. For gases at atmospheric pressure the distributor pressure drop is in the range 1000 to 10,000 N/m² (100 to 1000 mm water gauge).

For small particles where the sedimentation velocity depends on viscous forces, the Kozeny–Carman equation gives an estimate of minimum fluidization velocity,

$$V_M = \frac{\Delta p D_p^2 \epsilon_M^3}{150 H_M (1 - \epsilon_M)^2 \mu} \tag{6.14}$$

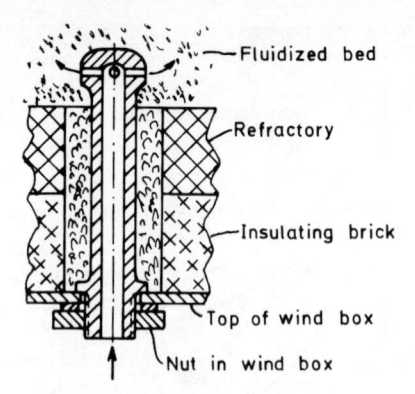

Figure 6.8 A club head gas distributor: the gas inlets are designed to prevent backflow of solids.

where V_M is the superficial velocity, i.e. the velocity referred to the total cross section. Combination with equation (6.13) gives

$$V_M = \frac{(\rho_p - \rho)gD_p^2\epsilon_M^3}{150\mu(1 - \epsilon_M)}. \tag{6.15}$$

Pneumatic conveying

Pneumatic conveying is used for granular and powder materials, such as, flour, seeds, grain, alumina, cement, and catalysts to and from fluidized beds. For calculations the reader is referred to the literature.[13,14,15]

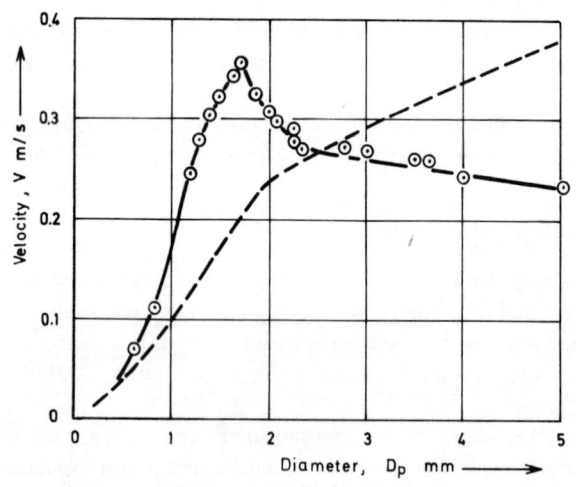

Figure 6.9 Solid curve, rising velocity of gas bubbles in water; broken curve, rising velocity of solid spheres according to equations (6.7), (6.8), and (6.9).

Drops and bubbles

Drops and bubbles can behave differently from solid spheres. This is due to deformation from spherical shape, circulation within the drops or bubbles, and (for bubbles) pulsations. Figure 6.9 gives experimentally determined rising velocities of gas bubbles in water.[16] Rising velocities for bubbles and bubble swarms in sea water, acetic acid, ethyl acetate, and glycerine solutions are given by Houghton et al.[17]

Decanting

Decanting can be carried out continuously, for example, in the decanter shown in Figure 6.10. For other types of decanters the reader is referred to literature in the field.[1,18] Previously decanters used in the petroleum industry were designed for one half to one hour residence time, while 5 to 10 minutes is sufficient for most low-viscosity liquids not showing a tendency to form stable emulsions.

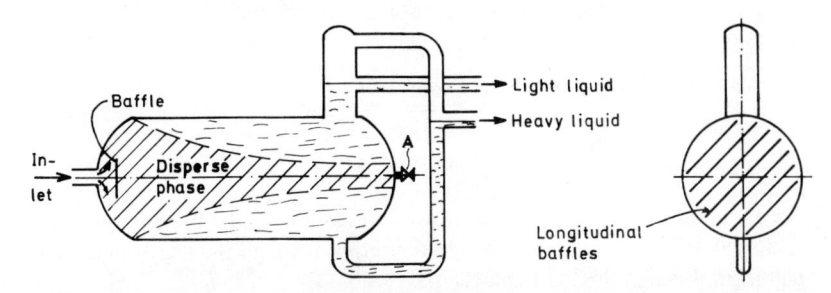

Figure 6.10 Horizontal continuous decanter with inclined baffles to secure laminar flow and decrease the distance of sedimentation before coalescence of the drops. The valve A is opened periodically to allow withdrawal of dust and dirt particles that accumulate at the interface.

Centrifuges, cyclones, and hydrocyclones

Centrifugal separators are centrifuges, cyclones, and hydrocyclones. Four types of centrifuges are shown schematically in Figures 6.11–6.14. Table 6.2 gives some characteristic data.[19]

Figure 6.15 shows the effluence of light phase at radius r_1 and the effluence of heavy phase at the adjustable radius r_2. A reduction in radius r_2 causes a reduction in the radius r_n of the 'neutral zone' indicated as a liquid/liquid interface. Small values of r_n should be used when it is important to have a low content of the light liquid in the heavy liquid, while some of the heavy liquid in the light liquid is less critical, as in separation of milk into skimmed milk and fat.

The location of the neutral zone can be derived from the centrifugal force dF acting on a cylinder with inner radius r and outer radius $r + dr$ (Figure 6.16):

$$dF = 2\pi r L \rho \omega^2 r dr.$$

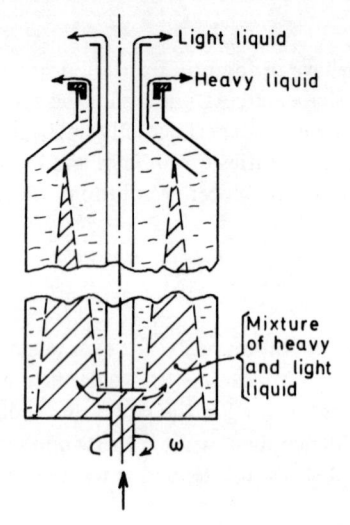

Figure 6.11   A tubular centrifuge.

The corresponding increase in pressure over dr is

$$dp = \frac{dF}{2\pi r L} = \rho \omega^2 r \, dr.$$

With the two phases and the symbols shown in Figure 6.15 the pressure on the outer cylinder wall, p_0, is given both by integration through the heavy liquid and by integration through the light and the heavy liquid,

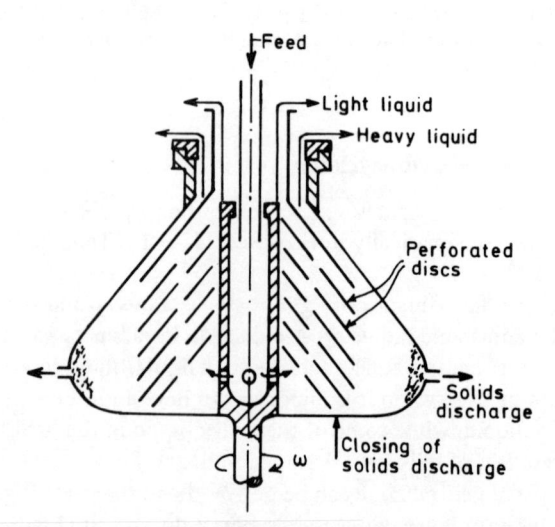

Figure 6.12   A disc centrifuge with discontinuous discharge of the concentrated solid phase.

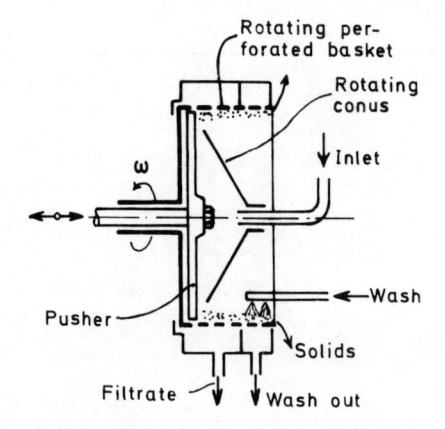

Figure 6.13 A pusher centrifuge.

$$p_0 = \rho_1 \omega^2 \int_{r_1}^{r_n} r\,dr + \rho_2 \omega^2 \int_{r_n}^{r_y} r\,dr = \rho_2 \omega^2 \int_{r_2}^{r_y} r\,dr$$

or

$$\frac{r_n}{r_1} = \sqrt{\frac{(r_2/r_1)^2 - (\rho_1/\rho_2)}{1 - (\rho_1/\rho_2)}}. \tag{6.16}$$

Cyclones are the least expensive and the most widely used means of separating solid particles or drops of diameter larger than 5 μm from a gas or vapour. As a result of agglomeration, they can also be used efficiently for particles smaller than 3 μm, in some cases with particle concentration higher than 200 grams per cubic meter of gas.

The main dimensions of a cyclone investigated by Lapple *et al.*[20] are given in Figure 6.17. The dust-laden gas enters tangentially and gets a rotating motion. The

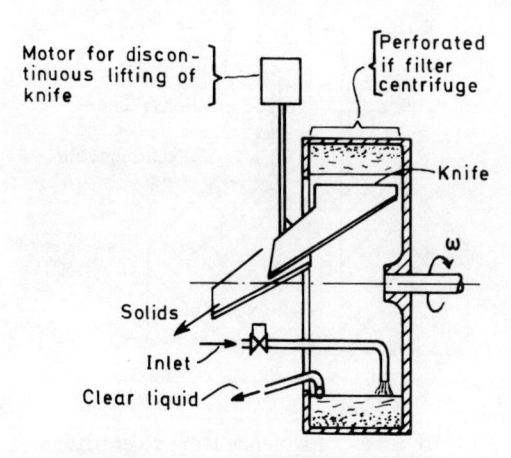

Figure 6.14 A peeler centrifuge.

Table 6.2 Characteristic data for four types of centrifuge

			Perforated basket	
Type of centrifuge Figure	Tubular 6.11	Disc 6.12	continuous 6.13	discontinuous 6.14
Suitable for:				
Breaking emulsions	X	X		
Liquid-liquid separation	X	X		
Classification	X	X	X	
Clarification	X	X	X	
Sludge concentration	X	X		
Filtering, washing, and/or drying			X	X
Capacity:				
Liquid, m³/h	0.02–6	0.12–70		
Solids, kg/h	0	9000	500–28,000	1000–12,000
Diameter, cm	4.5–15	25–120	40–250	
Height, cm	150	≈ diameter		
Motor size, kW	0.25–2.5	0.7–90	2–300	
Number of g	13,000–62,000	700–9500	100–2000	

vortex so formed develops a centrifugal force 5 to 2500 times the force of gravity, and causes the particles to settle towards the wall where they slide down to the dust outlet. The efficiency of the cyclone for particles with diameter D_p is

$$\eta = \frac{\text{mass of collected particles with diameter } D_p}{\text{mass of particles with diameter } D_p \text{ in the incoming gas}}. \quad (6.17)$$

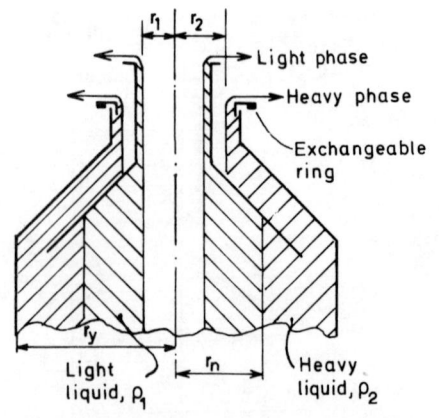

Figure 6.15 Effluence from a centrifuge: r_n = radius of the neutral zone.

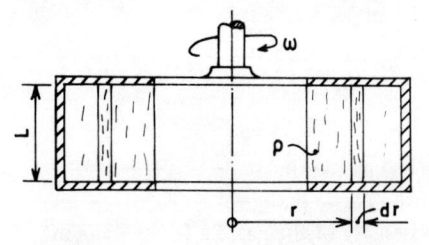

Figure 6.16 Rotating liquid-filled bowl where the gravitational force is negligible.

Cyclone calculations are based on the *cut size* D_{pc}, which is the diameter of particles of which 50 per cent are collected. For a cyclone with the dimensions as given in Figure 6.17 and low dust load, the total pressure drop through the cyclone is

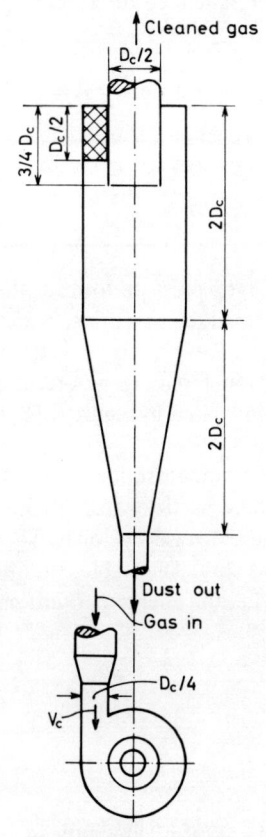

$$\Delta p_c \approx 8\rho V_c^2/2 \qquad (6.18)$$

Figure 6.17 The cyclone described by Lapple.[20]

and the cut size

$$D_{pc} \approx 0.27\sqrt{\mu D_c/V_c(\rho_p - \rho)} \tag{6.19}$$

where

ρ = density of the gas, kg/m³
ρ_p = density of the particles, kg/m³
μ = dynamic viscosity of the gas, kg/(m s) = N s/m²
V_c = gas velocity at the inlet (cross section $D_c^2/8$), usually between 6 and 20 m/s. Velocities $V_c >$ 20–25 m/s give reduced efficiency.[21]

Efficiencies for different particle sizes are given in Table 6.3.

Equations (6.18) and (6.19) show that several small cyclones in parallel ('multiclone') have a higher efficiency than one large cyclone with the same pressure drop. The equations are valid for low dust concentrations. High dust concentration gives less pressure drop due to less turbulence and some times higher efficiencies due to agglomeration in the cyclone.

Soft linings (rubber) are recommended for abrasive particles and particles larger than 200 μm.

Table 6.3 Cyclone efficiency η for different ratios of particle diameter to cut size, D_p/D_{pc}

D_p/D_{pc}	0.2	0.3	0.5	0.7	1.0	1.5	2.0	3.0	5.0	7.0	10
η	0.04	0.08	0.20	0.33	0.50	0.69	0.80	0.90	0.96	0.98	0.99

Note The vortex reduces the static pressure towards the centre. Hence, there can be sub-atmospheric pressure in a small diameter dust outlet for cyclones exhausted to the atmosphere.

In drop-collecting cyclones van Tengbergen[22] recommends the installation of a plate above the liquid level as indicated in Figure 6.18, to prevent re-entrainment from the liquid surface.

Hydrocyclones (Figure 6.19) were introduced after World War II by the Dutch State Mines for separation of solid particles from their suspensions by centrifugal force. The principle is the same as for gas cyclones. To obtain a high efficiency the hydrocyclone should have a gas core. The stable condition of a frictionless fluid in the cyclone is the free vortex, i.e. the energy per unit mass of fluid is constant over the cross section,

$$p + \rho V^2/2 = \text{constant}$$

or

$$\frac{\partial p}{\partial r} + \rho V \frac{\partial V}{\partial r} = 0.$$

The centrifugal force produces a pressure gradient,

$$\frac{\partial p}{\partial r} = \rho \frac{V^2}{r}.$$

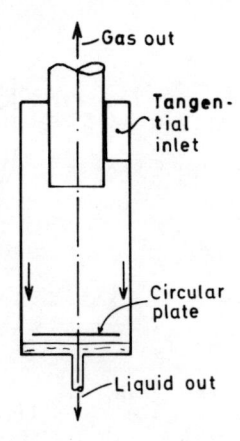

Figure 6.18 A cyclone for liquid drops.

Combination of the two equations gives

$$\frac{\partial V}{\partial r} + \frac{V}{r} = 0,$$

i.e., Vr = constant. (6.20)

Equation (6.20) corresponds to the outer broken velocity curve in Figure 6.20. Due to friction, however, the inner part of the fluid will approach the forced vortex where the angular velocity is constant. The free vortex gives constant centrifugal

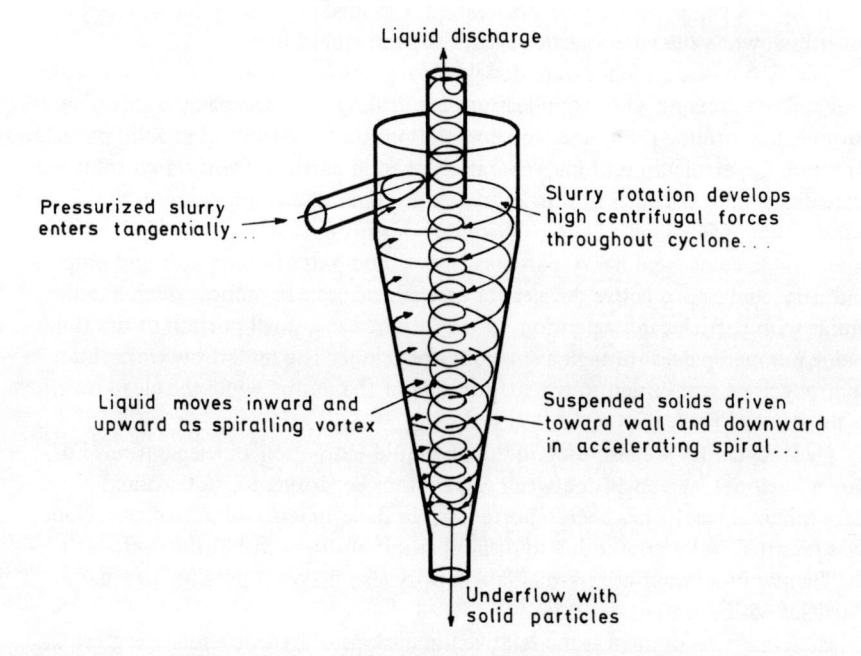

Figure 6.19 Spiral flow in a hydrocyclone.[23]

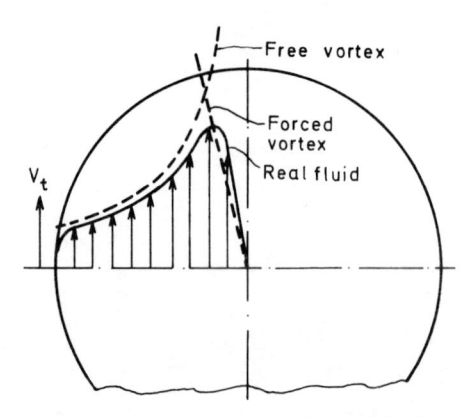

Figure 6.20 Tangential velocities in free vortex, forced vortex, and real fluids.

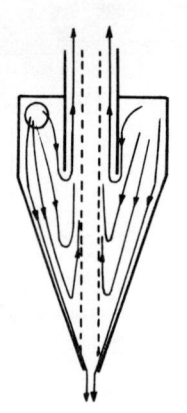

Figure 6.21 Axial and radial flow pattern: broken lines indicate the surface of a gas core.[24]

force on a particle, while the centrifugal force in the forced vortex decreases with the radius. With a gase core in the hydrocyclone (Figure 6.21) the tangential velocity approaches that of the free vortex with constant centrifugal acceleration. In small hydrocyclones the acceleration can reach values as high as 1000 times that of gravity.

In hydrocyclones used for classification, the smaller particles leave with the overflow, while the larger particles leave with the underflow.

The first hydrocyclones were designed for particle separation from slurries in mineral ore dressing. Other applications are drilling mud degassing, solids removal from 'clear' drilling fluids and well completion fluids, separation of solid particles from oil in petroleum refining, separation of solid particles from starch solutions, classification of particles in slurries of wet-ground raw material used in the production of Portland cement, thickening of slurries from crystallizers, removing undesirable coarse and heavy particles from dilute pulps in the paper and pulp industry, and as protective devices for mechanical seals in pumps which handle fluids with particles in suspension. In the latter case, a small portion of the fluid being pumped passes through a small hydrocyclone. The underflow from this hydrocyclone is returned to the suction side of the pump, while the clean overflow is returned to flush the mechanical seal.[26]

Hydrocyclones are also used in liquid–liquid extraction and separation. For larger cyclones, agreement between calculations according to Rietema and experimental results has been reported, while the efficiency of a smaller cyclone was reported to be poorer due to disintegration of drops within the cyclone.[27]

The use of a liquid of intermediate density also makes it possible to separate particles with different densities.[28]

Rietema[25] recommends the relative dimensions of hydrocyclones as given in

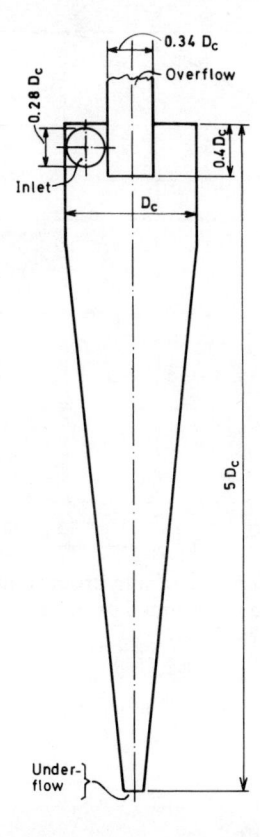

Figure 6.22 Hydrocyclone according to Rietema.[25]

Figure 6.22. He also recommends operation with approximately 10% of the liquid leaving as underflow and 90% as overflow, and that a gas core be provided as indicated by the broken lines in Figure 6.21. At very low particle concentrations the underflow can be reduced to well below 10%. The gas core can be formed either by release of gas dissolved in the slurry, or by air entering from the atmosphere. Air intake will occur automatically when the outlet is open to the atmosphere.

Figure 6.23 gives performance data for the hydrocyclone of Figure 6.22, where η is the mass fraction of input particles which appear in the underflow, D_p is the particle diameter, and D_{pc} is the diameter of particles, the efficiency of recovery of which is 50%.

Rietema[25] produced functions for D_{pc} and $\dot{Q}$ in terms of dimensionless groups for the hydrocyclone in Figure 6.22, which are well approximated by the equations

$$\left[\left(\frac{D_{pc}}{\mu}\right)^2 (\rho_p - \rho)\Delta p\right] = 0.177 \left[\frac{D_c\sqrt{\rho \Delta p}}{\mu}\right]^{0.85} \tag{6.21}$$

and

$$\left[\dot{Q}\left(\frac{\rho}{\mu}\right)^2 \sqrt{\frac{\Delta p}{\rho}}\right] = 0.2 \left[\frac{D_c\sqrt{\rho \Delta p}}{\mu}\right]^{1.85} \tag{6.22}$$

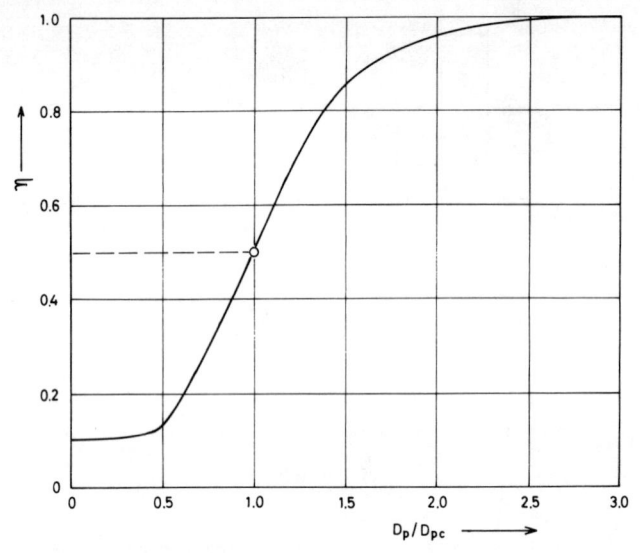

Figure 6.23 Efficiency of separation of a hydrocyclone plotted against particle size ratio determined experimentally by Rietema, with 10% underflow.[25]

where

$\dot{Q}$ = liquid flow rate, m³/s
D_c = diameter of hydrocyclone, m
V_c = liquid velocity at the inlet, m/s
ρ = liquid density, kg/m³
$\rho_p - \rho$ = particle density minus liquid density, kg/m³
μ = dynamic viscosity of the fluid, kg/(m s) = N s/m²
Δp = pressure drop, N/m².

Figure 6.24 shows a hydrocyclone design used in mineral ore dressing, and Figure 6.25 gives its corresponding efficiency.[29] Range of particle size and other characteristic data as reported by one manufacturer are given in Table 6.4.

Batteries of small hydrocyclones are often made in plastic, while larger hydrocyclones may have a rubber lining to reduce abrasion.

The feed to a hydrocyclone usually contains particles with a distribution of sizes. Figure 6.26 shows the cumulative weight fraction x of particles suspended in an oil as a function of particle diameter.

The weight of particles with diameters between D_p and $D_p + \Delta D_p$ (Figure 6.27) in the feed is $(\Delta x)(w_{in})$ where w_{in} is the total weight of particles in the feed. The fraction of $(\Delta x)(w_{in})$ that is not collected but leaves with the overflow, is $(1 - \eta)$. Hence

$$\Delta w_{out} = w_{in}(1 - \eta)\Delta x.$$

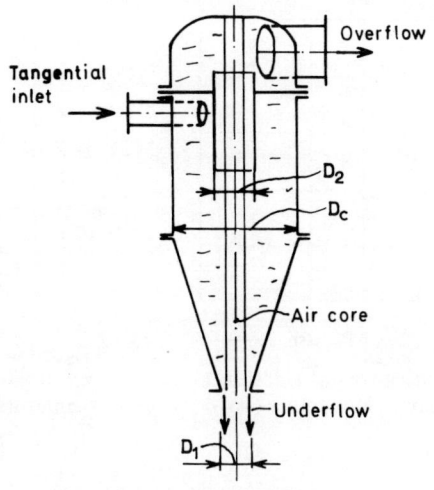

Figure 6.24 Hydrocyclone with area of overflow outlet four times that of inlet.

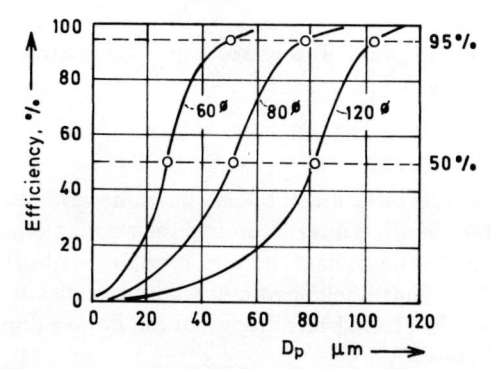

Figure 6.25 Efficiency of recovery (per cent) for the hydrocyclone of Figure 6.24 plotted against particle diameter. Cyclone diameter $D_c = 1200$ mm, $\Delta p = 0.25$ bar, sleeve diameters $D_2 = 60$, 80, and 120 mm.[29]

Table 6.4 Characteristic data for hydrocyclones used as classifiers[29]

Cyclone diameter, D_c mm	Nominal capacity per unit, m^3/h	Pressure drop, Δp MN/m^2	Nominal classification size range, D_p μm
76	2.3	0.28	5– 20
200	23	0.20	20– 40
350	50	0.20	100–325
610	240	0.14	Coarse separation or scalping

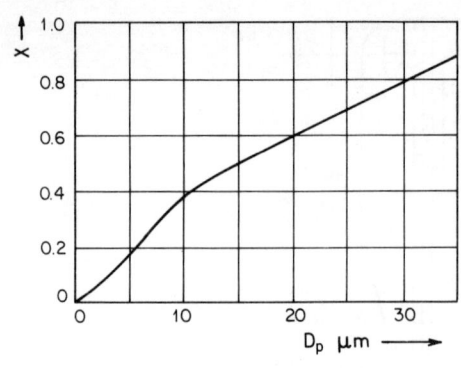

Figure 6.26 Cumulative weight fraction x of particles in an oil having diameters less than D_p.

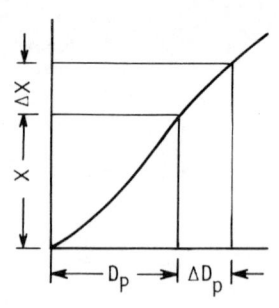

Figure 6.27 Weight fraction Δx of particles having diameters between D_p and $D_p + \Delta D_p$.

Integration gives

$$w_{\text{out}} = w_{\text{in}} \int_{x=0}^{x=1} (1 - \eta)\, dx. \tag{6.23}$$

Both η and x are functions of D_p. This makes it possible to express η as a function of x as shown in Example 6.9.

Impingement separators

Impingement separators make use of the inertia of the particles. An obstruction is placed in the gas flow which is diverted around the obstruction. Due to inertia, the particles will cross the stream lines and they may hit the obstruction as indicated in Figures 6.28 and 6.29. Small, solid particles will usually stick to the obstruction, while drops will collect and form larger drops that can flow or drip off without being carried away by the gas.

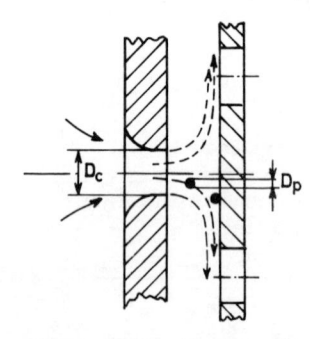

Figure 6.28 Gas-jet with diameter D_c that hits a wall in distance D_c; $D_p =$ particle diameter.

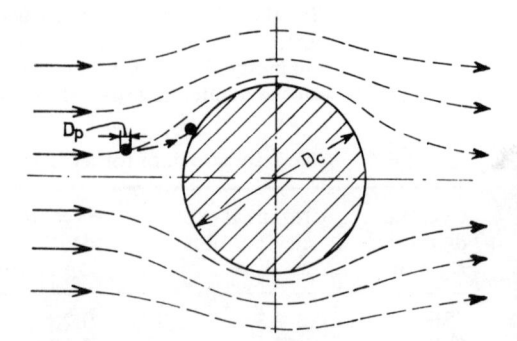

Figure 6.29 Gas flowing around a cylinder or a sphere with diameter D_c. The collector may be a fibre in a filter or a liquid drop in a scrubber.

The efficiency of impingement separators is conveniently given as a function of a dimensionless number Ψ which is the distance a particle with the velocity of the gas, V_0, will penetrate into stagnant gas, divided by the collector diamter D_c.

For spherical particles the drag force due to friction in the laminar region [equations (6.1) and (6.5)], is

$$F = 3\pi\mu D_p V$$

or with force equal to mass times acceleration,

$$3\pi\mu D_p V = -\left(\frac{\pi}{6} D_p^3 \rho_p\right)\left(\frac{dV}{dt}\right) \quad \text{or} \quad V dt = -\frac{\rho_p D_p^2}{18\mu} dV.$$

The distance a particle moves is $s = \int V dt$. Hence

$$s = -\frac{\rho_p D_p^2}{18\mu} \int_{V=V_0}^{0} dV = \frac{\rho_p D_p^2 V_0}{18\mu} \tag{6.24}$$

and

$$\Psi = \frac{s}{D_c} = \frac{C}{18} \frac{\rho_p D_p^2 V_0}{\mu D_c}. \tag{6.25}$$

The factor C is 1.0 for spherical particles and Reynolds number between 0.00001 and 2.0. For small particles with Reynolds numbers less than 0.00001, C is equal to the Cunningham correction given in square brackets in equation (6.10). The C-values for other shapes are given by Ranz and Wong.[30]

Vane type demisters, wire mesh demisters, fibre mats, tower packings, and the drops in scrubbers are common forms of impingement separators.

The efficiency for square and circular jets shown in Figure 6.28 is given in Figure 6.30. Data for the simplest vane type separator shown in Figure 6.31 are not available in the literature. The order of magnitude of the efficiency may be estimated from the curve for a square jet in Figure 6.30 with sides of the jet twice the distance between the vanes. With the length L in Figure 6.31 at least three times

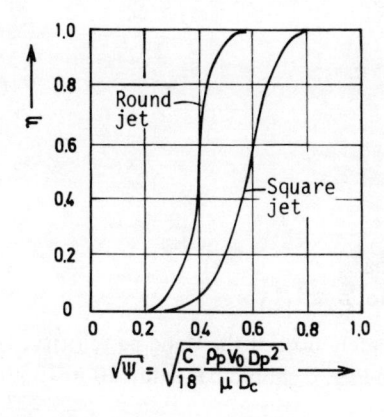

Figure 6.30 Efficiency η for an impingement separator depending on the target number Ψ. The curves are determined for jets hitting a wall at a distance D_c to $3D_c$ from the nozzle.

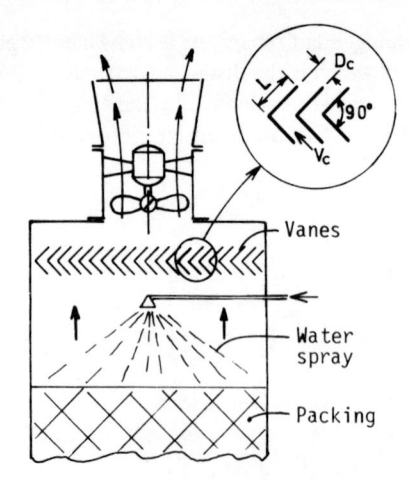

Figure 6.31 The top section of a cooling tower with a vane type demister.

the distance between the vanes, the pressure drop may be estimated as 1.3 velocity heads, or

$$\Delta p_f = 1.3 \rho \frac{V_c^2}{2} \text{ N/m}^2. \tag{6.26}$$

Wire mesh demisters

Wire mesh demisters are effective entrainment separators in evaporators, absorption towers, distillation columns, etc. The mesh is available in materials, such as, acid resistant steel, nickel, copper, aluminium, tantalum, hastelloy and Teflon, with thread diameters usually in the range from 0.075 mm to 0.4 mm. Some characteristic data are given in Table 6.5. The efficiency, η, given in the table refers to low viscosity liquids and the gas velocities calculated by equation (6.27).

If two mats are used in series, a high efficiency mat first with a more open mat above it at a distance 3/4 times the diameter of the mats is recommended.[32] Recommended gas velocity, referred to the total cross section of the mat not blanked by supports, is

$$V_s = C \sqrt{\frac{\rho_1 - \rho}{\rho}} \text{ m/s} \tag{6.27}$$

where

ρ_1 = density of the liquid, kg/m^3
ρ = density of the gas, kg/m^3
C = 0.09 for gas pressures $p > 1.2$ bar
   = 0.045 + 0.035 p with p in bar for $p < 1.2$ bar.

These values of C correspond to approximately 60% of the flooding velocity. To obtain high efficiency without risking flooding, C should be within 20 to 135% of

Table 6.5   Data for wire mesh demisters with thickness 100 mm and thread diameter 0.25 mm.[3,31]

Type	Weight kg/m^3	Free volume %	Surface area m^2/m^3	η %	Notes
Knitted, high efficiency mats	210–220	97.5	400	99.9+	
High efficiency mats	190	97.5	330–380	99.9+	relatively clean
Standard mats	150–160	98	300–330	99.5+	general purpose
Fishbone-woven mats and mats for high liquid contents	90–100	99	160–230	90.0+	services containing solids or 'dirty' materials

the value given above and never exceed 0.125. According to Campbell[33] a *C*-value of 0.107 has given good results for separation of oil and natural gas under high pressure.

The figures quoted apply to horizontally mounted wire mesh demisters placed at least 250 mm above the liquid surface as shown in Figure 6.32. For vertical and inclined mats a velocity approximately 1/3 of the value calculated by equation (6.27) is recommended.

Wire mesh demisters have a low pressure drop,

$$\Delta p_f = f_d \rho\, V_s^2 / 2 \qquad (6.28)$$

where V_s is the superficial velocity referred to the total cross-section, m/s, and f_d is the friction factor.

Based on measurements with air and water by Metex Mist Eliminators,[34]

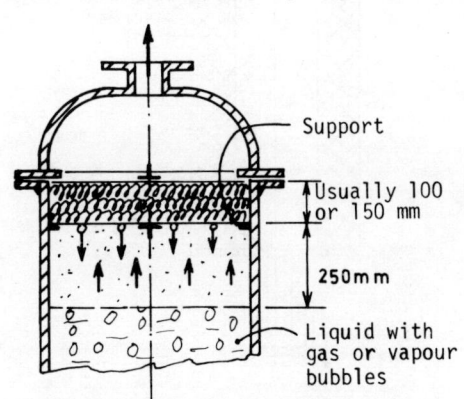

Figure 6.32   Horizontal wire mesh demister: liquid separated from the gas drips down.

$f_d = 40$ for 150 mm mats with large free volume as fishbone-woven mats, and $f_d = 67$ for standard 100 mm mats are recommended.

The friction factor increases considerably with increasing liquid load. Data given by Knit Mesh Ltd.[35] can be reproduced by the equation,

$$f_d = 104 w_0^{0.14} \pm 30\% \qquad (6.29)$$

and data given by Ludwig,[31]

$$f_d = 82 w_0^{0.3} \pm 15\% \qquad (6.30)$$

where w_0 is the liquid load, kg/(m² s).

Wire mesh demisters are sensitive to solids. If solids can deposit, nozzles for flushing of the mat should be installed.

Fibre mist eliminators

Fibre mist eliminators with silicone treated fibre glass or polyester fibres with diameters down to 1.3 μm, are used to catch acid fogs in exit gases. The mats are installed as cylinders kept in position by screens (Figure 6.33).[36] Soluble particles are removed with sprays of water or another solvent (Figure 6.34).

Drops larger than 3 μm are reported to be removed 100% in fibre mist eliminators. Data for drops 3 μm and smaller are given in Table 6.6.

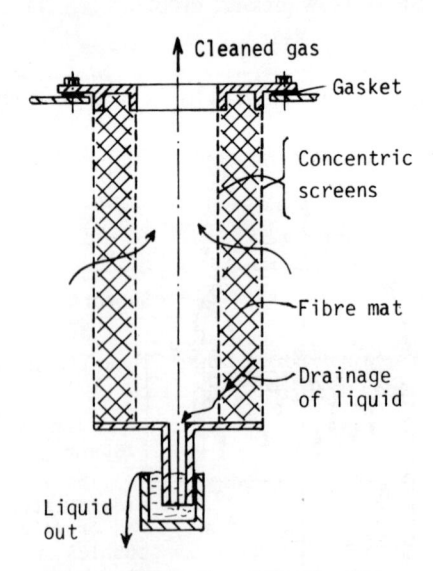

Figure 6.33 Cartridge of a fibre mat demister.

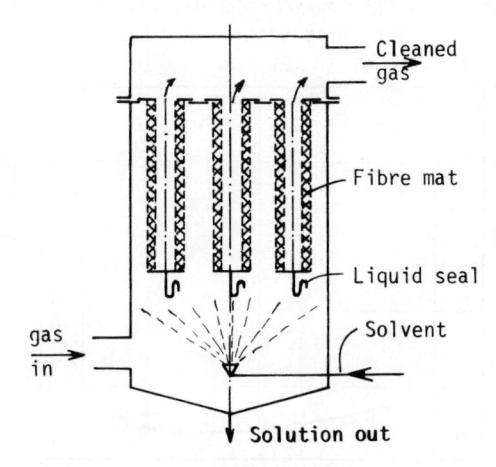

Figure 6.34 Cylindrical fibre cartridges for removal of soluble particles.[36]

Table 6.6 Characteristic data for fibre mist eliminators as used in sulphuric acid plants.[36]

Type of fibre mist eliminator	High efficiency	High velocity	Spray catcher
Controlling mechanism for mist collection	Brownian movement	impaction	impaction
Superficial velocity, m/s	0.07–0.2	2–2,5	2–2,5
Efficiency for particles <3 μm, %	95–99+	90–98	15–30
Pressure drop, N/m^2	1250–4000	1500–2000	125–250

Bacteria eliminating filters

Fibre glass filters are used to remove bacteria from air streams. Figure 6.35 shows efficiencies obtained with a 3 mm thick sheet of glass fibre impregnated with furfural resin binder.[37] The pressure drop through this filter was

$$\Delta p = 136 V^{1.22} \tag{6.31}$$

where Δp is in N/m^2 and V in m/s.

Both wire mesh demisters and fibre filters must fit close to the walls in order to prevent short circuiting.

Bag filters

Bag filters consist of cylindrical or flat filter elements made of woven cloth. Particle-containing gas passes through the cloth and particles are collected by inertial impaction on the fabric, and on the particles that gradually build up a filter cake.

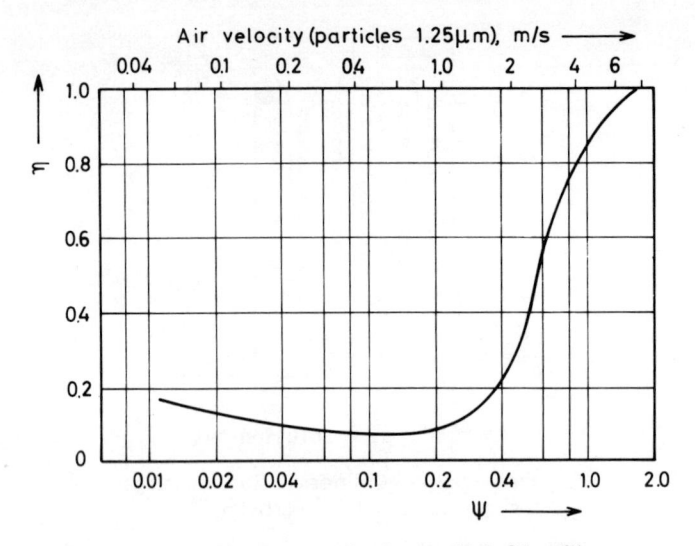

Figure 6.35 Efficiency η of a 3 mm thick fibre filter as a function of the inertial parameter Ψ: fibre diameter 16 ± 3.5 μm. The upper abscissa gives air velocity when spores of *B. subtilis* are collected.

For the smallest particles Brownian motions also helps to increase the efficiency. Electrostatic charges may also interfere. The collected particles are removed at intervals either by shaking the bags or by reverse flow of gas. For most applications the efficiency is very high, up to 99.99+%.

Many types of fibres are in use,[3] cotton being cheapest available. However, it should only be used at temperatures below 80°C. Wool has finer fibres than cotton, and the upper temperature limit for continuous use is 95°C. Both cotton and wool are sensitive to alkaline and oxidizing conditions. Polyacrylonitrile fibre cloths are used successfully at 135°C and polyester fibre cloth at 140°C. Glass fibre cloth is used in the temperature range 150–300°C.

In most cases satisfactory results are obtained with superficial gas velocities in the range 0.6 to 0.8 m/min. An exception is graphite dust where the velocity should be below 0.45 to 0.6 m/min. In some cases where the bags are cleaned with reverse gas flow, considerably higher velocities are used, from 2 m/min for fine smoke to 7 m/min for coarse dust.[38]

In general, the pressure drop through dust-free bags is negligible. The flow through the layer of dust is laminar. It gives a pressure drop in N/m²,

$$\Delta p = C_d \mu m_s V_0 = C_d \mu c_s V_0^2 t \tag{6.32}$$

or

$$\Delta p = C_d \mu c_s V_0^2 (t_1 + t) \tag{6.33}$$

Table 6.7 Some values of C_d in equation (6.32) and (6.33).[1]

Dust source	Various industrial dusts	Roasting of zinc ore	Talc	Dust from cellulose acetate	Coal dust <200 mesh
Particle range, μm	<10	1–5	1–4	5–20	7–50
C_d 10^{-10}	1.0–2.8	1.2	1.3	0.14–0.18	0.6–1.6

where

μ = dynamic viscosity of the gas, N s/m^2
m_s = particle concentration on the cloth, kg/m^2
c_s = particle concentration in the gas, kg/m^3
V_0 = velocity referred to the total cross section, m/s
t = time the filter has been in operation, s
t_1 = time to collect the amount of particles left on the cloth after cleaning, s
C_d = factor to be determined experimentally. Lacking experimental data, Table 6.7 may be a guide.

Bag filters are usually arranged in two or more groups with one group being cleaned while the others are in operation. This requires either reduced gas flow during cleaning or a fan that can force the total amount of gas through the smaller cloth area.

Scrubbers

Scrubbers for removal of liquid or solid particles use the same mechanisms of collection as gas filters: inertial impaction, interception, and diffusion.

Spray towers are the simplest type of gas scrubbers. The most effective drop size in the spray is reported to be in the range 400–1000 μm.[39] Tall spray towers

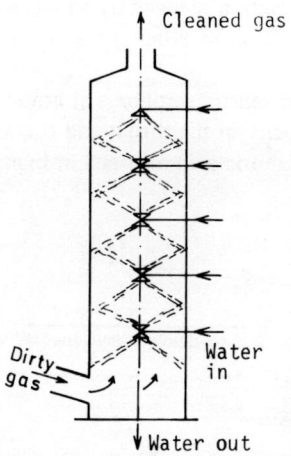

Figure 6.36 Spray tower with 9 nozzles, diameter 5 m, height 16 m.[40]

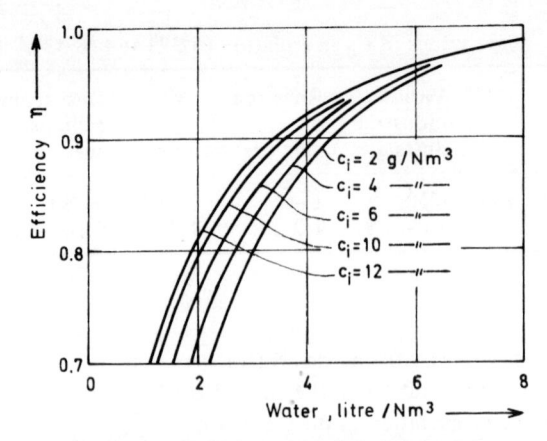

Figure 6.37 Efficiency η of the spray tower of Figure 6.36 as a function of the water consumption in litres per standard cubic metre of gas at different particle concentrations, c_i, grams per standard cubic metre, in gas from a steel works.[40]

should have spray nozzles at several levels (Figure 6.36). Figure 6.37 gives some efficiencies.[40]

Venturi scrubbers are used for removing smoke particles, fine dust, and acid fog. They rely on very high velocity of the gas and disintegration of the liquid by the gas. In humid gas, condensation on the drops may produce an effect in addition to inertial impaction and interception. The high velocity produces a reduced temperature causing condensation which then gives a net flow towards the drop and increases the drop size.

Figure 6.38 shows a Pease–Anthony venturi scrubber with the liquid introduced through radial holes in the throat. To obtain high efficiency at reduced gas load, either the liquid rate can be increased or the cross section reduced by an adjustable cone as indicated in Figure 6.38. Figures 6.39 and 6.40 show other types of venturi scrubbers.

Measured efficiency and pressure drop for venturi scrubbers at atmospheric pressure plotted as a function of the gas velocity in the throat and the water rate are given in Figure 6.41 for collection of sulphuric acid mist and in Figure 6.42 for

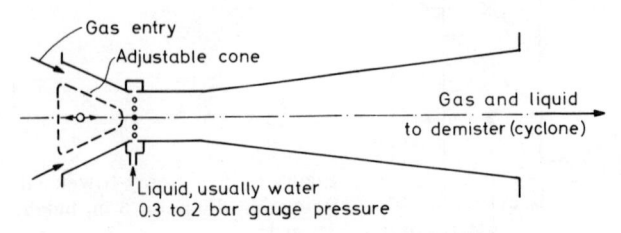

Figure 6.38 Pease–Anthony scrubber.

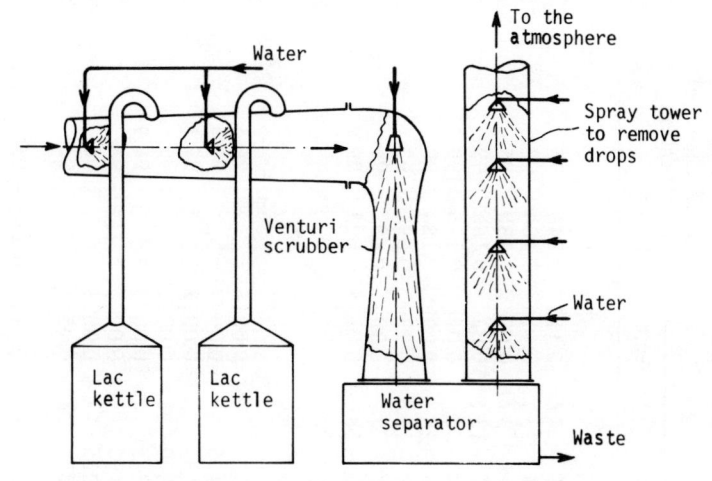

Figure 6.39 Venturi scrubber for lac-kettles. The water sprays into the suction line prevent accidental fires spreading to the neighbouring kettle.[43]

salt cake fume from kraft mill gases.[41] Table 6.8 gives performance data from various installations.[42]

For scrubbers with good mixing of liquid drops and gas without channelling or additional effects due to condensation, vaporization, or diffusion, Semrau[45] has suggested the following equation for scrubber efficiency,

$$\ln[1/(1-\eta)] = \alpha P_T^\beta \tag{6.34}$$

where P_T is the power consumption to overcome pressure loss for the gas in the

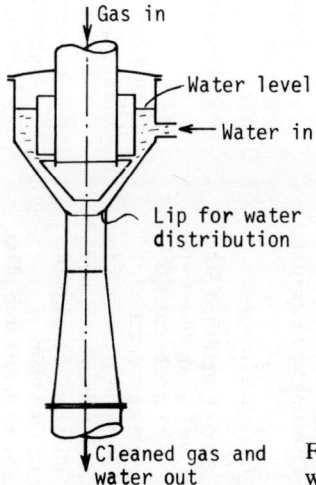

Figure 6.40 Venturi scrubber without pressure atomization.[44]

Table 6.8 Performance data for Pease–Anthony scrubbers. The efficiencies are average values for a particular plant or group of installations operating under a specific set of conditions.

Source of gas	Dust or mist	Particle size μm	Loading g m^{-3} inlet	Loading g m^{-3} exit	Average efficiency of collection %
Iron and Steel Industry					
Gray iron cupola	Iron, coke, silica dust	0.1 – 10	2.3 – 4.6	0.115 – 0.346	95
Oxygen steel converter	Iron oxide	0.5 – 2	18.4 – 23	0.115 – 0.184	98.5
Steel open-hearth furnace (scrap)	Iron and zinc oxide	0.08 – 1.00	1.15 – 3.46	0.69 – 0.138	95
Steel open-hearth furnace	Iron oxide	0.02 – 0.50	2.3 – 13.8	0.02 – 0.16	99
Blast furnace (iron)	Iron ore and coke dust	0.5 – 20	6.9 – 55	0.018 – 0.115	99
Electric furnace	Ferro-manganese fume	0.1 – 1	23.0 – 28	0.092 – 0.184	99
Electric furnace	Ferro silicon dust	0.1 – 1	2.3 – 11.5	0.23 – 0.69	92
Rotary kiln–iron reduction	Iron, carbon	0.5 – 50	6.9 – 23	0.23 – 0.69	99
Crushing and screening	Taconite iron ore dust	0.5 – 100	11.5 – 58	0.0115 – 0.023	99.9
Chemical Industry					
Acid–humidified SO$_3$	Sulphuric acid mist				
(a) scrub with water		—	0.30	0.0016	99.4
(b) scrub with 40% acid		—	0.405	0.00275	99.3
Acid concentrator	Sulphuric acid mist	—	0.133	0.0032	97.5
Copperas roasting kiln	Sulphuric acid mist	—	0.200	0.0021	99
Chlorosulphonic acid plant	Sulphuric acid mist	—	0.745	0.0078	98.9
Phosphoric acid plant	Orthophosphoric acid mist	—	0.193	0.0038	98
Dry ice plant	Amine fog	—	0.025	0.0021	90
Wood distillation plant	Tar and acetic acid	—	1.07	0.0575	95
Titanium chloride plant, titanium dioxide dryer	Titanium dioxide, hydrogen chloride fumes	0.5 – 1	2.3 – 11.5	0.115 – 0.23	95
Spray dryers	Detergents, fume, and odour	—	—	—	
Flash dryer	Furfural dust	0.1 – 1	4.6 – 13.8	0.115 – 0.345	95

Non-ferrous Metals Industry

Blast furnace (sec. lead)	Lead compounds	0.1 — 1	4.6 —13.8	0.115 −0.345	99
Reverberatory lead furnace	Lead and tin compounds	0.1 — 0.8	2.3 — 4.6	0.276	91
Ajax furnace—magnesium alloy	Aluminium chloride	0.1 — 0.9	6.9 —11.5	0.046 −0.115	95
Zinc sintering	Zinc and lead oxide dusts	0.1 — 1	2.3 —11.5	0.115 −0.23	98
Reverberatory brass furnace	Zinc oxide fume	0.05— 0.5	2.3 −18.4	0.23 −1.15	95

Mineral Products Industry

Lime kiln	Lime dust	1 — 50	11.5 −23	0.115 −0.345	99
Lime kiln	Soda fume	0.3 — 1	0.46−11.5	0.023 −0.115	99
Asphalt stone dryer	Limestone and rock dust	1 — 50	11.5 −34.5	0.115 −0.345	98
Cement kiln	Cement dust	0.5 — 55	2.3 − 4.6	0.115 −0.23	97

Petroleum Industry

Catalytic reformer	Catalyst dust	0.5 — 50	0.207	0.0115	95
Acid concentrator	Sulphuric acid mist	—	0.136	0.0032	97.5
TCC catalyst regenerator	Oil fumes	—	0.760	0.0080	98

Fertilizer Industry

Fertilizer dryer	Ammonium chloride fumes	0.05— 1	0.23− 1.15	0.115	85
Superphosphate den and mixer	Fluorine compounds	—	0.308	0.0055	98

Pulp and Paper Industry

Lime kiln	Lime dust	0.1 — 50	11.5 −23	0.115 −0.345	99
Lime kiln	Soda fume	0.1 — 2	4.6 −11.5	0.023 −0.115	99
Black liquor recovery boiler	Salt cake	—	9.3 −13.8	0.92 −1.38	90

Miscellaneous

Pickling tanks	Hydrogen chloride fumes	—	0.0253	0.0023	90
Boiler flue gas	Fly ash	0.1 — 3	2.3 − 4.6	0.115 −0.184	98
Sodium disposal incinerator	Sodium oxide fumes	0.1 — 0.3	1.15− 2.3	0.046	98

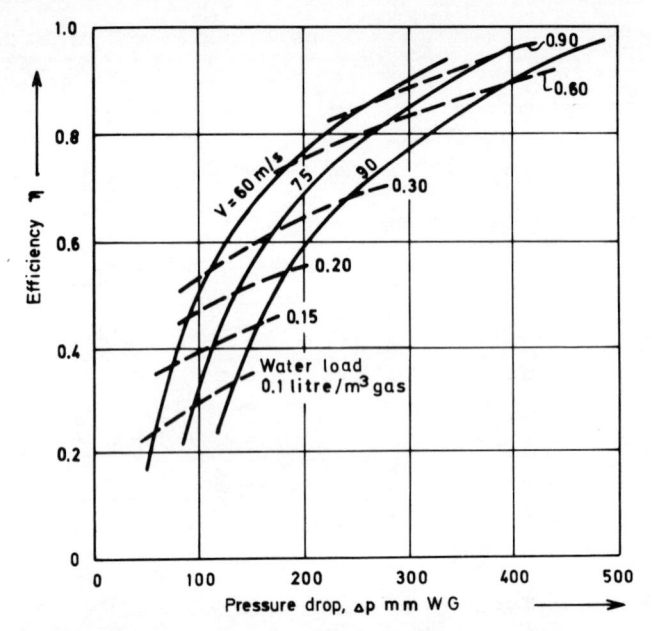

Figure 6.41 Efficiency of Pease–Anthony scrubber for sulphuric acid mist plotted as a function of pressure drop, and of liquid load and gas velocity in the throat.[41]

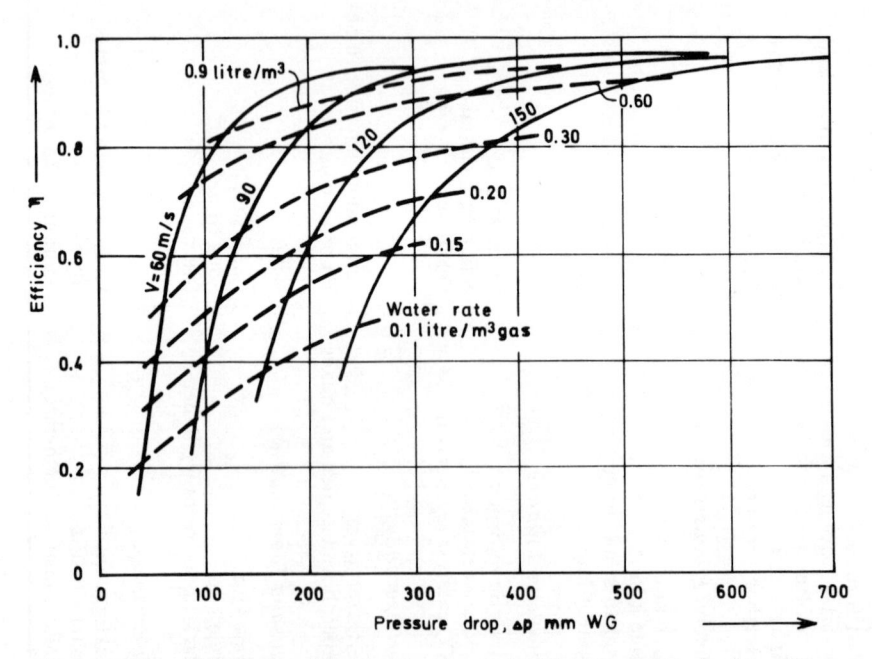

Figure 6.42 Efficiency of Pease–Anthony scrubber for salt cake fume from kraft mill gases plotted as a function of pressure drop in mm water gauge, water rate, and gas velocity in the throat.[41]

Table 6.9 Scrubber efficiency coefficients in equation (6.34).

Aerosol	Scrubber	Correlation parameter α	β
Lime kiln dust and fume (kraft mud kiln)			
Raw gas (lime dust and soda fume)	Venturi and cyclonic spray	1.47	1.05
Pre-washed gas (soda fume)	Venturi, pipe line and cyclonic spray	0.915	1.05
Talc dust	Venturi	2.97	0.362
	Orifice and pipe line	2.70	0.362
Black liquor recovery furnace fume.			
Cold scrubbing water humid gases	Venturi and cyclonic spray	1.75	0.620
Hot fume solution for scrubbing (humid gases)	Venturi, pipeline and cyclonic spray	0.740	0.861
Hot black liquor for scrubbing (dry gases)	Venturi evaporator	0.522	0.861
Phosphoric acid mist	Venturi	1.33	0.647
Foundry cupola dust	Venturi	1.35	0.621
Open hearth steel furnace fume	Venturi	1.26	0.569
Talc dust	Cyclone	1.16	0.655
Copper sulphate	Solivore		
	(A) with mechanical spray generator	0.390	1.14
	(B) with hydraulic nozzles	0.562	1.06
Ferrosilicon furnace fume	Venturi and cyclonic spray	0.870	0.459
Odorous mist	Venturi	0.363	1.41

scrubber and to atomize the liquid, kWh/m^3 gas, and α and β are coefficients characteristic of the dust being collected. Values for some dusts are given in Table 6.9.

The Peabody scrubber (Figure 6.43), may have a section for humidification of the gas before it enters the agglomeration section, followed by one or more plates with holes, or slots and reflectors, or impingement baffles above the holes or slots. The gas velocity through the openings is 4.5 to 6 m/s.

Electrostatic precipitators

In electrostatic precipitators, particles (solid or liquid) are charged electrically and precipitated in a high voltage electric field. Electrostatic precipitators are used where particles in the size range 0.5 to 20 μm can be removed from large volumes of gases without an explosion risk.

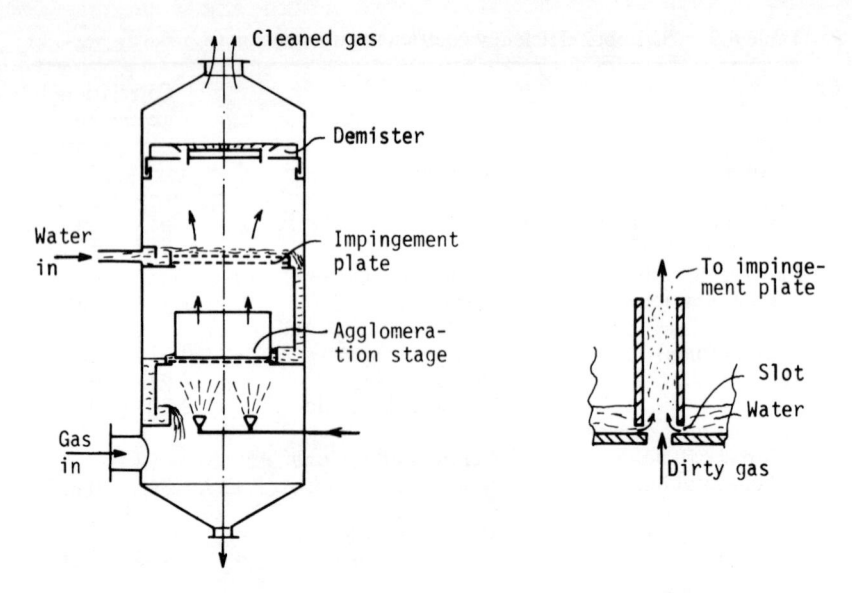

Figure 6.43 Peabody impingement plate scrubber. The inset on the right shows details of the agglomeration stage.

The principle of a two stage electrostatic precipitator is indicated in Figure 6.44. The first set of electrodes gives a corona and charges the particles, while the collection of particles takes place in the second set of electrodes.

One stage electrostatic precipitators have only two sets of electrodes. The first set creates both the electric field and the corona. They can be wires with sharp edges, either with a star-shaped cross section, or similar to barbed wire. The other set of electrodes is the collector, usually circular (Figure 6.45) or hexagonal tubes or

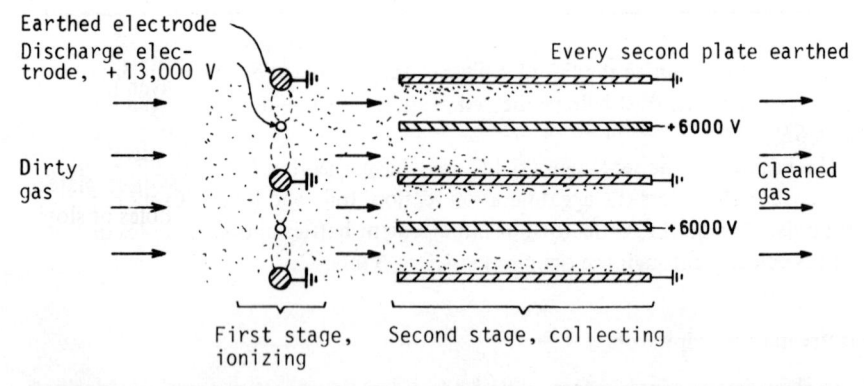

Figure 6.44 Two-stage electrostatic precipitator. Common electric fields are 5 kV/cm in the first and 4 kV/cm for sedimentation in the second stage. Negative discharge electrodes give the more stable corona. Positive discharge electrodes are used if formation of ozone must be avoided.

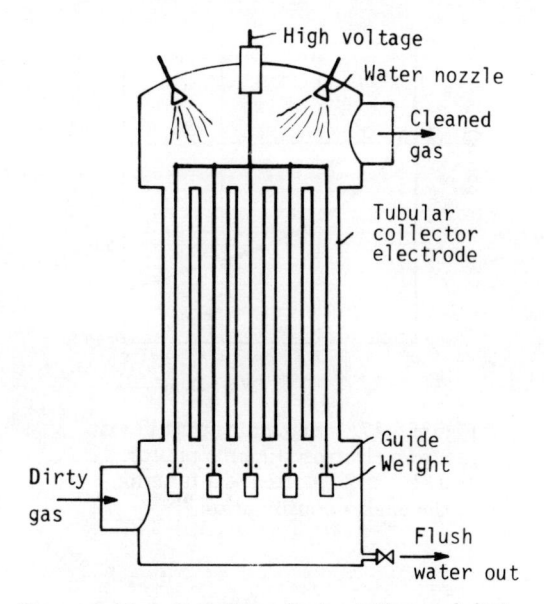

Figure 6.45 One stage electrostatic precipitator with tubular electrodes; dust is flushed down periodically.

plates (Figure 6.46). The particles are removed periodically, they are either shaken off by some kind of rapping, or flushed off.

The efficiency of an electrostatic precipitator is strongly dependent on the electrical conductivity of the particles, which is influenced by the humidity of the gas. Figure 6.47 gives the efficiency as a function of energy consumption.

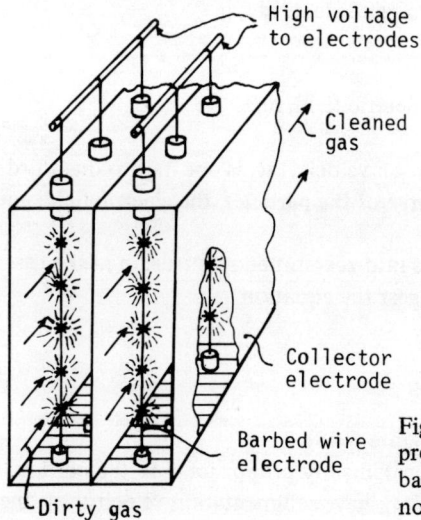

Figure 6.46 One stage electrostatic precipitator with corona around barbed wires. (Dust hoppers underneath not shown.)

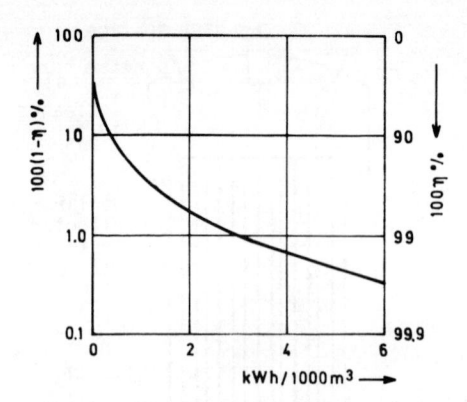

Figure 6.47 Remaining dust in exit gas from electrostatic precipitator, $100(1-\eta)\%$, plotted as a function of the energy consumption.[46]

Precipitators with plate collectors have an efficiency,[47]

$$\eta = 1 - e^{-(Lw/sV)} \qquad (6.35)$$

and with turbular collectors,

$$\eta = 1 - e^{-(2Lw/RV)} \qquad (6.36)$$

where

L = length of filter in direction of flow, m
s = distance from wire electrode to plate electrode, m
R = radius of tube, m
V = average gas velocity, m/s
w = sedimentation velocity in the electric field, m/s.

In industrial precipitators the sedimentation velocity, w, is one half to one third the velocity calculated from the electric charge of the particles, the electric field, and Stokes' law.[47]

A somewhat different approach takes into account eddy diffusion in the gas stream, and Williams and Jackson[48] suggest the equation

$$\eta = f(\tau, \varphi) \qquad (6.37)$$

where $\tau = 7.41 \, L/s$ and $\varphi = 6.67 \, w/V$ (Figure 6.48).

The sedimentation velocity, w, is approximately proportional to the particle diameter D_p. If particles with diameter D_{po} have sedimentation velocity w_o, the

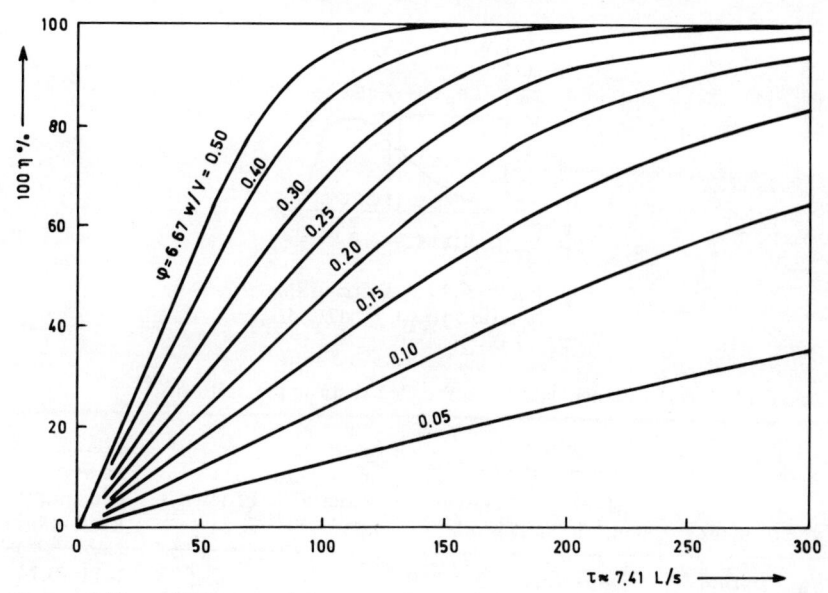

Figure 6.48 Efficiency of electrostatic precipitators plotted as a function of τ and ϕ.[48]

Table 6.10 Particle sizes and concentrations for some industrial dispersoids.[47]

Industry	Dispersoid	Particle size by weight 0–1 μm %	0–5 μm %	0–10 μm %	Particle loading g/m^3
Electrical power	Fly ash from pulverized coal	1	25	50	7
Electrical power	Fly ash from cyclone furnace	6	60	85	1.6
Cement	Kiln dust	1	20	40	23
Steel	Blast furnace after dry-dust catcher	5	30	60	7
Steel	Open-hearth fume	90	98	99	2.3
Non-ferrous smelters	Copper roaster dust			20	23
Non-ferrous smelters	Converter furnace dust			30	11
Non-ferrous smelters	Reverberatory furnace dust			60	7
Chemical	H_2SO_4 acid fume	99			0.11
Chemical	H_3PO_4 acid fume	15	99		46

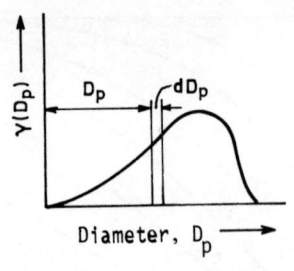

Figure 6.49 Distribution function $\gamma(D_p)$, $\int_0^\infty \gamma(D_p)dD_p = 1.0$.

Table 6.11 Performance data for some electrostatic precipitators.[49,50]

Type of plant	Entering gas 1000 m³/h	Dust g/m³	Gas velocity m/s	Efficiency %	Power consumption kW/1000 m³
Power station	85–400	25–250	–	96.6–99.4	0.11–0.14
Coke from peat	4.5	80	–	99.85	0.7
Coke from coal	2.4–3.1	250–370	–	99.9	0.9–1.6
Paper industry					
pyrite roaster	9.6–14.1	18–63	–	98.3–99.7	0.55–0.95
acid from burning of sulphur	2.5–5.0	105–190	1.2–1.5	99.1–99.4	0.8–0.95
burning black liquor	134	43	–	95.3	0.26
Cement industry					
rotary kiln, dry process	128–135	96–315	0.45–1.0	98.8–99.1	0.03–0.09
rotary kiln, wet process	146	170–325	0.45–1.5	98.2–99.7	0.08–0.10
raw material dryer, mill and packing machine	15–36	580–780	–	99.7–99.8	0.45–0.47
Non-ferrous metal industry, blast furnaces and kilns, lead, zinc, tin	3.5–45	75–600	–	98.7–99.8	0.16–0.27
Miscellaneous					
concentration of sulphuric acid	21	200	–	99.6	0.36
sulphur from combustion of hydrogen sulphide	4.3	390	–	99.2	1.7

sedimentation velocity for particles with diameter D_p will be $w \approx w_0 D_p/D_{p0}$ and the collection efficiency according to equations (6.35) and (6.36),

$$\eta = 1 - e^{-C_0 D_p/D_{p0}} \tag{6.38}$$

where $C_0 = w_0 L/sV$ for plate and $w_0 2L/RV$ for tubular precipitators.

The number of particles with diameter between D_p and $D_p + dD_p$ is $\gamma(D_p)dD_p$ where $\gamma(D_p)$ is the particle distribution function, for example, as given in Figure 6.49. The fraction of particles with diameter between D_p and $D_p + dD_p$ that passes through the precipitator is

$$dW_{out} = (1-\eta)\gamma(D_p)dD_p = e^{-C_0 D_p/D_{p0}}\gamma(D_p)dD_p.$$

The total fraction that passes through is

$$W_{out} = \int_0^\infty \gamma(D_p) e^{-C_0 D_p/D_{p0}} \, dD_p$$

and the efficiency,

$$\eta = 1 - \int_0^\infty \gamma(D_p) e^{-C_0 D_p/D_{p0}} \, dD_p. \tag{6.39}$$

Typical particle sizes and concentrations are listed in Table 6.10, and performance data from some industrial installations in Table 6.11.

EXAMPLE 6.1. TERMINAL VELOCITY IN A SPRAY DRYER

Figure 6.50 is a spray dryer for detergent containing 45 weight% water.

Calculate the terminal velocity of the largest drops, $D_p = 1.3$ mm, in the upper part of the dryer when the density of the humid air is $\rho = 0.88$ kg/m³ and its viscosity $\mu = 0.000021$ N s/m². The density of the drops is $\rho_p = 930$ kg/m³.

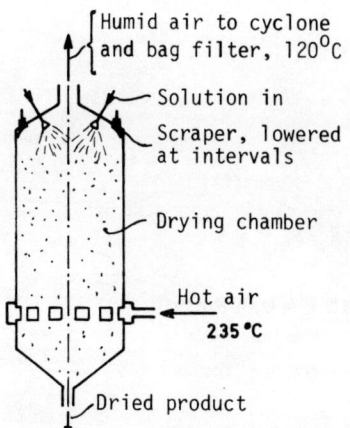

Figure 6.50 The spray dryer for Example 6.1

Solution

With drop diameter 1.3 mm, the settling is probably in the transient or the turbulent region. Equation (6.8) for the transient region gives the terminal velocity,

$$V_t = \left[(0.072)(0.0013^{1.6})\frac{930-0.88}{(0.000021^{0.6})(0.88^{0.4})}9.82\right]^{1/1.4}$$

$$= 5.4 \text{ m/s}.$$

Reynolds number can be checked,

$$Re_D = \frac{(0.88)(5.4)(0.0013)}{0.000021} = 294.$$

This is between 2 and 500, i.e. equation (6.8) applies.

EXAMPLE 6.2. HINDERED SETTLING

The terminal velocity of a fraction of fine quartz particles in water at 13°C (density $\rho = 1000$ kg/m^3, viscosity $\mu = 0.0012$ N s/m^2) is measured to be approximately 0.015 m/s. The density of quartz is $\rho_p = 2650$ kg/m^3.

Calculate the terminal velocity in hindered settling with 50 kg particles per 100 kg water.

Solution

1 500 kg of suspension contains 1 m^3 water and 500/2650 = 0.189 m^3 particles, i.e. the liquid fraction is

$$\epsilon = 1/(1 + 0.189) = 0.84.$$

Equation (6.12) with $n = 3.65$ given by Scholl, gives

$$V_h = (0.84^{3.65})(0.015)$$

$$= 0.008 \text{ m/s}.$$

The value of n from Maude and Whitmore's experiments is determined by trial and error. Equation (6.9) together with the terminal velocity for dilute suspensions, gives the equivalent particle diameter D,

$$0.015 = 0.0556 D^2 \frac{2650-1000}{0.0012} 9.82,$$

$$D = 0.00014 \text{ m}.$$

Reynolds number, can again be checked,

$$Re_D = \frac{(1000)(0.015)(0.00014)}{0.0012} = 1.75,$$

i.e. equation (6.9) applies. First trial with $V = 0.008$ m/s,

$$Re_D = \frac{(1000)(0.008)(0.00014)}{0.0012} = 0.93$$

which is only slightly outside the region $1 < \text{Re} < 2500$.

$$n \approx (4.4)(0.93^{-0.08}) = 4.42$$
$$V_h = (0.84^{4.42})(0.015) = 0.007 \text{ m/s}$$
$$\text{Re}_D = \frac{(1000)(0.007)(0.00014)}{0.0012} = 0.82$$
$$n \approx (4.4)(0.82^{-0.08}) = 4.47$$
$$V_h = (0.84^{4.47})(0.015)$$
$$= 0.007 \text{ m/s}.$$

EXAMPLE 6.3. SEPARATION ACCORDING TO DENSITY

Figure 6.51 shows diagramatically a double-cone apparatus for separating particles according to size and/or density. Particles with a terminal velocity exceeding the liquid flow velocity between the adjustable cone in the centre and the outer wall will end up in the underflow.

The apparatus can be used to separate particles with density $\rho_a = 4000 \text{ kg/cm}^3$ from particles with density $\rho_b = 2700 \text{ kg/m}^3$. The particles are crushed and passed through screens to give certain size fractions. The water has density $\rho = 1000 \text{ kg/m}^3$ and viscosity $\mu = 0.0016 \text{ N s/m}^2$ (1.6 cP).

Estimate the minimum diameter, D_2, that the smallest particles can have if the separation according to density shall be 100% and the diameter of the largest particle is (a) $D_1 = 0.6$ mm or (b) $D_1 = 0.12$ mm.

It is assumed that the suspension is dilute.

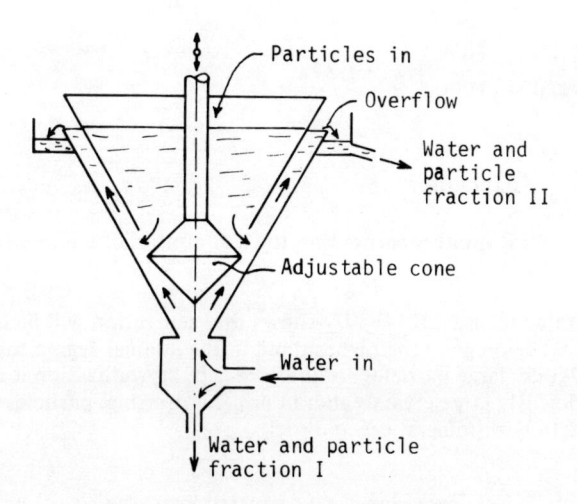

Figure 6.51 Double-cone apparatus.

Solution

For (a) a trial using equation (6.8) with a check of Reynolds number gives flow in the transition region for both types of particles with diameter 1.6 mm. Assuming transition region for all particles, equation (6.8) gives

$$\left[0.072 D_1^{1.6}\, \frac{2700-1000}{\mu^{0.6} \rho^{0.4}}\, g\right]^{1/1.4} < \left[0.072 D_2^{1.6}\, \frac{4000-1000}{\mu^{0.6} \rho^{0.4}}\, g\right]^{1/1.4}$$

$$\frac{D_1}{D_2} < \left(\frac{4000-1000}{2700-1000}\right)^{1/1.6} = 1.426$$

or

$$D_2 > 0.6/1.426 = 0.42 \text{ mm}.$$

Check of Reynolds number: the terminal velocity for particles b with diameter D_2 is

$$V_t = \left[(0.072)(0.00042^{1.6})\, \frac{2700-1000}{(0.0016^{0.6})(1000^{0.4})}\, 9.82\right]^{1/1.4} = 0.048 \text{ m/s}$$

$$\text{Re}_D = \frac{(1000)(0.048)(0.00042)}{0.0016} = 12.6.$$

The corresponding data for the largest particles a are $V_t = 0.108$ m/s and Re = 40.5, i.e. the assumption $2 < \text{Re}_D < 500$ was correct.

In case (b) flow will probably be laminar around all particles, and equation (6.9) gives,

$$0.0556 D_1^2\, \frac{2700-1000}{\mu}\, g < 0.0556 D_2^2\, \frac{4000-1000}{\mu}$$

$$\frac{D_1}{D_2} < \left(\frac{4000-1000}{2700-1000}\right)^{0.5} = 1.33$$

or

$$D_2 > \frac{0.12}{1.33} = 0.09 \text{ mm}.$$

A check of Reynolds number shows that the assumption of laminar flow was correct ($\text{Re}_D < 2$).

Note A high value of the ratio D_1/D_2 shows that separation will be easy. Separation in the turbulent region gives the highest, and in the laminar region the lowest, value of D_1/D_2, i.e. large particles are preferred. In crystallization it may be difficult to obtain sufficiently large crystals and, in mineral dressing, particles that are too large will contain both mineral ore and rock.

EXAMPLE 6.4. FLUIDIZED BED

Particles of mineral ore from a flotation process are to be roasted at 975°C and at atmospheric pressure ($p = 1$ bar) in a fluidized bed. The bed should be at least 1.0 m high. The particle size varies between 0.04 mm and 0.18 mm with most of the particles in the range from 0.08 to 0.16 mm. The density of the particles $\rho_p = 2800$ kg/m^3. The average molecular weight of the gas $M = 33.3$ and its viscosity at 975°C, $\mu = 0.000049$ N s/m^2.

(a) Estimate the pressure drop in the fluidized bed.

(b) Estimate the pressure in the wind box (Figure 6.5).

(c) Estimate the minimum gas velocity necessary to produce a fluidized bed.

(d) Calculate the gas velocity at which particles with diameter 0.04 and 0.08 mm respectively are carried away with the gas.

Solution

(a) The density of the gas

$\rho = Mp/RT = (33.3)(100,000)/(8314)(1248) = 0.32$ kg/m^3.

According to Figure 6.7 the minimum porosity of fluidized sand with the same particle size as most of the particles (0.08 to 0.16 mm) $\epsilon_M = 0.55$ to 0.59. The smaller particles, however, have a tendency to fill in voids between the larger particles and thus reduce the porosity. Values of ϵ_M from 0.53 to 0.57 are used in the following. Equation (6.13) gives with

$\epsilon_M = 0.53$, $\Delta p = (1.0)(1.0 - 0.53)(2800 - 0.3)(9.82) = 12,920$ N/m^2

with

$\epsilon_M = 0.57$, $\Delta p = (1.0)(1.0 - 0.57)(2800 - 0.3)(9.82) = 11,820$ N/m^2.

The highest value is used, to be on the safe side, i.e.,

$\Delta p = 13,000$ N/m^2.

(b) The pressure drop through the nozzles of the distributor plate is usually in the range from 100 to 1000 mm water gauge or 1000 to 10,000 N/m^2. The lowest value is less than 10% of the pressure drop in the fluidized bed and the highest value almost 80%. The lowest value is probably too low to give a reasonably good gas distribution over the cross section, while the highest value may be unnecessarily high. Senecal[51] states, as a rule of thumb, that the pressure drop through the distributor should exceed 10 velocity heads at the inlet to the space under the distributor. Assuming the density of the air in the wind box, $\rho_{air} = 1.3$ kg/m^3, and the air velocity at the inlet to the wind box, $V = 16$ m/s (Figure 1.10), Senecal's rule gives,

$$\Delta p > (10)(1.3)\frac{16^2}{2} = 1664 \text{ N/m}^2$$

or more than 13% of the pressure drop through the fluidized bed. Twice this value is proposed, giving a pressure drop in the nozzles,

$\Delta p = (2)(1664) \approx 3300$ N/m^2

or gauge pressure in the wind box,

$p = 13,000 + 3300 = 16,300$ N/m^2.

(c) The largest particles have to be fluidized, i.e. $D_p = 0.00018$ m has to be inserted in equation (6.15),

$$V_M = \frac{(2800 - 0.3)(9.82)(0.00018^2)\epsilon_M^3}{(150)(0.000049)(1 - \epsilon_M)} = 0.1212 \frac{\epsilon_M^3}{1 - \epsilon_M}$$

with

$$\epsilon_M = 0.53, V_M = 0.038 \text{ m/s}$$

and with

$$\epsilon_M = 0.57, V_M = 0.052 \text{ m/s}.$$

Minimum fluidization velocity is probably in the range 0.038 to 0.052 m/s.

(d) Equation (6.9) for laminar flow gives,

$$V = 0.0556 D_p^2 \; \frac{2800 - 0.3}{0.000049} \; 9.82 = (31.2)(10^6) D_p^2$$

with

$D_p = 0.04$ mm $= 0.00004$ m,

$V = 0.05$ m/s

and with

$D_p = 0.08$ mm $= 0.00008$ m,

$V = 0.20$ m/s.

A check of Reynolds number shows that the assumption of laminar flow is correct $(0.00001 < \text{Re}_D < 2)$.

EXAMPLE 6.5. DECANTING A LIQUID WITH VARYING TEMPERATURE

The cooling water from a batch process varies in temperature between 20°C and 80°C. The hot water is collected, and the cold water discarded to the sewer. Figure 6.52 shows the decanter to be used.

(a) Calculate the minimum height z_1 in Figure 6.52, when $z_1 - z_2$ is at least 20 mm and the temperature in the decanter is assumed to increase linearly from 20°C at the bottom to 80°C at the top. Pressure drop through the cold water outlet can be neglected.

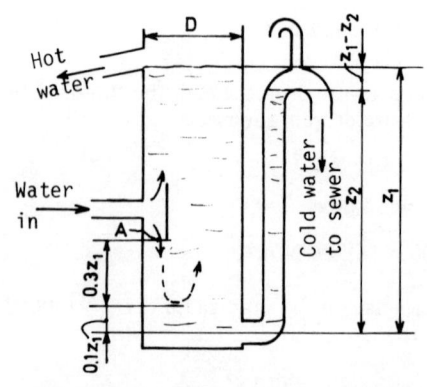

Figure 6.52   Decanter for separation of hot and cold water.

(b) Estimate maximum velocity at point A (Figure 6.52) ($0.4z_1$ above the bottom), when incoming water at $50°C$ penetrates to a maximum of $0.3z_1$ down before it flows up again as indicated by the broken arrow. Heat exchange with the ambient air can be neglected, and the temperature gradient in the vessel is the same as in (a).

Density variation for water is shown below.

Temperature, $\theta°C$	20	30	40	50	60	70	80
Density, ρ kg/m^3	998.3	995.7	992.3	998.0	983.1	977.6	971.5

Solution

(a) The static pressure at the bottom of the vessel, p_1, equals the pressure exerted by the liquid column in the outlet for cold water with height $z_2 = z_1 - 0.02$ m and density $\rho = 998.3$ kg/m^3,

$$p_1 = (z_1 - 0.02)(998.3)g.$$

The same pressure is obtained by integration through the liquid in the vessel,

$$p_1 = \int_0^{z_1} \rho g \, dz,$$

or

$$(998.3)(z_1 - 0.02) = \int_0^{z_1} \rho \, dz. \qquad (a)$$

Equation (a) can be solved graphically as shown in Figure 6.53, where the liquid density is plotted against the distance above the bottom. In this case, a good

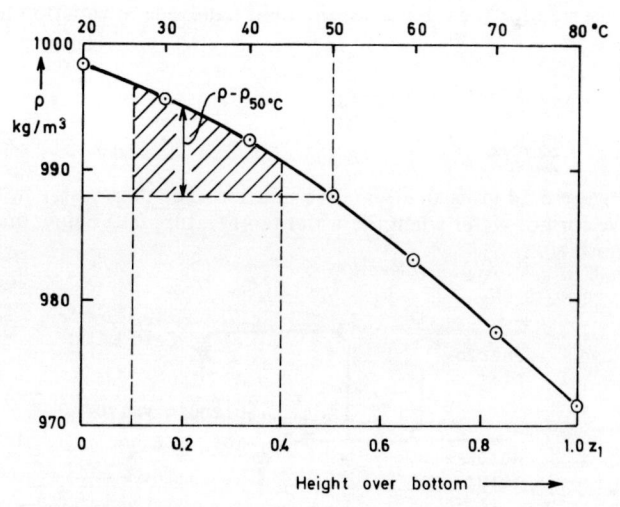

Figure 6.53 Density of water plotted against distance above the bottom of the tank in Figure 6.52.

approximation is obtained with the curve between the points replaced by straight lines,

$$\int_0^{z_1} \rho \, dz = \frac{z_1}{6}\left(\frac{998.3}{2} + 995.7 + 992.3 + 988 + 983.1 + 977.6 + \frac{971.5}{2}\right)$$

$$= 986.9 \, z_1$$

$$(z_1 - 0.02)(998.3) = 986.9 \, z_1$$

$$z_1 = 1.75 \text{ m}.$$

(b)   The bouyancy per unit volume is $(\rho - 998)g$ where 998 is the water density at 50°C. The bouyancy equals mass times acceleration,

$$(\rho - 998)(9.82) = 998 \, \frac{dV}{dt}. \tag{b}$$

In the time dt the liquid moves a distance $-dz = V dt$, i.e. $dt = -dz/V$. Insertion into equation (b) and integration gives,

$$\int_{V_A}^0 V \, dV = \frac{9.82}{988} \int_{0.4z_1}^{0.1z_1} (\rho - 988) \, dz. \tag{c}$$

The integral on the right hand side in equation (c) is the hatched area in Figure 6.53. The average height of the area is

$$[(995.7 - 988) + (992.3 - 988)]/2 = 6 \text{ kg/m}^3$$

and the area itself,

$$6(0.1 - 0.4)z_1 = -(6)(0.3)(1.75)$$

$$= -3.15 \text{ kg/m}^2.$$

This value, inserted for the integral on the right hand side in equation (c), gives

$$-\frac{V_A^2}{2} = \frac{9.82}{988}(-3.15),$$

$$V_A = 0.25 \text{ m/s}.$$

Comment Figure 6.54 gives an alternative solution. The thermostat closes the solenoid valve for hot water when the water temperature falls below the set-point of the thermostat.

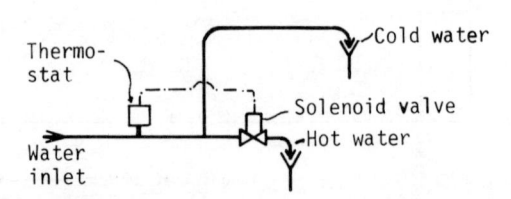

Figure 6.54 An alternative to the arrangement in Figure 6.52.

EXAMPLE 6.6. DECANTING

The decanter shown in Figure 6.10 is to be used to separate 2 m³/h of a mixture containing 40 vol% of a light liquid dispersed in a heavy liquid.

	Heavy liquid	light liquid
Density, ρ kg/m³	1000	870
Viscosity, μ cP	1.00	0.60

The following table gives the results of a sedimentation test with a graduated cylinder (Figure 6.55).

Time, minutes		2	4	6	10
Approximate height	clear, heavy liquid, z_1 cm	8.7	14.5	17.5	20.0
	transition layer, z_2 cm	21.8	12.5	7.3	2.5
	clear, light liquid, z_3 cm	4.5	8.0	10.2	12.5

The decanter is designed to give a transition layer at the outlet (z_2 in Figure 6.55) not exceeding 30% of the diameter of the decanter, and located in the middle of the decanter as indicated in Figure 6.56.

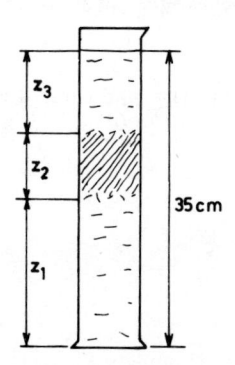

Figure 6.55 Sedimentation test.

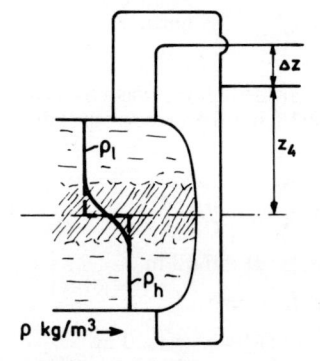

Figure 6.56 Density variation at the outlet.

(a) Estimate the distance from the centre of the decanter to the outlet for the heavy liquid, z_4 in Figure 6.56, based on a distance $\Delta z = 100$ mm between the two outlets.

(b) Estimate the diameter and length of the decanter with the ratio length to diameter between 2.5 and 3.0. Is it necessary to have baffles to prevent turbulence?

Solution

(a) The approximation that the density changes from $\rho_h = 1000$ kg/m³ to $\rho_1 = 870$ kg/m³ in the centre, gives

$$1000 z_4 = 870(z_4 + 0.1), \quad z_4 = 0.67 \text{ m}$$

$$z_4 + \Delta z = 0.77 \text{ m}.$$

(b) If the diameter of the decanter is approximately equal to the height of the graduated cylinder, the test results can be used directly. The height of the transitional layer, z_2, is plotted in Figure 6.57 as a function of time.

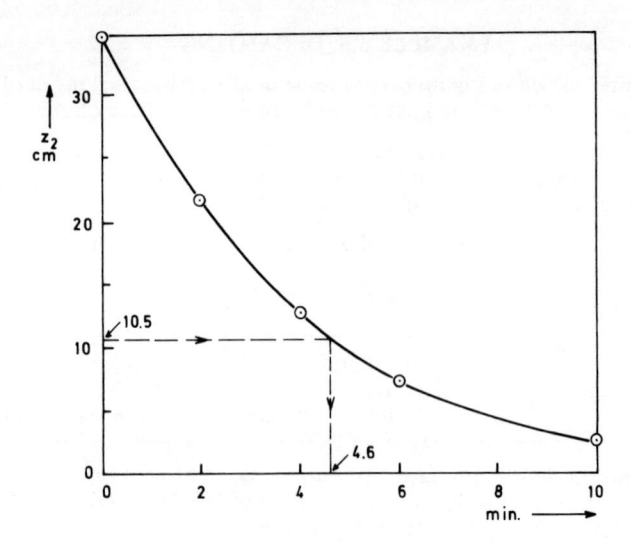

Figure 6.57 Height of the transition layer, z_2, as a function of time in minutes.

The transition layer was reduced to 30% or $(0.3)(35) = 10.5$ cm in 4.6 minutes (Figure 6.57), corresponding to a volume, $(2)(4.6)/60 = 0.153$ m^3. With $L = 3D$,

$$3D \frac{\pi}{4} D^2 = 0.153,$$

$$D = 0.40 \text{ m}$$

and effective length,

$$L = (3)(0.40) = 1.20 \text{ m}.$$

The velocities are calculated as if the two liquids have an interface at the centre line. Velocity of light liquid,

$$V_1 = \frac{(0.4)(2)}{(3600)\left(\frac{1}{2}\right)\left(\frac{\pi}{4}\right)(0.40^2)}$$

$$= 0.0035 \text{ m/s}$$

heavy liquid,

$$V_h = \frac{(0.6)(2)}{(3600)\left(\frac{1}{2}\right)\left(\frac{\pi}{4}\right)(0.40^2)}$$

$$= 0.0053 \text{ m/s}.$$

Eddies in one of the phases will influence the other phase across the transition layer. Hence, to be on the safe side, the diameter of the decanter is inserted as hydraulic diameter in the Reynolds number calculation:

$$\text{light phase, } Re_l = \frac{(870)(0.0035)(0.040)}{0.0006}$$

$$= 2030$$

$$\text{heavy phase, } Re_h = \frac{(1000)(0.0053)(0.040)}{0.001}$$

$$= 2120.$$

Minor disturbances, such as variations in the flow rate about the design value, can give turbulence, and it is recommended that a few longitudinal baffles be used as indicated in Figure 6.10.

EXAMPLE 6.7. CYCLONES IN PARALLEL AND IN SERIES

Cyclones are to be used to remove particles with density $\rho_p = 1100$ kg/m^3 from 0.40 m^3/s air with density $\rho = 1.2$ kg/m^3 and viscosity $\mu = 0.018$ cP. Maximum allowable pressure drop is 130 mm water gauge. A sedimentation test gave the particle diameter as 4.5 μm.

What fraction of the particles will be removed by cyclones with the geometrical ratios given in Figure 6.17, with

(a) one cyclone,

(b) five cyclones in series,

(c) five cyclones in parallel.

Solution

(a)   From equation (6.18),

$$(0.13)(1000)(9.82) = (8)(1.2)\, V_c^2/2,$$

$$V_c = 16.3 \text{ m/s}.$$

Using the inlet cross-sectional area (Figure 6.17), $D_c^2/8$,

$$\frac{D_c^2}{8}\, 16.3 = 0.40, \quad D_c = 0.443 \text{ m}.$$

The cut size, according to equation (6.19),

$$D_{pc} \approx 0.27\sqrt{(0.000018)(0.443)/(16.3)(1100-1.2)}$$

$$= (5.7)(10^{-6}) \text{ m}$$

$D_p/D_{pc} = 4.5/5.7 = 0.79.$

Interpolation from Table 6.3 gives $\eta = 0.38$ or 38% removed.

(b)   Neglecting the pressure drop in the connections between the cyclones gives a pressure drop per cyclone, $\Delta p = (0.13)(1000)(9.82)/5 = 255$ N/m^2.

From equation (6.18),

$$255 = (8)(1.2) V_c^2/2,$$

$$V_c = 7.3 \text{ m/s}$$

$$\frac{D_c^2}{8} 7.3 = 0.40,$$

$$D_c = 0.622 \text{ m}$$

$$D_{pc} \approx 0.27\sqrt{(0.000018)(0.662)/(7.3)(1100-1.2)}$$

$$= (10.4)(10^{-6}) \text{ m}$$

$$D_p/D_{pc} = 4.5/10.4 = 0.43.$$

Interpolation from Table 6.3 gives $\eta = 0.16$. The fraction leaving the first cyclone is $(1 - 0.16)$, and leaving the fifth cyclone $(1 - 0.16)^5 = 0.42$, i.e. 58% removed.

(c) Air flow rate per cyclone,

$$0.40/5 = 0.08 \text{ m}^3/\text{s}$$

$$\frac{D_c^2}{8} 16.3 = 0.08,$$

$$D_c = 0.198 \text{ m}$$

$$D_{pc} \approx 0.27\sqrt{(0.000018)(0.198)/(16.3)(1100 - 1.2)}$$

$$= (3.8)(10^{-6}) \text{ m}$$

$$D_p/D_{pc} = 4.5/3.8 = 1.18$$

or

$$\eta = 0.57, \text{ i.e. } 57\% \text{ removed.}$$

Note The efficiency is practically the same for cyclones in series and in parallel. Cyclones in parallel are smaller and cheaper than those in series.

EXAMPLE 6.8. CYCLONE FOR PARTICLES OF VARYING SIZE

2.11 m³/s of gas from roasting mineral ore passes through a cyclone followed by a scrubber and an electrostatic precipitator. Gas density $\rho = 0.78$ kg/m³ and viscosity $\mu = 0.0265$ cP. Particle density $\rho_p = 1800$ kg/m³ and particle size distribution is given in Figure 6.58.

Estimate the weight fraction of the particles removed in a cyclone with geometrical ratios as in Figure 6.17 and inlet velocity $V_c = 20$ m/s.

Solution

The weight of collected particles with diameter between D_p and $D_p + dD_p$ (Figure 6.59), is ηdw_{in}, where η is the cyclone efficiency for particles with diameter D_p. The fraction leaving the cyclone, is

$$\int dw_{out} = \int (1 - \eta) \, dw_{in}. \tag{a}$$

To integrate equation (*a*), the term $(1 - \eta)$ has to be determined as a function of w_{in}.

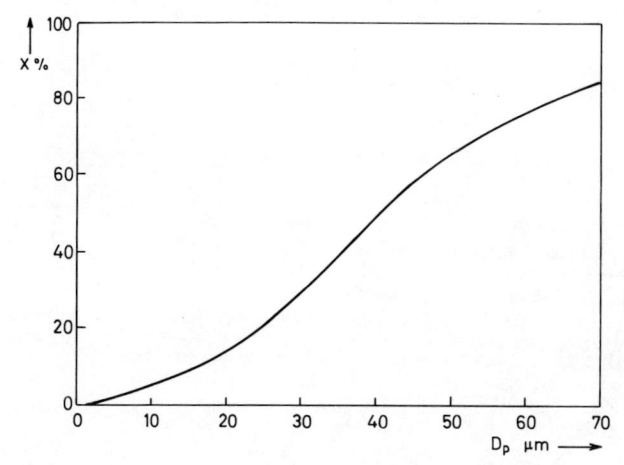

Figure 6.58 Weight % particles x, with diameter less than D_p.

The equation of continuity and inlet cross section $D_c^2/8$, gives

$$\frac{D_c^2}{8} 20 = 2.11,$$

or cyclone diameter $D_c = 0.92$ m. Cut size, according to equation (6.19),

$$D_{pc} \approx 0.27\sqrt{(0.0000265)(0.92)/(20)(1800 - 0.78)} = (7)(10^{-6}) \text{ m}.$$

A cut size of 7 µm and efficiencies from Table 6.3 correspond to:

D_p/D_{pc}	0.5	1.0	1.5	2.0	3.0	5.0	7.0	10.0	Notes
D_p µm	3.5	7.0	10.5	14	21	35	49	70	$7D_p/D_{pc}$
η	0.20	0.50	0.69	0.80	0.90	0.96	0.98	0.99	Table 6.3
$1 - \eta$	0.80	0.50	0.31	0.20	0.10	0.04	0.02	0.01	
w_{in}* kg	0.8	2.5	5	7.5	15	38	64	85	from Figure 6.58

*w_{in} is kg of particles with diameter less than D_p, per 100 kg particles entering (from Figure 6.58).

The figures from the last two rows in the table are plotted in Figure 6.60, where the hatched area corresponds to the integral of the right hand side of equation (a). The hatched area is 9.3% of the total area, i.e. the cyclone removes $100 - 9.3 = 90.7$ weight % of the entering dust.

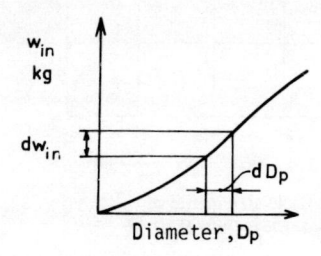

Figure 6.59 Cumulative weight of particles in entering gas, w_{in}, plotted as a function of particle size.

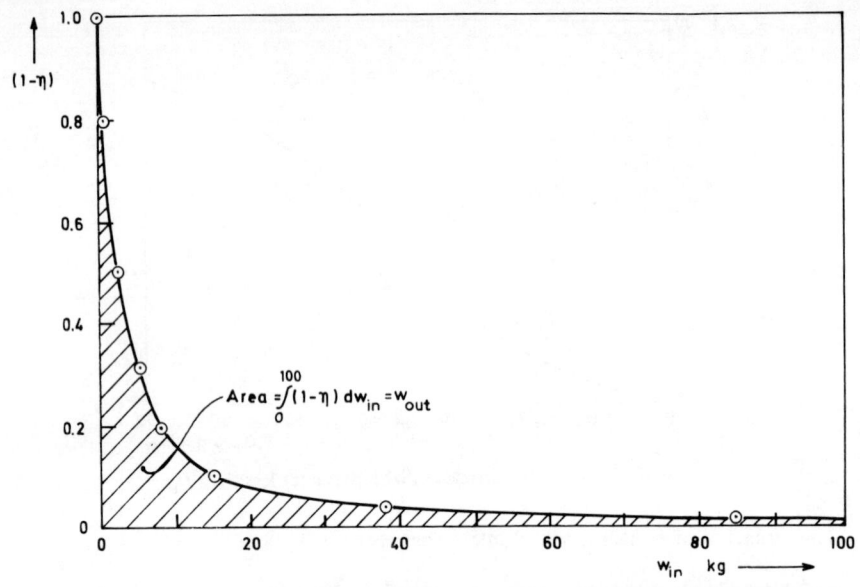

Figure 6.60 Diagram for integration of the right hand side of equation (*a*) in Example 6.8.

EXAMPLE 6.9. HYDROCYCLONE

An oil refinery has a continuous flow of 1000 t/day of oil at a temperature of 240°C. The oil contains 0.1 weight % solid particles with a cumulative particle size as given in Figure 6.61. Particle content is to be reduced to 0.02 weight % by means of hydrocyclones operating with 4 bar pressure drop. Estimate

(*a*) suitable cut size D_{pc}, and

(*b*) the main dimensions of the hydrocyclones.

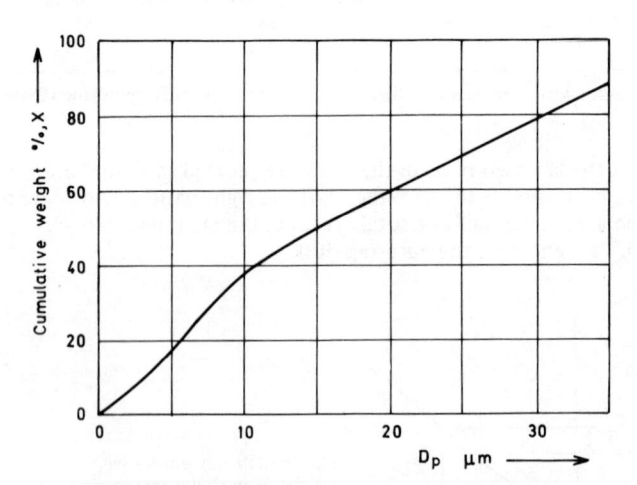

Figure 6.61 Cumulative particle size in the oil:[25] x = weight % particles with diameter less than D_p.

Density of the oil, $\rho = 760$ kg/m^3 and viscosity $\mu = 0.7$ cP. Density of the particles (wetted with oil), $\rho_p = 1800$ kg/m^3.

Solution

(a) Cut size D_{pc} is adjusted until a size giving the desired separation is found. The calculation is the same as in example 6.8. The amount of particles leaving the cyclones with the overflow, x_{out}, is determined by integration of equation (a) in example 6.8 except that w is replaced by x,

$$x_{out} = \int_{x=0}^{x=100} (1-\eta)\, dx \, \% \qquad (b)$$

The integration is carried out graphically. A trial with $D_{pc} = 4$ μm gives the data below:

D_p/D_{pc}	0.25	0.5	0.75	1.0	1.5	2.0	3.0	Notes
D_p μm	1.0	2.0	3.0	4.0	6.0	8.0	12.0	$D_p = 4 D_{pc}$
η	0.10	0.13	0.30	0.50	0.86	0.96	1.00	from Figure 6.23
$1-\eta$	0.90	0.87	0.70	0.50	0.14	0.04	0.00	
$x\,\%$	2.7	5.3	8.7	12.7	21.5	30	43	from Figure 6.61

The figures from the two last rows in the table are plotted in Figure 6.62, where the hatched area represents the integral on the right hand side of equation (b). The hatched area is 6.7 squares, while $\int_{x=0}^{x=100} dx$ corresponds to 50 squares (Figure 6.62 stops at $x = 50\%$ and shows only 25 of the squares). Thus, $D_{pc} = 4$ μm corresponds to $(0.1)(6.7)/50 = 0.0134$ weight % particles in the overflow. This is well below the required 0.02 weight %, and a reasonable safety margin is included if the hydrocyclones are designed for cut size

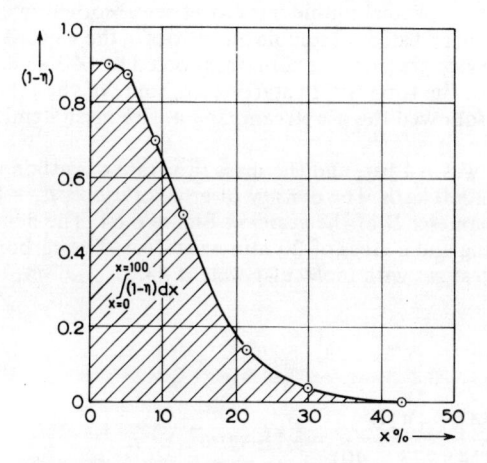

$$D_{pc} = 4 \text{ μm}$$

Figure 6.62 Diagram for graphical integration of equation (b) in Example 6.9.

(b) From equation (6.21),

$$\left[\left(\frac{0.000004}{0.0007}\right)^2 (1800 - 760)(4)(10^5)\right] = 0.177 \left[\frac{D_c\sqrt{(760)(400,000)}}{0.0007}\right]^{0.85}$$

$$D_c = 0.0224 \text{ m}$$
$$= 22.4 \text{ mm}.$$

Other dimensions are determined from the ratios in Figure 6.22.

Diameter of overflow, (0.34)(22.4) =   7.6 mm.
Diameter of inlet,   (0.28)(22.4) =   6.3 mm.
Height of cyclone,   (5)(22.4) = 112   mm.

Equation (6.22) gives the capacity $\dot{Q}$ of each cyclone,

$$\left[\dot{Q}\left(\frac{760}{0.0007}\right)^2 \sqrt{\frac{(4)(10^5)}{760}}\right] = 0.2 \left[\frac{0.0224\sqrt{(760)(4)(10^5)}}{0.0007}\right]^{1.85}$$

$$\dot{Q} = 0.00032 \text{ m}^3/\text{s or 21,000 kg/day}$$

which requires

$$\frac{1,000,000}{21,000} = 48 \text{ cyclones in parallel.}$$

Note Rietema reports[25] that a battery of 54 hydrocyclones was built to remove particles from the oil specified in this example. They had 6 groups of 9 cyclones each to give flexibility. The oil contained some gas bubbles which escaped towards the centre and provided the gas core. Due to the low particle concentration, 3 to 5% underflow was sufficient. Very hard material was used for construction to reduce erosion. A test after several hundred hours operation showed 0.01 to 0.02 weight % particles in the overflow.

EXAMPLE 6.10. WIRE MESH DEMISTER

Figure 6.63 shows an arrangement for purifying carbon dioxide released by fermentation of beer.[52] Water soluble impurities are absorbed in cold water in the bottom of the column, and alcohol soluble impurities in the 95 vol% ethanol on the top tray. Carbon dioxide gas from the column is cooled to $-40°C$, and condensed ethanol is refluxed to the top tray. In starting up, some of the ethanol condensate formed a mist that followed the gas stream, and a wire mesh demister was installed (Figure 6.64).

The gas pressure was 1.4 bar, and the mass flow rate of carbon dioxide varied between 1200 and 2000 kg/h. The density of ethanol drops, $\rho_1 = 830 \text{ kg/m}^3$.

Determine the diameter D of the vessel in Figure 6.64. The demister is supported by a 25 mm wide ring and a cross of 20 mm wide Tees. The carbon dioxide is assumed to be an ideal gas with molecular weight 44.

Solution

Gas density,

$$\rho = \frac{Mp}{RT} = \frac{(44)(1.4)(10^5)}{(8314)(273 - 40)} = 3.18 \text{ kg/m}^3.$$

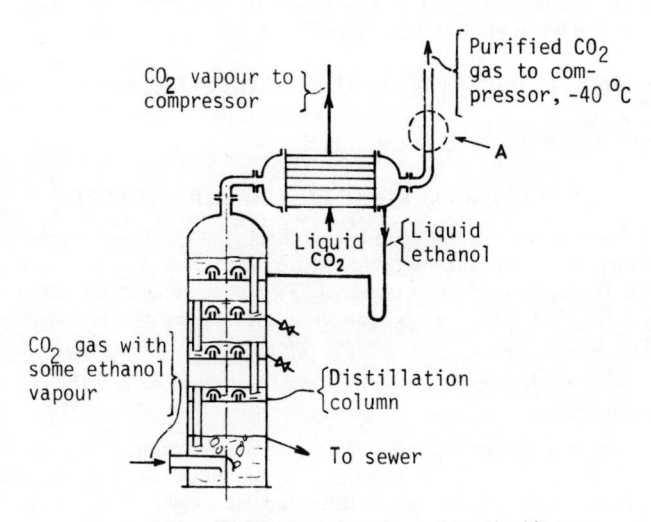

Figure 6.63 Purification plant for carbon dioxide gas.

Equation (6.27), with $C = 0.09$, gives

$$V_s = 0.09 \sqrt{\frac{830 - 3.18}{3.18}} = 1.45 \text{ m/s},$$

corresponding to a cross-section area,

$$A = \frac{2000}{(3600)(3.18)(1.45)} = 0.123 \text{ m}^2.$$

Neglecting the Tees gives

$$\frac{\pi}{4}(D - 0.05)^2 = 0.123,$$

$$D = 0.45 \text{ m}.$$

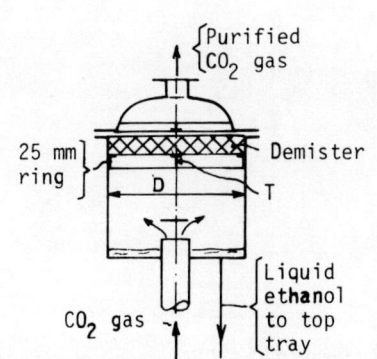

Figure 6.64 Wire mesh demister installed at A in Figure 6.63.

Introducing $D = 0.48$ m gives the free cross-section

$$A = \frac{\pi}{4}(0.48 - 0.05)^2 - (2)(0.02)(0.48 - 0.05) = 0.128 \text{ m}^2$$

i.e. standard 100 mm mat with $D = 480$ mm.

EXAMPLE 6.11. FIBRE FILTER FOR BACTERIA

The air for penicillin fermentation passes through a fibre glass bacteria filter. The air pressure is 1.2 bar and the temperature 30°C. The efficiency calculated from the data in Figure 6.35, should be 99.99% under normal operation, and at least 99.8% if the flow rate of air is reduced due to failure of one of the air compressors.

(*a*) Estimate the thickness of the fibre filter in order to satisfy the requirement for reduced flow rate.

(*b*) Determine a reasonable air velocity under normal operation with fibre diameter $D_c = 16$ μm, particle diameter $D_p = 1$ μm and density $\rho_p = 1000$ kg/m³. The air is assumed to be an ideal gas with molecular weight 29 and viscosity $\mu = 0.018$ cP.

Solution

(*a*) From Figure 6.35 $\eta_{min} = 0.08$ for the 3 mm thick filter at $\Psi = 0.09$. The fraction of bacteria that passes point 1 in Figure 6.65 three mm from the inlet, is $(1 - \eta)$, at point 2 $(1 - \eta)^2$, and at point n $(1 - \eta)^n$, i.e.

$$1 - 0.998 = (1 - 0.08)^n \text{ or } n = 74.5.$$

Filter thickness,

$$L = (74.5)(0.003)$$
$$= 0.22 \text{ m}.$$

Figure 6.35 is valid for furfural resin bound mats. In a bed of packed fibres, some channelling may occur in certain parts of the bed. It is therefore recommended that 2–3 times the thickness be used for a packed bed, i.e.

$$L \approx (3)(0.22) = 0.66 \text{ m}.$$

(*b*) An efficiency of 99.99% requires an efficiency η_3 per 3 mm thickness,

$$1 - 0.9999 = (1 - \eta_3)^{74.5},$$

$$\eta_3 = 0.13.$$

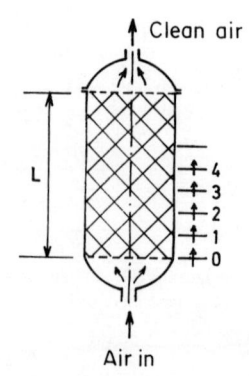

Figure 6.65 Fibre filter: points 0, 1, 3, etc., are 3 mm apart.

Figure 6.35 gives the corresponding inertial parameter $\Psi = 0.28$.
Air density,

$$\rho = \frac{Mp}{RT} = \frac{(29)(1.2)(10^5)}{(8314)(303)} = 1.38 \text{ kg/m}^3$$

Mean free path [equation (6.11)],

$$\lambda = 3 \frac{0.000018}{1.38} \sqrt{\frac{29\pi}{(8)(8314)(303)}}$$
$$= (0.083)(10^{-6}) \text{ m}.$$

The factor C in equation (6.25) equals the Cunningham correction factor in equation (6.10),

$$C = 1 + \frac{0.083}{1}(1.644 + 0.552e^{-0.656/0.083}) = 1.14.$$

Equation (6.25) with $\Psi = 0.28$,

$$0.28 = \frac{1.14}{18} \frac{(1000)V_0(10^{-6})^2}{(0.000018)(0.000016)},$$

$V_0 = 1.27$ m/s.

To be safe, the filter is designed for

$V_0 = 1.4$ m/s.

EXAMPLE 6.12. BAG FILTER

Approximately 1.5 kg/h valuable dust is lost in 2000 m³/h exit air from a spray dryer for detergent after the air has passed a cyclone. Investigate the possibility of recovering dust by means of a bag filter as indicated in Figure 6.66.

(a) Estimate the filter area.

(b) Estimate the pressure drop if the filter is cleaned after 8 hours operation, and 12% of the dust remains on the bags after cleaning. The viscosity of the air $\mu = 0.019$ cP.

Solution

(a) Particles from a spray dryer are almost spherical. Therefore a relatively high gas velocity is assumed, i.e. 0.8 m/min. The area

$$A = \frac{2000}{(60)(0.8)}$$
$$= 42 \text{ m}^2.$$

(b) The factor C_d in equation (6.33) is estimated as average for industrial dusts, $(1.9)(10^{10})$ (Table 6.7), and $t_1 + t$ is taken as $1.12t$. After 8 hours operation, equation (6.33) gives

$$\Delta p = (1.9)(10^{10})(0.000019)\frac{1.5}{2000}\left(\frac{0.8}{60}\right)^2 (1.12)(8)(3600)$$

$$= 1550 \text{ N/m}^2$$

or 158 mm water gauge.

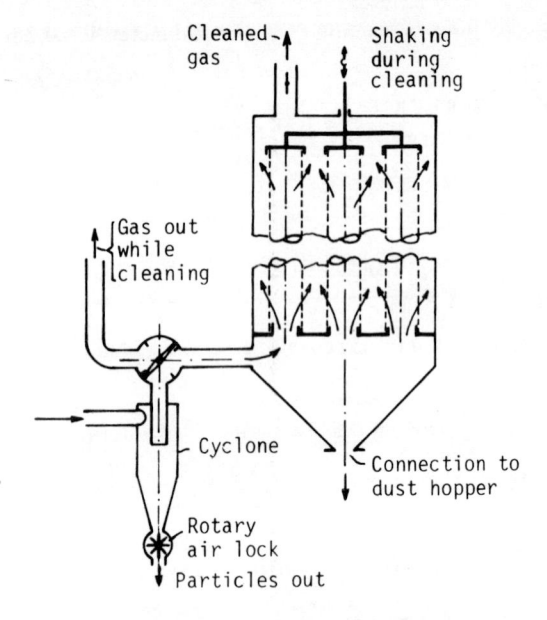

Figure 6.66 Cyclone and bag filter.

Note The values are too uncertain as a basis for design. Tests must be carried out to determine the constants and the possibility of gradual clogging of the filter cloth.

EXAMPLE 6.13. FIBRE FILTER, VENTURI SCRUBBER AND ELECTROSTATIC PRECIPITATOR FOR SULPHURIC ACID MIST

Approximately 16,000 m^3/h air with pressure 1 bar and temperature 120°C, contains some sulphur trioxide that reacts with water vapour, giving sulphuric acid mist in the air. 97% of the mist must be recovered. The hot air passes through a spray chamber before mist recovery in a fibre filter, a venturi scrubber, or an electrostatic precipitator. The injection of 20°C water into the spray chamber is adjusted to give an exit gas temperature of 80°C. A fan with isothermal efficiency η_{is} = 0.65 is used to overcome the pressure drops of the fibre filter and the venturi scrubber. The gas has molecular weight 31 and specific heat capacity c_p = 0.98 kJ/(kg °C).

(*a*) Estimate the size of a fibre filter and the power consumption of the fan.

(*b*) Estimate the throat diameter, water consumption, pressure drop, and power consumption of a venturi scrubber.

(*c*) Estimate the power consumption of an electrostatic precipitator.

Solution

An enthalpy balance around the spray chamber with gas and water at 0°C as reference state, gives kilograms of evaporated water per kilogram of gas, *W*:

$$(0.98)(120) + (4.19)(20)W = (0.98)(80) + 2644W$$

where 2644 kJ/kg is the enthalpy of water vapour at 80°C. This gives *W* = 0.0153 kg water vapour/kg gas, and a total volume after the spray chamber,

$$v_0 = \frac{nRT}{Mp} = \frac{16{,}000}{3600}\left(\frac{1}{31} + \frac{0.0153}{18}\right)\frac{(8314)(353)}{10^5}$$

$$= 4.32 \text{ m}^3/\text{s}.$$

(a) According to Table 6.6, the specified efficiency may be obtained with a fibre mist eliminator based on impaction and velocity ~2.5 m/s and pressure drop 2000 N/m² (alternative 1), or a filter based on Brownian movement with velocity 0.10–0.15 m/s and pressure drop probably in the range 2000 to 3000 N/m² (alternative 2).

ALTERNATIVE 1

Area,

$$A = 4.32/2.5$$
$$= 1.73 \text{ m}^2$$

Power,

$$P = \frac{v_0 \Delta p}{1000 \eta_{is}} = \frac{(4.32)(2000)}{(1000)(0.65)}$$
$$= 13.3 \text{ kW}.$$

ALTERNATIVE 2

With velocity 0.125 m/s and estimated pressure drop 2500 N/m²:
Area,

$$A = 4.32/0.125$$
$$= 35 \text{ m}^2.$$

Power,

$$P = \frac{(4.32)(2500)}{(1000)(0.65)}$$
$$= 16.6 \text{ kW}.$$

(b) Figure 6.41 gives efficiency $\eta = 0.97$ with velocity $V = 90$ m/s, water load approximately 0.8 litre/m³ gas, and a pressure drop $\Delta p = 490$ mm water gauge:
Cross section in throat,

$$\frac{\pi}{4}D^2 = \frac{4.32}{90}, \quad D = 0.25 \text{ m}.$$

Water consumption,

$$\frac{0.8}{1000}(4.32)(3600) = 12 \text{ m}^3/\text{h}.$$

Power consumption,

$$\frac{(4.32)(490)(9.82)}{(1000)(0.65)} = 32 \text{ kW}.$$

(c) Table 6.11 reports efficiency 99.6%, using 0.35 kW/1000 m^3 gas per hour, or

power consumption = (0.36)(4.32)(3600)/1000

= 5.6 kW.

Figure 6.47 gives an energy consumption of 1.25 kW/1000 m^3 gas per hour in order to obtain 97% efficiency, or

power consumption = (0.36)(4.32)(3600)/1000

= 5.6 kW.

EXAMPLE 6.14. ELECTROSTATIC PRECIPITATOR

Dust-laden gas from roasting of pyrites passes through a cyclone, a scrubber, and finally through two electrostatic precipitators with tubular electrodes (Figure 6.45). Remaining dust, after the precipitators, deposits in the catalyst beds of the reactor in which oxidation to sulphur trioxide occurs and reduces the lifetime of the catalyst.

The precipitators are in parallel, and have an efficiency of 98%. Every 12 hours both precipitators are bypassed for 12 minutes while they are flushed for cleaning. The tubular electrodes are 2.6 m long with internal diameter 0.25 m.

(a) Estimate the reduction in amount of dust to the catalyst beds if the precipitators are flushed at different times.

(b) A third, identical precipitator may be installed to increase the overall efficiency. Consider whether it should be in series or in parallel with the previous two which remain in parallel.

Solution

(a) The dust in front of the precipitators is W kg/day.

Filtration time,

24 − (2)(12)/60 = 23.6 h/day.

Dust discharged during filtration,

$$0.02 \frac{23.6}{24} W = \qquad 0.0197 \, W \text{ kg/day.}$$

Dust discharged during flushing,

$$\frac{0.4}{24} W = \qquad \underline{0.0167 \, W \text{ kg/day.}}$$

Present emission,   0.0364 W kg/day.

Efficiency $\eta = 0.98$ inserted into equation (6.36), gives

$$0.98 = 1 - e^{-(2Lw/RV)}, \quad \text{or} \quad 2Lw/RV = 3.91.$$

Efficiency at doubled velocity, $\eta = 1 - e^{-3.91/2} = 0.858$.

Flushing at different times allows 23.2 h/day with both filters in operation.

Amount discharged in 23.2 hours,

$$0.02 \frac{23.2}{24} W = 0.0193 \, W \text{ kg/day.}$$

Amount discharged during flushing,

$$(1 - 0.858)\frac{0.8}{24} W = 0.0047 \, W \text{ kg/day}.$$

Emission, $(0.0193 + 0.0047)W = 0.0240 \, W$ kg/day

or

$$100 \, \frac{0.0364 - 0.0240}{0.0364} = 34\% \text{ reduction in emission}.$$

Alternative calculation using equation (6.37) and Figure 6.48. For tubular electrodes L/s is replaced by $2L/R$, i.e.

$$\tau = (7.41)(2)(2.6)/0.125 = 308$$

Figure 6.48 with $\tau = 308$ and $\eta = 0.98$, gives $\varphi = 0.25$.
A doubled gas velocity gives $\varphi = 0.25/2 = 0.125$.
This value and $\tau \approx 308$ gives (Figure 6.48) $\eta = 0.75$.

Amount discharged in 23.2 hours, $0.0193 \, W$ kg/day

Amount discharged during flushing,

$(0.25)\dfrac{0.8}{24} W =$ $\underline{0.0083 \, W \text{ kg/day}}$

     Emission, $0.0276 \, W$ kg/day

or

$$100 \, \frac{0.0364 - 0.0276}{0.0364} = 24\% \text{ reduction in emission}.$$

The estimates give between 24 and 34% reduction of the dust load in emission.

(b)   Emission when *all three precipitators are in parallel* and the filters are flushed at different times can be calculated as follows. All precipitators in operation $24 - (2)(3)(12)/60 = 22.8$ h/day. Efficiency with 3 filters in operation [equation (6.36)],

$$\eta = 1 - e^{-(3)(3.91)/2} = 0.996.$$

Amount discharged in 22.8 hours,

$0.004 \dfrac{22.8}{24} W =$ $0.0038 \, W$ kg/day.

Amount discharged in 1.2 hours,

$0.02 \dfrac{1.2}{24} W =$ $\underline{0.0010 \, W \text{ kg/day}}$

     Emission, $0.0048 \, W$ kg/day.

A similar calculation can be made for *two precipitators in parallel and the third in series*. According to equation (6.36), the dust load to the third precipitator will be only $(100)(2)(0.024) = 4.8\%$ of the dust arriving in each of the two precipitators in parallel. This indicates that flushing of the third precipitator can be neglected in the estimate.

Amount discharged with all three precipitators in operation,

$$(1 - 0.98)(1 - 0.858)\frac{23.2}{24} W = 0.0027 \ W \ \text{kg/day}.$$

Amount discharged with two in series

$$(1 - 0.858)^2 \frac{0.8}{24} W = \qquad 0.0007 \ W \ \text{Kg/day}$$

Emission, 0.0034 W kg/day.

Note The third precipitator reduces the dust load to approximately 1/5 of what is obtainable with the two precipitators. The calculated emissions for precipitators in parallel and in series are of the same order of magnitude. Therefore, the choice may depend on other considerations, such as simplicity of piping, easier service for the operators, and performance if one of the original precipitators is out of operation for repair.

Problems

6.1. Estimate the terminal velocity of a cylindrical catalyst particle with diameter 4 mm, length 5 mm and density 960 kg/m^3, falling in a gas with density 0.6 kg/m^3 and viscosity 0.025 cP.

6.2. A tubular centrifuge (Figure 6.11) with diameter 100 mm is used to separate liquids with densities ρ_1 = 920 kg/m^3 and ρ_2 = 1000 kg/m^3. The outlet for the light liquid has a radius r_1 12.5 mm. Calculate the radius r_2 for the outlet of the heavy liquid to maintain the neutral zone in the middle of the liquid cylinder.

6.3. Estimate the efficiency of a second cyclone in series with the cyclone in Example 6.8 and with the same dimensions.

6.4. Figure 6.67 is a vane type demister for a cooling tower (dimensions in mm). The temperature of the humid air is 35°C.

	Mol. wt.	Partial pressure bar	Viscosity cP
Air	29	0.944	0.0183
Water vapour	18	0.056	0.0098

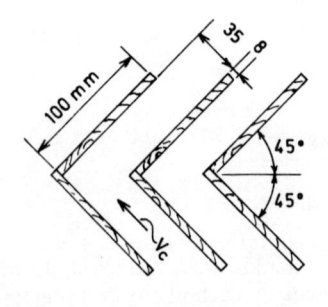

Figure 6.67 Vane type demister.

(*a*) Calculate the diameter of water drops with terminal velocity 2.8 m/s.

(*b*) Estimate the efficiency of the demister as a function of drop diameter.

(*c*) Estimate the pressure drop.

6.5. Estimate the pressure drop of the demister in Example 6.10 and comment on the results obtained from different calculations.

6.6. 4000 kmol ferrosilicon furnace fume is cooled to 160°C in a steam boiler before it is cleaned in a venturi scrubber and a cyclone.

(*a*) Estimate the theoretical power requirement of the scrubber in order to obtain 85, 90, and 95% efficiency.

(*b*) Estimate the pressure drop in the scrubber and cyclone if 3% of the total power is used for pumping water through the nozzles and the gas pressure is 1 bar absolute.

References

1   Perry, R. H. and C. H. Chilton: *Chemical Engineers' Handbook*, McGraw-Hill, New York, 5th Edn., 1973.
2   Batel, W.: *Entstaubungstechnik*, Springer-Verlag, Berlin, 1972.
3   Strauss, W.: *Industrial Gas Cleaning*, Pergamon Press, Oxford, 2nd Edn., 1975.
4   Marchello, J. M. and J. J. Kelly: *Gas Cleaning for Air Quality Control*, Marcel Dekker, New York, 1975.
5   Brown, G. G. et al.: *Unit Operations*, Wiley, New York, 1950.
6   Wasser, E.: Das Wiederstandgesetz kleiner Kugeln in reibenden Medien, *Phys. Zeitschrift*, **34**, 257–279 (1973).
7   Bird, R., W. E. Stewart and E. N. Lightfoot: *Transport Phenomena*, Wiley, New York, 1960.
8   Gery, W. B.: Thickeners, *Chem. Eng.*, **62** (6), 228–233 (1955).
9   Scholl, K. H.: Der Einfluss der Konzentration auf die Sinkgeschwindigkeit von Partikeln, *Chem.-Ing.-Techn.*, **48**, 149–150 (1976).
10 Maude, A. D. and R. L. Whitmore: A generalized theory of sedimentation, *Brit. J. Appl. Physics*, **9**, 447–482 (1958).
11 Flood, H. W. and B. S. Lee: Fluidization, *Scientific American*, **219** (No. 1), 94–104 (1968).
12 Leva, M.: *Fluidization*, McGraw-Hill, New York, 1959.
13 Gluck, S. E.: Design tips for pneumatic conveyers, *Hydrocarbon Processing*, **47** (Oct. 1968).
14 Gerchow, F. J.: How to select a pneumatic-conveying system, *Chem. Eng.*, Feb. 17, 72–86 (1975).
15 Gerchow, F. J.: Specifying components of pneumatic-conveying systems, *Chem. Eng.*, March 31, 88–96 (1975).
16 Stuke, B.: Das Verhalten der Oberfläche von sich in Flüssigkeiten bewegenden Gasblasen, *Die Naturwissenschaften*, **39**, 325–326 (1952).
17 Houghton, G., A. M. McLean and P. D. Ritchie: Mechanism of formation of gas bubble-beds, *Chem. Eng. Sci.*, **7**, 40–50 (1957).
18 Treybal, R. E.: *Liquid Extraction*, McGraw-Hill, New York, 2nd Edn., 1963.
19 Bingeman, J. B.: Preliminary selection of centrifugal equipment, *Chem. Eng. Progress*, **51**, 272–277 (1955).
20 Lapple, C. E. et al.: *Fluid and Particle Mechanics*, Chapter 14, University of Delaware, 1951.
21 Stern, A. S.: *Air Pollution*, Volume 3, Academic Press, New York and London, 2nd Edn., 1968.
22 van Ebbenhorst Tengbergen, H. J.: *The separation of liquids from gases by cyclones*, Chapter 7 in reference 24.

23 Day, R. W.: The hydrocyclone in process and pollution control, *Chem. Eng. Progress*, **69**, 67–72 (1973).
24 Rietma, K. and C. C. Verver: *Cyclones in Industry*, Elsevier, Amsterdam, 1961.
25 Rietema, K.: Performance and design of hydrocyclones, *Chem. Eng. Sci.*, **15**, 298–325 (1961).
26 API Standards No. 610: *Centrifugal Pumps for General Refinery Services*, Am. Petroleum Inst., Washington D.C.
27 Johnson, R. A., W. E. Gibson and D. R. Libby, Performance of liquid-liquid cyclones, *Ind. Eng. Chem. Fundamentals*, **15**, No. 2, 1976.
28 Tangel, O. F. and R. J. Brison: Wet cyclones, *Chem. Eng.*, **62** (6), 234–238 (1955).
29 Neumann, K.: Der Einsatz von Klassierzyklonen in der Mesaufbereitung Calbecht, *Erzmetall*, 7, H.12 (1954).
30 Ranz, W. E. and J. B. Wong: Impact of dust and smoke particles, *Ind. Eng. Chem.*, **44**, 1371–1381 (1952).
31 Ludwig, E.: *Applied Process Design for Chemical and Petrochemical Plants*, Volume 1, Gulf, Houston, Texas, 1964.
32 Massey, O. D.: How well do filters trap stray stack mist, *Chem. Eng.*, **66**, 13. July, 143–146 (1959).
33 Campbell, J. M.: *Gas Conditioning and Processing*, J. M. Campbell & Co., Norman, Oklahoma, 1972.
34 *Metex Mist Eliminators*, Bulletin ME-9-58, Metal Textile Corp., Roselle, N.J., 1958.
35 Knit-Mesh Ltd., 36 Victoria Street, London S.W.1.
36 Brink, J. A. et al.: Mist eliminators for sulfuric acid plants, *Chem. Eng. Progress*, **64** (No. 11), 82–86 (1968).
37 Humprey, A. E. and E. L. Gaden jr.: Air sterilization by fibrous media, *Ind. Eng. Chem*, **47**, 924–930 (1955) and **48**, 2172–2176 (1956).
38 Harris, W. B. and M. G. Mason: Operating economics of air-cleaning equipment utilizing the reverse jet principle, *Ind. Eng. Chem.*, **47**, 2423–2425 (1955).
39 Stairmand, C. J.: Dust collection by impingement and diffusion, *Trans. Instn. Chem. Engrs.*, **28**, 130–139 (1950).
40 Hausberg, G.: Die Nassentstaubung in der Hüttenindustrie, Entwicklungsmöglichkeiten, *Staub*, **21**, 418–425 (1961).
41 Johnstone, H. F. and F. O. Eckman: Collection of aerosols in a venturi scrubber, *Ind. Eng. Chem.*, **43**, 1358–1363 (1951).
42 Chemical Construction Corp., 525 West 43rd St., New York.
43 Webb, R. L., *Paint Industry Magazine*, Jan. and Oct. 1958.
44 Güntheroth, H.: Venturi-Scrubber zur Abscheidung von Rauch, Nebeln und Aerosolen, *Staub*, **21**, 430–434 (1961).
45 Semrau, K.: Neuere Erkenntnisse auf dem Gebiete der Nassentstauber, *Staub*, **22**, 184–188 (1962).
46 Lagarias, Paper 51–29, Air Pollution Control Association Meeting, 21–25 June, 1959.
47 White, H. J.: *Industrial Electrostatic Precipitation*, Addison-Wesley, Reading, Mass., 1963.
48 Williams, J. C. and R. Jackson: *The Interaction between Fluids and Particles*, Symposium, European Federation of Chemical Engineers, 22 June, 1962.
49 Heinrich, R. F. and J. R. Anderson: *Chemical Engineering Practice*, Ed. H. W. Cremer, Volume 3, Butterworths, London, 1957, p. 464.
50 Funke, G.: *Zement, Kalk, Gips*, **12**, 189 (1959).
51 Senecal, V. E.: Fluid Distribution in Process Equipment, *Ind. Eng. Chem.*, **49**, 993–997 (1957).
52 Hougen, O. A. and A. L. Lydersen: *Apparatus and Method of Removing Impurities from Highly Volatile Gas*, U.S. pat. 2,862,819.

CHAPTER 7
Liquid Filtration and Flotation

Filtration

Liquid filtration is used to remove solid particles from a liquid by means of a porous medium or septum, which retains the particles and allows the liquid to pass through.

After the first layer of particles is deposited on the septum, the filter medium is the *filter cake*, consisting of collected particles. In some cases a *filter aid* must be added to the feed to obtain sufficiently permeable filter cake, or to give a satisfactory clarity of the filtrate. Kieselguhr or diatomaceous earth with particles of diameter from 50 to 150 µm are the most common filter aids. Cellulosic and asbestos fibres are also used. Certain colloidal dispersions may require special treatment to produce agglomeration before filtration.[1]

A multitude of filter types has been developed,[2] each having its advantages and disadvantages, depending on characteristics of the suspension and the filtrate. Only some of the more common types are described in the following.

Sand filters are used for water clarification. They can be open gravity filters or closed filters for operation under pressure (Figure 7.1). The sand bed can be up to 2 m thick, particle diameter 0.5 to 1.5 mm, and the superficial water velocity can range from 5 m/h for open, to 10 or 12 m/h for closed designs.[3,4]

Equation (1.18) describes the pressure drop during filtration. The pressure drop

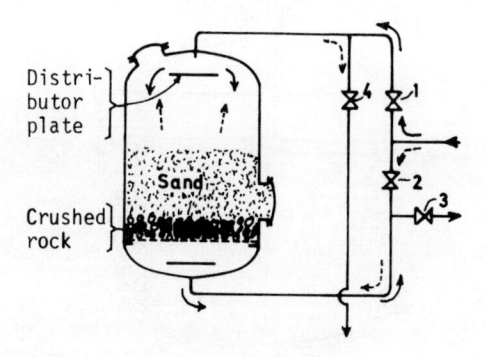

Figure 7.1 A closed sand filter.
Filtration: valves 1 and 3 open, 2 and 4 closed. Backwash (broken arrows): valves 2 and 4 open, 1 and 3 closed.

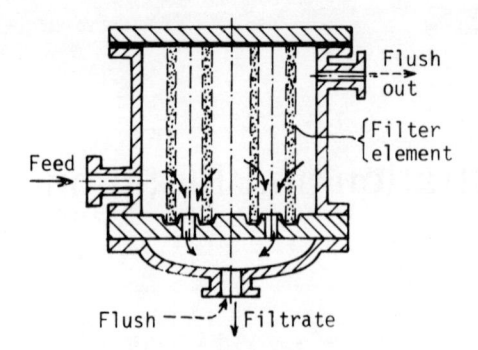

Figure 7.2 Nutsche with cylindrical filter elements of porous, ceramic material or springs. Springs must have a precoat of filter aid.

through the fluidized bed during the backwash corresponds to the weight of the particles as given by equation (6.13).

A *nutsche* is an open or closed tank with porous or perforated bottom that supports a filter medium. There are open vacuum nutsches similar to those used in laboratories, or closed pressure nutsches as shown in Figures 7.2 and 7.3. The diameter is usually less than 2 m and the cake thickness 0.1 to 0.3 m.

The *filter press*, Figure 7.4, is one of the most widely used filter types. The plates and frames alternate, with filter cloths over each plate. The slurry is pumped into the frames, and the filtrate leaves through grooves behind the filter cloth. Common sizes of square plates and frames are from 0.16 to 2.8 m² with approximately 85% of the area as effective filter area. Frame widths are from 25 to 125 mm.

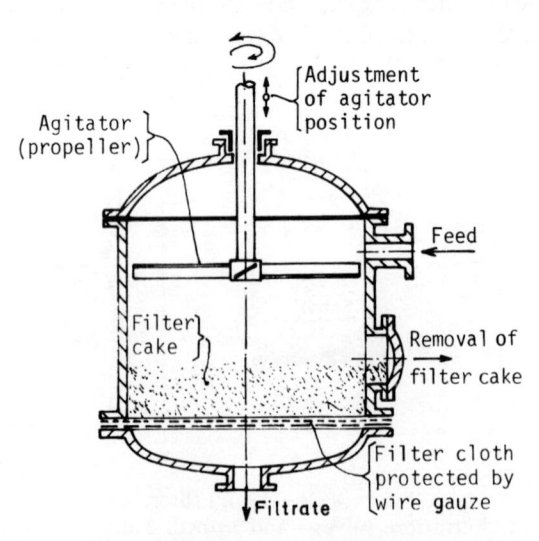

Figure 7.3 Pressure nutsche with agitator used to compress and to disperse the filter cake.

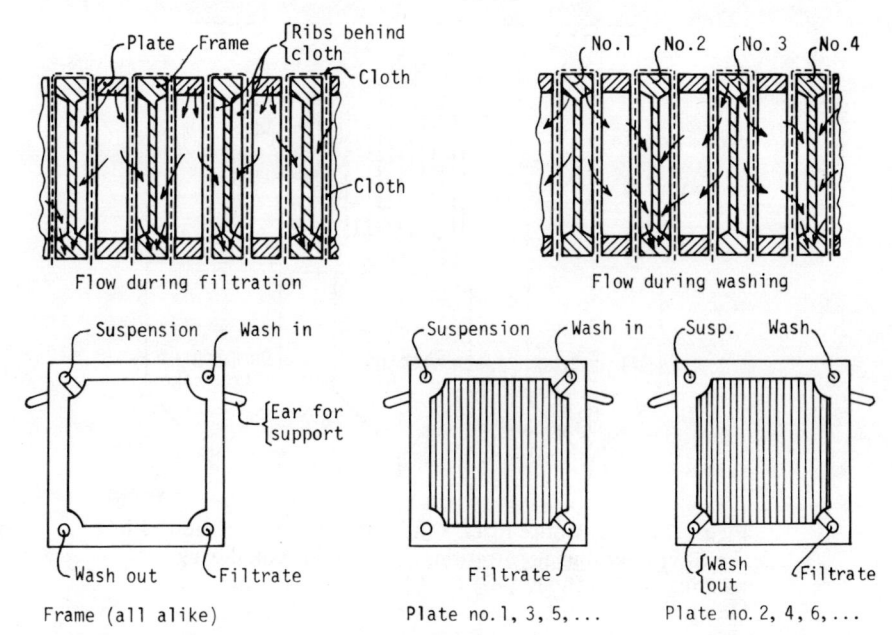

Figure 7.4   Filter press: upper left filtration, upper right washing period. Lower drawings show the conduits for feed, wash, and filtrate. The filter cloths have holes for the conduits.

Pressures range from 1 to 7 bar, in special cases up to 15 bar. During filtration and washing the filter is compressed by a mechanical or hydraulic arrangement, and the press is opened by retraction of one of the end plates.

Leaf filters are composed of flat filter elements that can be built of wire gauze (Figure 7.5). Pressure leaf filters are placed inside a pressure shell, while vacuum filters, such as the *Moore filter* in Figure 7.6, can be placed in open tanks.

Sweetland filters are pressure filters with circular leaves, while *Kelly filters*

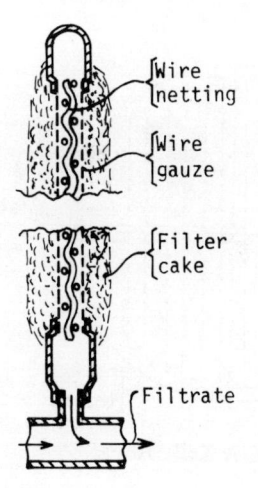

Figure 7.5   A wire filter leaf.

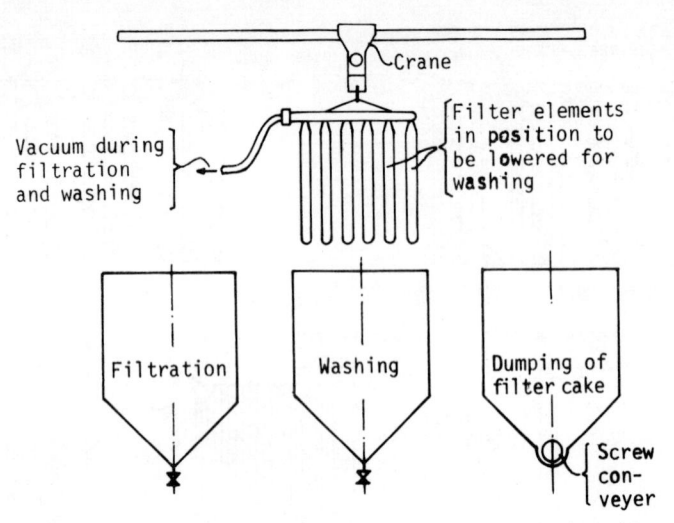

Figure 7.6 A Moore filter. Filter elements are submerged in the filter tank during filtration, in the wash tank during washing, and in the hopper for filter cake when the cake is blown off.

(Figure 7.7) have longitudinal leaves in a cylindrical pressure shell. Kelly filters are designed for pressures up to 17 bar. The shell can be opened and the filter leaves pulled out.

Next to filter presses, *rotating filters* are the most widely used filters in chemical process industries and in the mineral dressing industry.[5] The operation of a rotating filter can be seen from Figures 7.8, 7.9, and 7.10.

Easily settling slurries can be conveyed to the top of the filter as 'top feed', i.e. 'feed' instead of 'wash water' in Figure 7.8.

With particularly difficult filter cakes, such as mycelium from penicillin fermentation, a precoat of filter aid is often used, with or without filter aid in the feed.

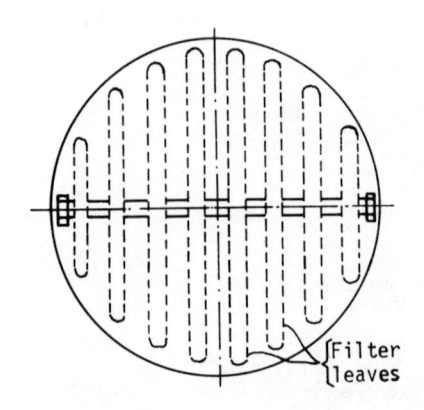

Figure 7.7 A Kelly filter.

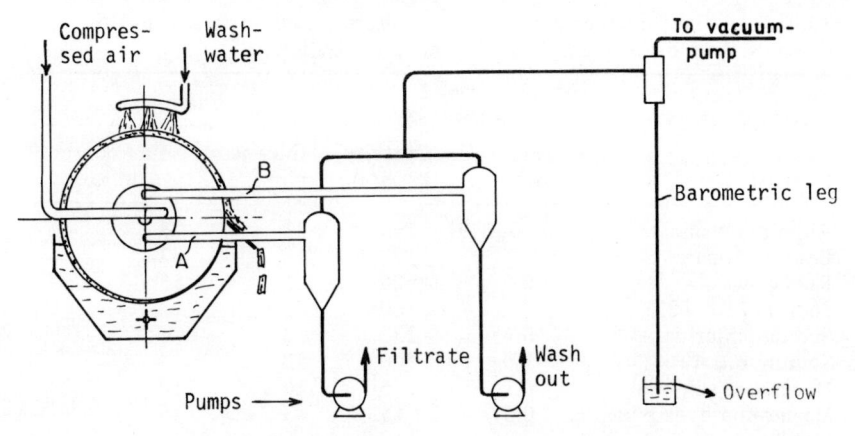

Figure 7.8   Rotating vacuum filter with suction line A for filtrate and B for wash water.

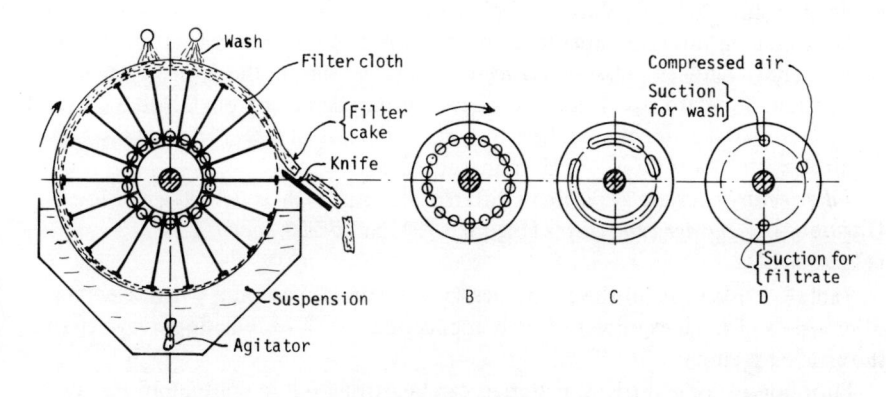

Figure 7.9   Details of rotating vacuum filter (Figure 7.8). Left: section through rotating drum. Right: valve plates taken apart. The cover D is also seen in Figure 7.8. C is behind D. C and D are stationary, and C slides against the rotating valve plate B with openings connected to the compartments of the rotating drum.

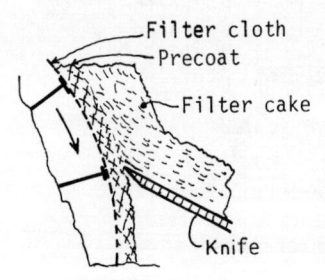

Figure 7.10   Detail of rotating filter with precoat. The knife moves slowly inwards.

Table 7.1 Performance data for rotary filters (based on data from Dorr Oliver N.V.; additional empirical data are given by Sperry[6]).

			Filter cake	
Particles	Rev. per min.	Capacity, kg/m² h	thickness, mm	liquid content, weight %
Aluminium silicate	0.25–1.0	245	43	73
Lead carbonate		122	12	62
Rubber wastes	0.3	20	5	55
Sodium carbonate		760	75	18
Sodium chloride	0.45	235	12	6
Sodium nitrate	1.0	2400	50	4
Manganese sulphate	0.1	176	16	10
Magnesium hydroxide	0.67	15	3	25
Starch	0.25	112	9	40

The outer layer of the precoat is removed as the knife (Figure 7.10) moves forwards a fraction of a millimetre for each revolution of the drum. At intervals new precoat is added by filtration of a slurry of filter aid.

Most vacuum filters are operated with atmospheric pressure in front of the filter cake and have reduced pressure between 1/3 and 2/3 bar on the downstream side of the septum. Standard size drum filters have a drum diameter from 0.3 to 3.0 m and a length from 0.3 m for the smallest to 4 m for the largest size. Performance data obtained with rotary filters are given in Table 7.1.

Filter centrifuges are available in different designs, such as *basket centrifuges*, (Figure 7.11), *pusher centrifuges* (Figure 6.13), and *peeler centrifuges* (Figure 6.14).

Table 7.2 lists some of the properties to be taken into account when selecting filter type and some examples of their application. For a systematic survey refer to the article by Purchas.[7]

Filtration of some difficult materials can be carried out in continuous *screw*

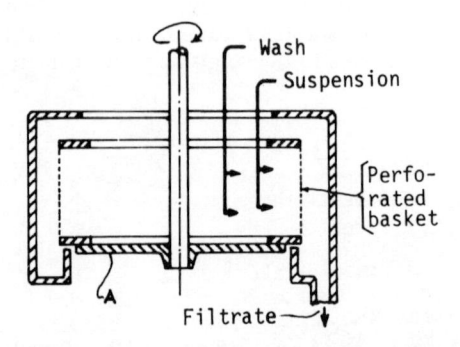

Figure 7.11 Discontinuous basket centrifuge. The bottom A is lowered for removal of filter cake during stops.

Table 7.2 Application of some filter types (see also Table 6.2).

Filter type	Figure	Advantages and important properties	Examples of applications
Sand	7.1	Inexpensive	Almost only in water purification.
Nutsche	7.2, 7.3	Possibilities of excellent washing. Closed design for toxic solvents. Can be made of corrosion-resistant materials. Unsuitable for large solid volumes.	Filtering of synthetic dyes, clarification of fruit juice, wine and lacquer.
Filter press	7.4	Can have large filter area, easy change of filter cloth. Less expensive than continuous filters. Takes much labour if large volumes of filter cake. Washing medium good.	Removal of bleaching earth from oils, cold clarification of cod liver oil, filtration of liquid for fibre production.
Moore	7.6	Can have large filter area and moderate labour requirement.	Removal of magnesium hydroxide produced from brine.
Kelly and Sweetland	7.7	Suitable for hot suspensions and high filter pressures (Kelly filters up to 35 bar).	Filtering of certain pigments, cracking residues in oil refining, cane sugar solutions.
Rotating	7.8, 7.9	Continuous, therefore little labour. Suitable for easy filtering operations (expensive per m^2 filter area) and filtering with precoat.	Filtering of pulp for flotation in mineral dressing, clarification of liquids in the food processing industry, removal of paper fibres and mycelium in the pharmaceutical industry.
Centrifuge	7.11, 6.13, 6.14	Gives the driest possible filter cake. Continuous types with little labour. The only continuous filter type without vacuum. Expensive per m^2 filter area.	Separation of crystals from mother liquor (sugar, ammonium sulphate), centrifugation of wool and fibres.

presses. These consist of one or two rotating screws fitting closely inside a curb. The curb can be cylindrical or conical. It is provided with perforations or slots through which the liquid escapes. They have been used for a long time in the animal and vegetable oil industries. Today they are also used to reduce the liquid content of some filter cakes and for dewatering of alkali cellulose, synthetic rubber, paunch manure, garbage, and waste from sugar and whisky production.

Ultrafiltration and hyperfiltration are methods for obtaining a high concentrate on the high pressure side and a lower concentration in the liquid that passes through a semipermeable film or membrane, without a filter cake on the high pressure side. The membranes are made of polymers such as cellulose acetate, PVC, polyamide and polyimide, with different qualities depending on temperature, pH resistance, and capacity.[8]

Ultrafiltration is carried out with pressures from 5 to 15 bar. The membranes are fairly coarse with a cut-off value in the macromolecular range with molecular weights from 2000 to 100,000. Water, salts, and low molecular compounds pass through the membrane.

Hyperfiltration or *reversed osmosis* is carried out with pressures from 40 to 60 bar or higher with the more dense membranes. The hyperfiltration membranes retain all dissolved solids, and only water, dissolved gases, volatile acids and, to a limited extent, ions pass through. The method is used in desalination of brackish water, for concentration of lignosulphonate from spent sulphite liquor, and in the dairy industry.

Filter calculations are semi-empirical. The assumptions have to be checked and the constants determined experimentally. Flow is almost invariably laminar through the fine pores in the filter cake, i.e. the friction factor in equation (1.8) is $f = 16/\mathrm{Re}$ (Figure 1.8) and the pressure drop,

$$\Delta p_\mathrm{f} = 32 \frac{\mu L V}{D_\mathrm{h}^2}. \tag{7.1}$$

The hydraulic diameter of a channel, equation (1.10), can be written as

$$D_\mathrm{h} = 4 \left(\frac{\text{cross-sectional area}}{\text{wetted circumference}} \right) \left(\frac{\text{length of channel}}{\text{length of channel}} \right)$$

$$= 4 \frac{\text{channel volume}}{\text{surface area of channel}}.$$

In a filter cake, the ratio of pore volume to surface are is $\epsilon/S_0(1-\epsilon)$ where S_0 is the specific surface area of the particles, m² area/m³ particle volume, and ϵ the porosity, i.e. (pore volume)/(filter cake volume). This leads to the assumption that the hydraulic diameter of the pores in a filter cake is proportional to $\epsilon/S_0(1-\epsilon)$, or

$$D_\mathrm{h} = \mathrm{const.} \frac{\epsilon}{S_0(1-\epsilon)}. \tag{7.2}$$

The velocity in the pores is proportional to the superficial velocity referring to the total cross section of the filter cake, V_0, and probably also to the reciprocal of the porosity,

$$V = (\mathrm{const.}) V_0/\epsilon. \tag{7.3}$$

Equations (7.2) and (7.3) inserted in equation (7.1) give Kozeny's equation,[9]

$$V_0 = C_1 \frac{\Delta p_\mathrm{f}}{L} \left[\frac{1}{\mu S_0^2} \frac{\epsilon^3}{(1-\epsilon)^2} \right] \tag{7.4}$$

where C_1 is a constant depending on the character of the filter cake.

For a filter cake with a certain porosity and specific surface area, and a given filtrate viscosity, the product of C_1 and the term in brackets in equation (7.4) can be substituted by a constant, C_2, and equation (7.4) simplifies to,

$$V_0 = C_2 \Delta p_f / L$$

Even if C_2 is treated as a constant, it should be kept in mind that it changes if the viscosity changes due to changed temperature, or if the porosity changes due to addition of filter aid or compression of the filter cake. The volume of filtrate per unit time is,

$$dQ/dt = V_0 A = C_2 (\Delta p_f / L) A \ \mathrm{m^3/s} \qquad (7.5)$$

where A is the filter cake area, m².

In an incompressible filter cake the pressure decreases linearly with the distance from the surface of the cake (Figure 7.12), while the pressure gradient in a compressible filter cake will increase with the distance from the surface of the cake (Figure 7.13). The following calculations are based on the assumptions of incompressible filter cake, constant porosity through the cake except at the filter cloth, and that the pressure drop through the filter cloth[10] in the calculations can be replaced by pressure drop through a superficial layer of the filter cake, L_c in Figure 7.12.

The length of the channels in an incompressible filter cake is proportional to the thickness of the filter cake, $L + L_c$ (Figure 7.12). L is also proportional to the amount of filtrate per unit area, Q/A, and L_c can be replaced by Q_c/A where Q_c is the amount of filtrate that would leave behind a filter cake with thickness L_c. Therefore L in equation (7.5) can be replaced by

$$C_3 \left(\frac{Q}{A} + \frac{Q_c}{A} \right),$$

and equation (7.5) becomes,

$$\frac{dQ}{dt} = C_4 \frac{\Delta p_f A^2}{Q + Q_c}$$

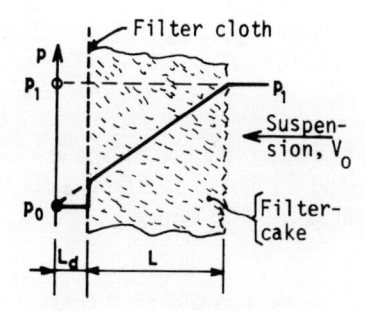

Figure 7.12 Pressure drop through incompressible filter cake. L_c corresponds to flow resistance through the cloth.

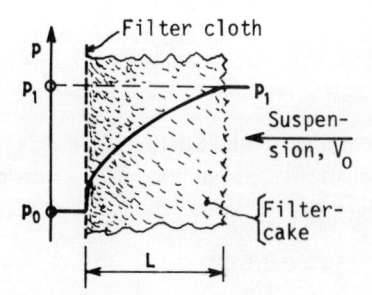

Figure 7.13 Pressure drop through compressible filter cake.

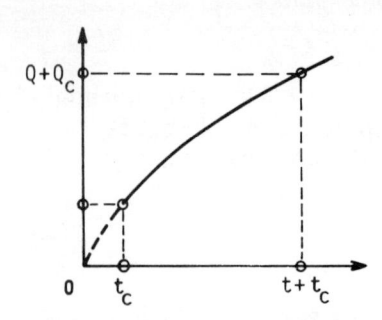

Figure 7.14 Filtrate Q plotted as a function of filtration time t.

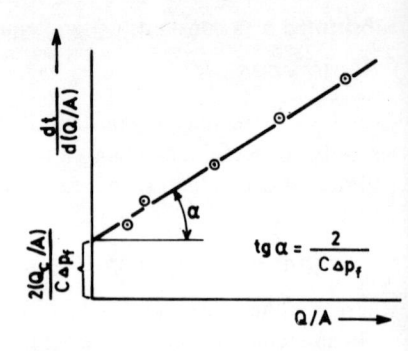

Figure 7.15 Experimental values of $dt/d(Q/A)$ plotted against Q/A.

or

$$(Q + Q_c)dQ = C_4 \Delta p_f A^2 dt. \tag{7.6}$$

If Δp_f and A are constants, an integration of equation (7.6) with the integration limits shown in Figure 7.14, gives,

$$\int_0^{Q+Q_c}(Q + Q_c)d(Q + Q_c) = C_4 \Delta p_f A^2 \int_0^{t+t_c} dt$$

or

$$\left(\frac{Q}{A} + \frac{Q_c}{A}\right)^2 = C\Delta p_f(t + t_c) \tag{7.7}$$

where t_c is the time required to produce a filter cake with the resistance of the cloth.

Equation (7.7) has three constants, Q_c, C, and t_c, which have to be determined experimentally. These can be obtained by measurements of Q for three values of filtration time t, but this gives no check of the validity of the assumptions made. A better procedure is differentiation of equation (7.7),

$$\frac{dt}{d(Q/A)} = \frac{2}{C\Delta p_f}\left(\frac{Q}{A}\right) + \frac{2}{C\Delta p_f}\left(\frac{Q_c}{A}\right). \tag{7.8}$$

According to this equation, $dt/d(Q/A)$ should be a linear function of Q/A as shown in Figure 7.15. Equation (7.7) is shown to be valid if a plot of $dt/d(Q/A)$ against Q/A is a straight line. The constant C is determined from the slope

$$\operatorname{tg}\alpha = 2/C\Delta p_f \tag{7.9}$$

and Q_c from the interception of the line with the ordinate,

$$\frac{2(Q_c/A)}{C\Delta p_f} = \left(\frac{Q_c}{A}\right)\operatorname{tg}\alpha. \tag{7.10}$$

The constant t_c is determined by insertion of C and Q_c into equation (7.7) and solving for t and Q equal zero.

Nearly all filter cakes show some deviation from the preceding theory. This is partly due to compressibility and partly to the tendency of smaller particles to be drawn forward in the cake. For practical purposes, however, many filter cakes can be regarded as incompressible. This is checked by filtration with different pressure drops, as compressible filter cakes give a decreasing constant C with increasing pressure drop Δp_f. As an empirical approximation, the product $C\Delta p_f$ in equation (7.7) may be substituted by $C\Delta p_f^n$ where $n < 1.0$,

$$\left(\frac{Q}{A} + \frac{Q_c}{A}\right)^2 = C\Delta p_f^n (t + t_c). \tag{7.11}$$

For more details of compressible filter cakes refer to Coulson and Richardson[11] and references therein.

For cylindrical filter cakes (Figure 7.16) equation (7.5) is written,

$$\frac{dQ}{dt} = C_2 \frac{dp}{dr} 2\pi r L$$

where L is the length of the filter element. Integration at time t with radius of the filter cake r gives,

$$\frac{dQ}{dt} \int_{r_1}^{r} \frac{dr}{r} = 2\pi L C_2 \int_{p_1}^{p_1 + \Delta p_f} dp$$

$$(\ln r - \ln r_1) dQ = 2\pi L C_2 \Delta p_f\, dt. \tag{7.12}$$

In terms of s (m³ filter cake)/(m³ filtrate),

$$s\, dQ = 2\pi r L\, dr.$$

Insertion in equation (7.12) and integration gives,

$$\frac{r_2^2}{2}\left(\ln \frac{r_2}{r_1} - \frac{1}{2}\right) + \frac{1}{4} r_1^2 = sC_2 \Delta p_f (t + t_c) \tag{7.13}$$

where t_c is the time taken to deposit a filter cake with thickness $r_0 - r_1$ (Figure 7.16). In the derivation of equation (7.7) from equation (7.5), the constant C in

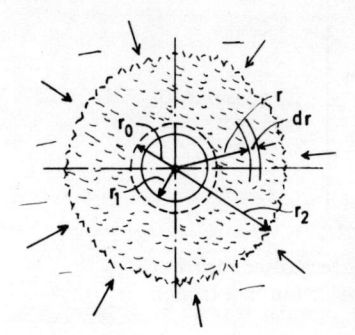

Figure 7.16 Cylindrical filter cake: r_0 = radius of filter cloth, $r_0 - r_1$ = thickness of cake with the resistance of the cloth.

equation (7.7) is $2C_2/s$, i.e. $C_2 = sC/2$. This value inserted in equation (7.13) gives,

$$t + t_c = \frac{r_2^2\left(2\ln\frac{r_2}{r_1} - 1\right) + r_1^2}{2s^2 C \Delta p_f}. \tag{7.14}$$

Equation (7.14) with $r_2 = r_0$ and $t = 0$ gives t_c. In addition,

$$Q = \pi(r_2^2 - r_0^2)L/s. \tag{7.15}$$

This equation can be used to determine either Q or the filter cake radius r_2 in equation (7.14).

Calculation of the time for *washing* is based on the filtration rate $d(Q/A)/dt$ at the end of the filtration period. This value is obtained from equation (7.8) with Q at the end of the filtration period, with C corrected for viscosity if the viscosity of the wash differs from the viscosity of the filtrate. For filter presses corrections must also be made for difference in flow area and in thickness of the filter cake. From Figure 7.4 it can be seen that the flow area is half and the cake thickness twice the values at the end of the filtration, i.e. estimates of washing are based on 1/4 the flow rate from equation (7.8) at the end of filtration.[1,2]

Laboratory tests to determine the empirical constants must be carried out under conditions comparable to the full-scale operation. The pressure drop through the filter cake must be the same in order to include the possible influence of compressibility. For rotating vacuum filters where the suspension is in a trough, the test must be carried out with the filter cloth pointing downwards (Figure 7.17). For rotating filters with precoat, the arrangement shown in Figure 7.18 can be used. The precoat is obtained by filtration of filter aid followed by removal of any excess filter aid outside the adjustable ring. To simulate filtration, the filter is kept in the suspension for a period corresponding to one revolution of the filter. Next, the adjustable ring is turned back the distance the knife will advance during one revolution of the drum. Then the filter cake outside the adjustable ring is removed

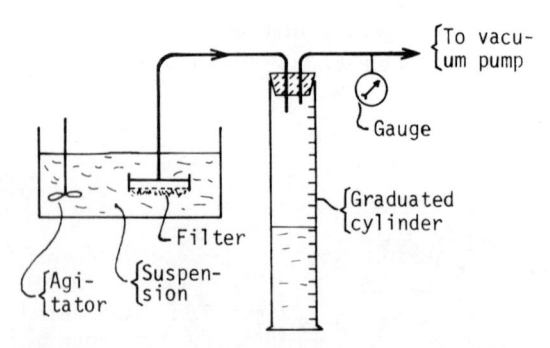

Figure 7.17 Laboratory test filter for rotating vacuum filter with the suspension in a trough under the filter.

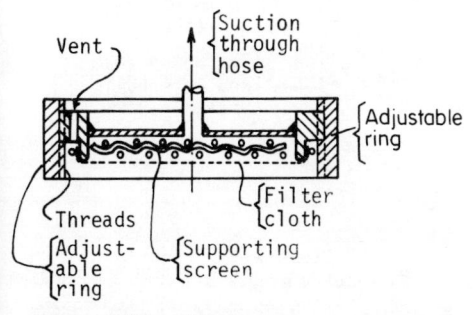

Figure 7.18 Laboratory filter for tests with precoat.

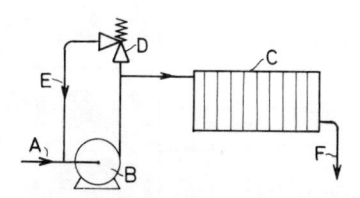

Figure 7.19 Filtration with constant flow rate initially and constant pressure later: A, slurry; F, filtrate.

with a knife that rests on the ring. This procedure is repeated until the precoat has to be renewed.

Installation of both flat and cylindrical filters with an effective filtration area 25 to 60% larger than determined by laboratory tests is recommended.

Filters can be operated at *constant pressure* or at *constant flow rate*. The first method can make particles penetrate into the filter cloth at the beginning of the filtration and give high cloth resistance. With colloidal particles present the cloth can be almost completely clogged. Constant flow rate, however, gives lower filtration rates over most of the filtration period. The best results may be obtained with constant flow rate in the beginning, and constant pressure after the pressure drop has reached a certain value (Figure 7.19). B is a pump with constant flow rate, e.g. a Mono-pump, pumping slurry to the filter C. The constant pressure valve D opens automatically for recycling through E, when the pressure reaches the set point for the valve.

Flotation

Flotation is a process in which particles suspended in a liquid are carried to the surface by air bubbles sticking to them. Separation according to composition can be carried out as particles well wetted by the liquid sink to the bottom. Figure 7.20 shows the principle of continuous flotation with particles separated according to their wetting properties. In most cases flotation agents must be added to obtain the desired effect.

Frothing agents or *frothers* prevent coalescence of the air bubbles before the particles are removed. Pine oil and heavy alcohols (5 to 8 carbon atoms) are common frothing agents.

Collectors and *promoters* are adsorbed on the surface of the particles. Collectors form films several molecule layers thick, while promoters form films one molecule thick. They make air bubbles stick to the solids. Pine oil acts both as a frother and a collector, while xanthates, such as sodium xanthate, act as a promoter with the hydrocarbon part of the molecule pointing outwards.

Activators are used to activate the adsorption of collectors or promoters onto

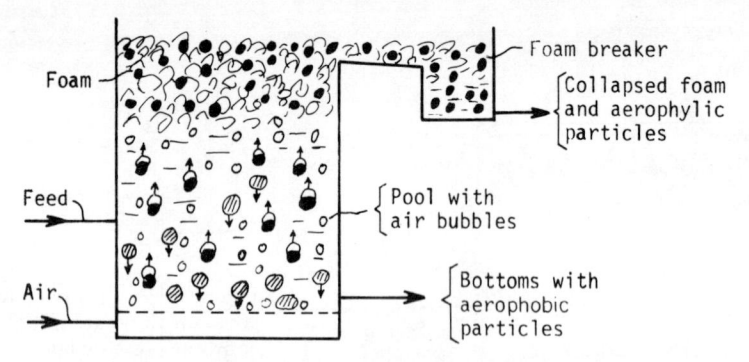

Figure 7.20　The principle involved in simple froth flotation.

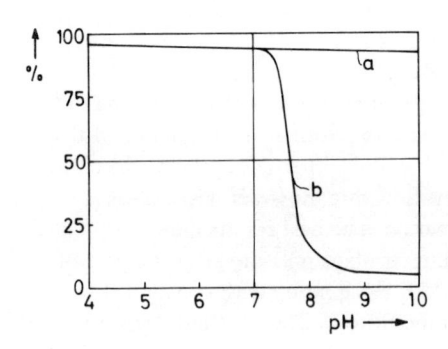

Figure 7.21　% Yield in flotation of chalcocite (curve a) and pyrite (curve b) with ethyl xanthate, pine-oil, and lime.

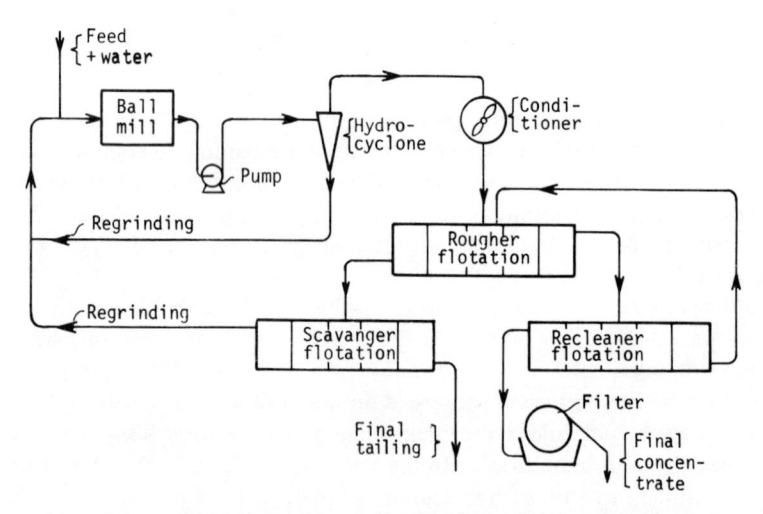

Figure 7.22　Flow sheet for a simple flotation plant with feed from a gyratory crusher and flotation reagents added in a conditioner.

particles, while *depressants* prevent adsorption of collectors or promoters on certain particles, thus increasing the selectivity in separation of particles of different composition. Figure 7.21 is an example of activator effect obtained by adjustment of pH, while sodium or calcium cyanide and zinc sulphate are examples of depressants used in flotation of other minerals.

The original and still by far the major application of flotation is in the field of mineral dressing. It accounts for approximately 80% of the concentration of minerals. The feed is usually ground to a particle diameter between 0.02 and 0.2 mm, to get particles small enough to be carried up by the air bubbles injected into the flotation cells, and also to reduce the amount of mixed grain (particles containing both mineral and gangue). Figure 7.22 is a flow sheet for a simple flotation plant, and Table 7.3 indicates some results obtained in concentration of minerals.

Table 7.3   Examples of reagents used and yields obtained in flotation of mineral ores.[16]*

Ore	Reagents, kg/ton	% recovery	assay
Zinc sulphide in carbonate gangue, 4% Zn, 5% pyrite	0.20 kg $CuSO_4 \cdot 5H_2O$ 0.03 kg dithiophosphate 0.06 kg NaCN 0.01 kg frother	98.4	63.5% Zn
Phosphate rock, 30% BPL with silica gangue	Phosphate flotation 0.5 kg tall oil 1.0 kg fuel oil 0.4 kg NaOH Silica flotation from phosphate conc. 0.05 kg fatty amine 0.2 kg kerosene		Phosphate rougher conc. 12% insol. 65% BPL Silica tailings (final phosphate conc.) 75–77% BPL 3% insol.
		85	
Fluorite, 40% CaF_2 in $CaCO_3$, SiO_2, iron oxide gangue	3.2 kg Na_2CO_3 0.7 kg Na_2SiO_3 0.36 kg refined fatty acid 0.05 kg alcohol frother 0.07 kg quebracho	95	99% CaF_2
Iron ore assaying, 30–35% Fe with siliceous gangue	1.4 kg H_2SO_4 0.9 kg petroleum sulphonate 0.3 kg fuel oil 0.2 kg tall oil 0.6 kg Na_2SiO_3	70	58–60% Fe 6–8% SiO_2
Copper sulphide ore, 0.9% Cu with pyrite and silicate gangue	2.7 kg lime 0.014 kg dithiophosphate 0.009 kg alcohol frother	91	24% Cu

*Reproduced by permission of McGraw-Hill Book Co.

Froth flotation and its modifications have applications in many other fields, such as recovery of oil from oil sands, separation of sodium and potassium salts, clarification of rayon spinning bath, deinking and dewaxing of waste paper, flotation of impurities from peas, and removal of cellulose particles from extracts of algaeic acids.

For calculations and experimental data refer to the literature.[13,14] A special application of kinetics in flotation has also been introduced.[15]

EXAMPLE 7.1. LABORATORY FILTRATION TEST

Laboratory filtrations conducted at constant pressure drop with a slurry of 23.5 g $CaCO_3$ per litre of water at 25°C gave the following filtration times in hours:

Test	I	II	III	IV	V
Pressure drop, Δp_f bar	0.463	1.12	1.95	2.52	3.39

Filtrate, $\dfrac{Q}{A} \dfrac{m^3}{m^2 \text{ cloth}}$

0.0114	0.0048	0.0019	0.0018	0.0014	0.0012
0.0227	0.0115	0.0053	0.0039	0.0032	0.0026
0.0341	0.0200	0.0096	0.0067	0.0055	0.0045
0.0455	0.0301	0.0148	0.0103	0.0084	0.0068
0.0568	0.0423	0.0211	0.0144	0.0118	0.0096
0.0682	0.0560	0.0283	0.0192	0.0158	0.0128
0.0795		0.0364	0.0247	0.0203	0.0164
0.0909		0.0453	0.0306	0.0253	0.0204
0.1023			0.0372	0.0308	0.0248
0.1136			0.0444	0.0369	0.0298
0.1250				0.0436	
0.1364				0.0507	

(a) Determine the constants in equation (7.7).

(b) Estimate necessary filter area for filtration of 30 m³ suspension per 6 hours, at a temperature of 18°C, particle density 2930 kg/m³, density of wet filter cake 1930 kg/m³, and:

 (α) the filter is a rotating vaccum filter with length equal to the diameter. 6% of the filter area is covered by supports and therefore ineffective. 30% of the circumference is submerged in the suspension. There is approximately 1.0 mm clearance between the knife and the filter cloth. The pressure difference during filtration is 0.6 bar.

 (β) The filter is a filter press with effective filtration area 85% of the frame area. The frame width is 50 mm. It takes 30 minutes to open the press, remove the filter cake, and close the press for the next cycle. The press will be operated with 2.0 bar pressure drop.

Solution

(a) For incompressible filter cakes constant C in equation (7.7) is independent of the pressure drop Δp_f. C is determined from equation (7.8) with differences,

$\Delta t/\Delta(Q/A)$, instead of from the differential $dt/d(Q/A)$. In test series I, $\Delta t/\Delta(Q/A)$ for the first interval is $0.0048/0.0114 = 0.421$ h/m, for the second interval $(0.0115 - 0.0048)/(0.0227 - 0.0114) = 0.593$ h/m, etc. Figures 7.23 and 7.24 give $\Delta t/\Delta(Q/A)$ calculated for all intervals as a step function of Q/A.

Figures 7.23 and 7.24 show the linear relationship for each value of Δp_f. The slopes of the lines and the intercepts with the ordinate give the constants C and Q_c/A for each test:

Test	I	II	III	IV	V	Notes
Δp_f MN/m^2	0.0463	0.112	0.195	0.252	0.339	observed from
tg α h/m^2	13.6	7.45	4.9	4.11	3.31	Figures 7.23 and 7.24
$C\Delta p_f$ m^2/h	0.147	0.268	0.408	0.487	0.604	$C\Delta p_f = 2/\mathrm{tg}\,\alpha$ [equation (7.9)]
C m^4/MN h	3.17	2.39	2.09	1.93	1.78	$C = C\Delta p_f/\Delta p_f$
$2(Q_c/A)/C\Delta p_f$	0.35	0.163	0.112	0.088	0.072	From Figures 7.23 and 7.24
Q_c/A m^3/m^2	0.0257	0.0219	0.0229	0.0214	0.0217	$\dfrac{Q_c}{A} = \left[\dfrac{2(Q_c/A)}{C\Delta p_f}\right]\dfrac{C\Delta p_f}{2}$

The last line shows the almost constant value $Q_c/A \approx 0.0227$ m^3/m^2.

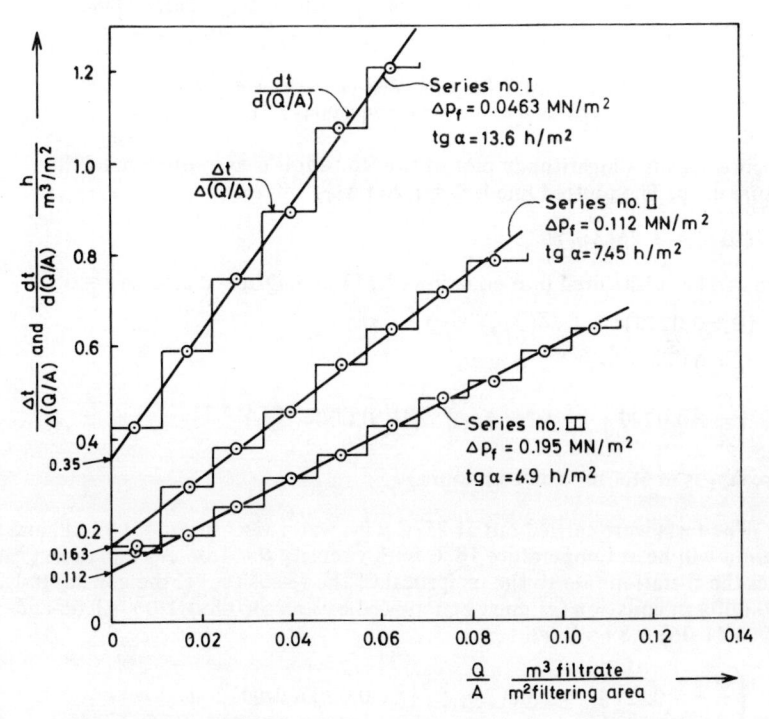

Figure 7.23 Diagram to check compressibility and to determine the constants in equation (7.7).

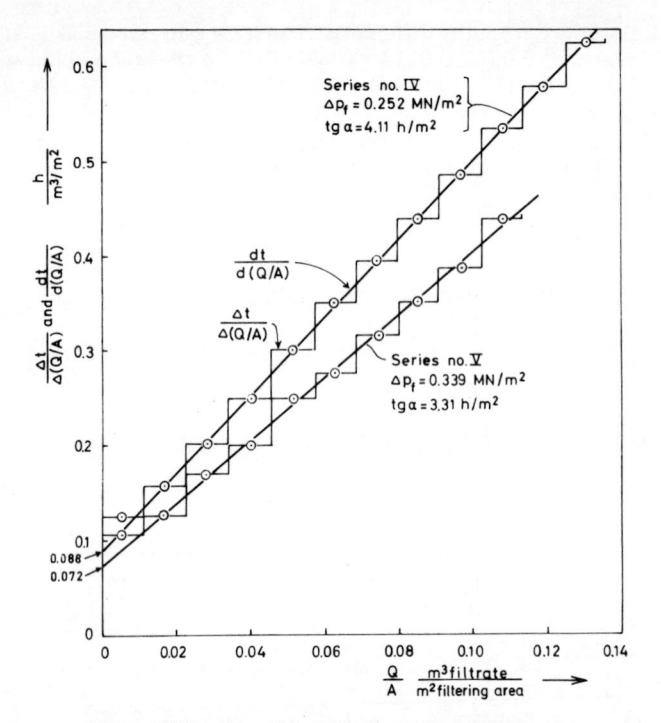

Figure 7.24 Diagram to check compressibility and to determine the constants in equation (7.7).

Figure 7.25 is a logarithmic plot of the 'constant' C as a function of the pressure drop. The plotted line is $C = 1.26\,(\Delta p_f)^{-0.3}$, or

$$C\Delta p_f^n = 1.26(\Delta p_f)^{0.7}$$

which can be substituted into equation (7.11) with $Q/A = 0$ at time $t = 0$:

$$(0 + 0.0227)^2 = 1.26(\Delta p_f)^{0.7}(0 + t_c),$$

$$t_c = 0.0004(\Delta p_f)^{-0.7} \text{ hours}$$

$$\left(\frac{Q}{A} + 0.0227\right)^2 = 1.26(\Delta p_f)^{0.7}[t + 0.0004(\Delta p_f)^{-0.7}] \quad (a)$$

where Δp_f is in MN/m² and t in hours.

(b)   The tests were carried out at 25°C with water viscosity $\mu = 0.89$ cP, and the filtration will be at temperature 18°C with viscosity $\mu = 1.06$ cP. Equation (7.4) relates the filtration rate to the reciprocal of the viscosity, i.e. the constants 1.26 and 0.0004 in equation (a) must be replaced by $(1.26)(0.89)/1.06 = 1.06$ and $(0.0004)(1.06)/0.89 = 0.0005$,

$$\left(\frac{Q}{A} + 0.0227\right)^2 = 1.06(\Delta p_f)^{0.7}[t + 0.0005(\Delta p_f)^{-0.7}]. \quad (b)$$

1 m³ filter cake contains x m³ particles,

$$2930x + 1000(1 - x) = 1930,$$

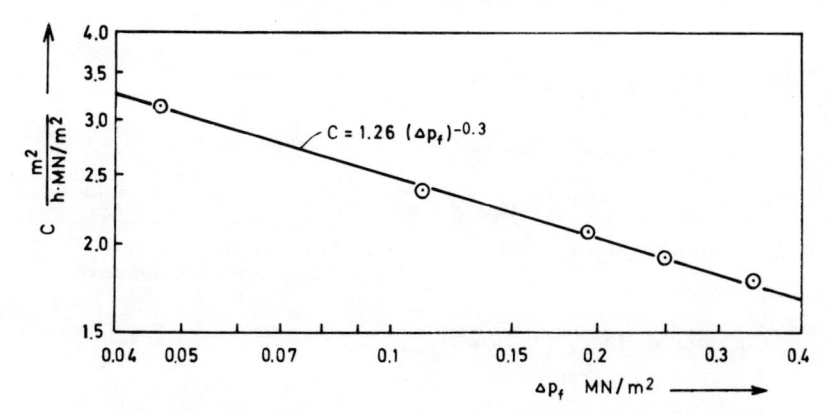

Figure 7.25 C in equation (7.7) plotted against the pressure drop Δp_f.

or
$$x = 0.48 \text{ m}^3$$
corresponding to $(0.48)(2930) = 1406$ kg particles

$$\text{or } 1406/23.5 = \quad 60 \, \frac{\text{m}^3 \text{ water}}{\text{m}^3 \text{ filter cake}}$$

$$60 - (1 - 0.48) = \quad 59 \, \frac{\text{m}^3 \text{ filtrate}}{\text{m}^3 \text{ filter cake.}}$$

(α) The knife is adjusted to leave 1 mm filter cake, i.e. the filter cake left corresponds to $Q/A = 0.059$ m^3 filtrate/m^2 cloth. This is added to the cloth resistance, giving Q_c/A in equation (b) equal to $0.059 + 0.0227 = 0.0817$ m^3 filtrate/m^2 cloth. Q and t equal to zero in equation (b) give $t_c = 0.0063 \, (\Delta p_f)^{-0.7}$ hours:

$$\left(\frac{Q}{A} + 0.0817\right)^2 = 1.06(\Delta p_f)^{0.7}[t + 0.0063(\Delta p_f)^{-0.7}]$$

or with $\Delta p_f = (0.1)(0.6) = 0.06$ MN/m^2,

$$\left(\frac{Q}{A} + 0.0817\right)^2 = (0.148)(t + 0.045). \hspace{2cm} (c)$$

The filter element dB in Figure 7.26 is submerged while it passes from C to D,

$$t = 0.3/(60\,n) \text{ hour.}$$

The length of the filter equals the drum diameter, d, and 94% of the filter cloth is effective filter area,

$$\therefore \quad A = (0.3)(0.94)\pi dd = 0.886 \, d^2.$$

The volumetric flow rate of filtrate is,

$$\frac{(30)(59/60)}{6} = 4.92 \text{ m}^3/\text{h}$$

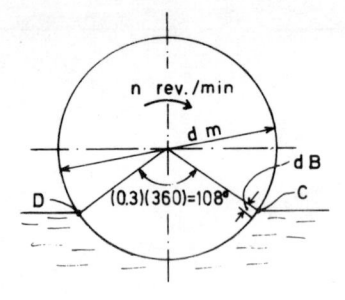

Figure 7.26 Rotating filter with 30% of the circumference submerged.

or in filtration time $t = 0.3/(60\,n)$,
$$Q = (4.92)(0.3)/(60\,n) = 0.0246/n \text{ m}^3.$$
Equation (c) becomes
$$\left(\frac{0.0246}{0.886\,d^2 n} + 0.0817\right)^2 = 0.148\left(\frac{0.3}{60\,n} + 0.045\right)$$
$$\left(\frac{0.0278}{nd^2} + 0.0817\right)^2 = \frac{0.00074}{n} + 0.0067. \qquad (d)$$

Equation (d) gives the filter drum diameter, d, as a function of speed in revolutions per minute, n. Usually, n is in the range 0.25 to 1.0 rev./min. The following table gives drum diameter d calculated from equation (d) for different values of n:

n rev./min.	0.25	0.5	0.75	1.0
d m	2.59	2.52	2.49	2.47

A standard size filter should be chosen. Rotational speed $n = 0.5$ rev./min. and 35% excess area give diameter
$$d = 2.52\sqrt{1.35}$$
$$= 2.93 \text{ m}.$$

(β) 30 m³ suspension contain y m³ water and $23.5\,y/2930$ m³ particles,
$$y + 23.5y/2930 = 30, y = 29.76 \text{ m}^3$$
Volume of filter cake, $(30 - 29.76)/0.48 = 0.5$ m³
Volume of filtrate, $30 - 0.5 = 29.5$ m³.
Equation (b) with $\Delta p_f = 0.2$ MN/m² and filtering time $t = 5.5$ hours, gives
$$\left(\frac{29.5}{A} + 0.0227\right)^2 = (1.06)(0.2)^{0.7}[5.5 + 0.0005(0.2)^{-0.7}]$$
$$A = 21.8 \text{ m}^2$$

or frame area $21.8/2 = 10.9$ m² and frame volume $(10.9)(0.05) = 0.55$ m³. With 35% added for uncertainty and two filter cloths per frame, the total frame area is
$$(1.25)(21.8)/(2)(0.85) = 17.3 \text{ m}^2.$$
It could be 21 frames of dimensions 0.9 × 0.9 m.

EXAMPLE 7.2. FILTER PRESS, FILTRATION, AND WASHING

A factory has a filter press with 25 m² effective filter area and 40 mm frames, and a suspension pump giving 0.12 MN/m² gauge pressure. This equipment is to be used for filtering an aqueous solution, at 20°C, containing 4 kg solid per 100 kg suspension. Filtration tests gave the empirical equation

$$(Q + 1.00)^2 = (62.5)(t + 0.016) \qquad (a)$$

where Q is amount of filtrate in m³ and t filtration time in hours.

The wet, incompressible filter cake contained 420 kg particles per m³. The density of the particles is 2100 kg/m³ and of the water 1000 kg/m³. The viscosity of water at 20°C μ = 1.005 cP and at 50°C 0.55 cP.

(a) Estimate the filtration time taken for the frames to fill with filter cake, at a filtration temperature of 20°C and pressure 0.12 MN/m².

(b) Estimate the time required for washing with a volume of water 4 times the volume of the frames, assuming unchanged temperature and pressure.

(c) Calculate maximum volume of filtrate per 24 hours. It takes half an hour to open the press, remove the filter cake and make the press ready for the next filtration.

Calculate volume of filtrate per 24 hours, if:

(d) the frames are replaced with new frames with width 80 mm;

(e) the pressure during filtration and washing is increased to 0.24 MN/m²;

(f) the temperature of suspension and wash water is increased to 50°C.

Solution

(a) The first step is to calculate the amount of filtrate per filtration.

Volume of filter cake, (0.04)(25)/2	= 0.5 m³.
Particles in the filter cake, (0.5)(420)/2100	= 0.1 m³.
or (0.1)(2100)	= 210 kg.
Suspension to be filtered, (100)(210)/4	= 5250 kg.
Weight of filter cake, 210 + (0.5 − 0.1)(1000)	= 610 kg.
Weight of filtrate per filtration,	= 4640 kg.
Volume of filtrate, 4640/1000	= 4.64 m³.

Equation (a),

$$(4.64 + 1.00)^2 = (62.5)(t + 0.016),$$
$$t = 0.49 \text{ h}$$
$$\approx 30 \text{ min.}$$

(b) Differentiation of equation (a) with Q = 4.64 m³ gives the final filtrate flow rate,

$$2(Q + 1.00)dQ = 62.5 \, dt$$

$$\frac{dQ}{dt} = \frac{62.5}{2(Q + 1.00)} = \frac{62.5}{(2)(5.64)} = 5.54 \text{ m}^3/\text{h}.$$

The wash water passes through two filter cakes (half the area and twice the distance of the filtration), giving a flow rate,

$$\left(\frac{dQ}{dt}\right)_{wash} \approx 5.54/4 = 1.39 \text{ m}^3/\text{h}.$$

Volume,

$$Q_{wash} = 4(0.5 - 0.1) = 1.6 \text{ m}^3$$

and time,

$$t_{wash} = 1.6/1.39 = 1.15 \text{ h}$$
$$\approx 70 \text{ min}.$$

(c) Time per cycle, $0.49 + 1.15 + 0.5 = 2.14$ hours and maximum amount of filtrate in 24 hours,

$$(4.64)(24)/2.14 = 52 \text{ m}^3/24 \text{ h}.$$

(d) Amount of filtrate per cycle, $(2)(4.64) = 9.28 \text{ m}^3$,

$$(9.28 + 1.00)^2 = 62.5(t + 0.016),$$
$$t = 1.67 \text{ h},$$

Final filtrate flow rate,

$$\frac{dQ}{dt} = \frac{62.5}{2(9.28 + 1.00)}$$
$$= 3.04 \text{ m}^3/\text{h}.$$

Wash water flow rate,

$$\left(\frac{dQ}{dt}\right)_{wash} \approx \frac{3.04}{4}$$
$$= 0.76 \text{ m}^3/\text{h}.$$

and time for washing,

$$t_{wash} = (2)(1.6)/0.76$$
$$= 4.21 \text{ h}.$$

Maximum amount of filtrate in 24 hours,

$$(9.28)(24)/(1.67 + 4.21 + 0.5) = 35 \text{ m}^3/24 \text{ h}.$$

(e) With incompressible filter cake the constant 62.5 in equation (a) is doubled and the constant 0.016 is halved,

$$(4.64 + 1.00)^2 = 125(t + 0.008),$$
$$t = 0.25 \text{ h}.$$

Wash water flow rate doubled and washing time halved,

$$t_{wash} = 1.15/2$$
$$= 0.58 \text{ h}.$$

Maximum amount of filtrate in 24 hours,

$$(4.64)(24)/(0.25 + 0.58 + 0.30) = 99 \text{ m}^3/24 \text{ h}.$$

(*f*) The viscosity is reduced from 1.005 cP to 0.55 cP, giving the constants on the right hand side of equation (*a*), (62.5)(1.005)/0.55 = 114 and (0.016)(0.55)/1.005 = 0.0088.

$$(4.64 + 1.00)^2 = 114(t + 0.0088),$$
$$t = 0.27 \text{ h}.$$

Wash water flow rate,

$$\left(\frac{dQ}{dt}\right)_{\text{wash}} \approx \frac{114}{(4)(2)(4.64 + 1.00)} = 2.53 \text{ m}^3/\text{h}.$$

and

$$t_{\text{wash}} = 1.6/2.53$$
$$= 0.63 \text{ h}.$$

Maximum amount of filtrate,

$$(4.64)(24)/(0.27 + 0.63 + 0.5) = 80 \text{ m}^3/24 \text{ h}.$$

EXAMPLE 7.3. CYLINDRICAL FILTER ELEMENTS

The laboratory filtrations in Example 7.1 could be reproduced by the equation

$$\left(\frac{Q}{A} + 0.0227\right)^2 = 1.06(\Delta p_f)^{0.7}[t + 0.0005(\Delta p_f)^{-0.7}] \tag{a}$$

where Q/A is m³ filtrate per m² filter area, Δp_f is the pressure drop in MN/m² and t the filtering time in hours. The filter cake volume is $s = 1/59$ m³/m³ filtrate.

A nutsche with cylindrical filter elements of a diameter $d_0 = 25$ mm is to be used to remove the particles from 14 m³ of the suspension per 24 hours. The filter can be flushed from the inside in 9 minutes at set intervals.

(*a*) Give the filtering equation for pressure drop $\Delta p_f = 0.16$ MN/m².

(*b*) Find the best time intervals for flushing.

(*c*) Estimate the total length of the filter elements.

Solution

(*a*) Equation (*a*) for a flat filter with $\Delta p_f = 0.16$ MN/m² becomes,

$$\left(\frac{Q}{A} + 0.0227\right)^2 = 0.294(t + 0.0018).$$

The constant 0.0227 m³ filtrate/m³ filter area corresponds to 0.0227 s = 0.0227/57 = 0.004 m filter cake, or (Figure 7.16) $r_1 - r_0 \approx 4$ mm and $d_1 = 24.2$ mm.

Equation (7.14) with $t = 0$, $r_2 = r_0 = 0.0125$ m, and $C_2 \Delta p_f = 0.294$ becomes,

$$t_c = \frac{0.0125^2 \left(2\ln\dfrac{25}{24.2} - 1\right) + \left(\dfrac{0.0242}{2}\right)^2}{2\dfrac{1}{59^2}0.294} = 0.0018 \text{ h}$$

$$t + 0.0018 = \frac{r_2^2 \left(2\ln\dfrac{r_2}{0.0121} - 1\right) + 0.0121^2}{2\dfrac{1}{59^2}0.294}$$

or the filtering equation,

$$t = 5920 r_2^2 \left(2 \ln \frac{r_2}{0.0121} - 1\right) + 0.865. \tag{a}$$

(b)   Equation (7.15) gives the amount of filtrate,

$$Q = \frac{\pi(r_2^2 - 0.0125^2)L}{1/59} = 185(r_2^2 - 0.000156)L \tag{b}$$

where L is the total length of the filter elements.
The following table and Figure 7.27 give the amount of filtrate per metre filter element per hour, $Q/(Lt_{cy})$, as a function of the filter cake radius, r_2.

Radius, r_2	m	0.014	0.016	0.018	0.020	0.025	Notes
Filtering time, t	h	0.043	0.196	0.471	0.877	2.535	equation (a)
Time per cycle, t_{cy}	h	0.193	0.346	0.621	1.027	2.685	$t + 9/60$
Q/L	m³/m	0.0074	0.0185	0.0311	0.0451	0.0868	equation (b)
$Q/(L\,t_{cy})$	m³/(m h)	0.0383	0.0534	0.0501	0.0440	0.032	

The curve of Figure 7.27 has a maximum at $r_2 \approx 0.016$ m with filtering time (table) 0.196 h or about 12 min.

(c)   Amount of filtrate per cycle of $12 + 9 = 21$ minutes or 0.35 hours,

$$Q = 14 \frac{59}{60} \frac{0.35}{24} = 0.20 \text{ m}^3,$$

which substituted in equation (b) gives,

$$0.20 = 185(0.016^2 - 0.000156)L,$$

$$L = 10.8 \text{ m}.$$

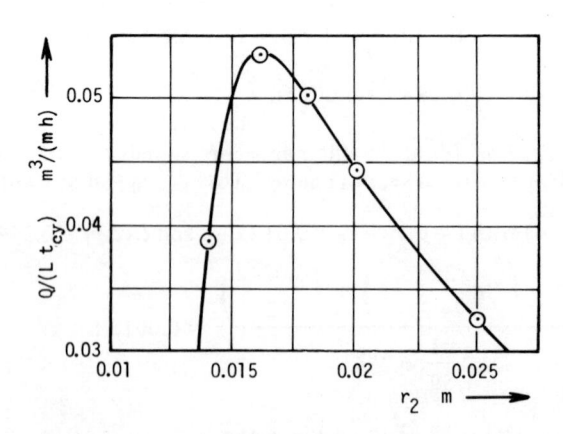

Figure 7.27   Filtrate in m³ per m filter element per hour, $Q/(Lt_{cy})$, plotted as a function of the filter cake radius, r_2.

With 35% extra length,

$$L \approx (1.35)(10.8)$$
$$= 14.6 \text{ m},$$

which could be made up of 12 elements each 1.2 m long.

Problems

7.1. Figure 7.28 shows a section of a sand filter for water, consisting of a 0.3 m layer of coarse stones at the bottom and a 0.5 m layer of sand above with particle diameter 0.3 mm and void fraction $\epsilon = 0.40$. The friction loss through the layer of coarse stones is negligible.

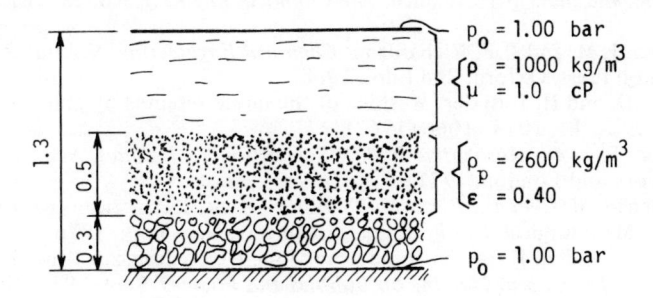

Figure 7.28 Sand filter with gravity flow of water.

(a) Calculate by means of data from chapter 1 the rate of flow through the filter in m^3 per m^2 filter area per hour.

(b) The water supply to the filter is cut off suddenly. Calculate how long it takes for the water surface to fall to the level of the top of the sand, and to fall to the bottom of the sand layer (neglect capillary forces).

7.2. A sample of a slurry filtered in a laboratory test filter gave the filtering equation,

$$\left(\frac{Q}{A} + 0.9\right)^2 = 30\Delta p_f \left(t + \frac{0.027}{\Delta p_f}\right)$$

where Δp_f is in bar and t in hours. The filter cake volume was 0.02 m^3/m^3 filtrate. A filter press with 18 frames, each 0.8 m^2 and 50 mm thick, is used for the same slurry. Estimate the filtering time before the frames are 90% filled with filter cake if the filter is operated with constant flow rate, $Q/A = 18$ m^3/(m^2 h), until the pressure difference is $\Delta p_f = 1.4$ bar. Then the pressure is kept constant for the rest of the filtration time. Incompressible filter cake is assumed.

References

1 Schuster, W. W. and L. K. Wang: Role of polyelectrolytes in the filtration of colloidal particles from water and wastewater, *Separation and Purification Methods*, **6** (1), 153–187 (1977).
2 Purchas, D. B.: *Industrial Filtration of Liquids*, Leonard Hill Books, London, 1971.
3 *J. Am. Water Works Ass.*, **42**, 687 (1950).

4 Ives, K. J., *Trans. Instn. Chem. Engrs.*, **43**, T238 (1965) and **48**, T94 (1970).
5 Roberts, E. J. et al.: Solids concentration, *Chem. Eng.*, **77**, No. 14, 52–68 (1970).
6 Sperry, D. R.: Analysis of filtration data, *Ind. Eng. Chem.*, **36**, 323–329 (1944).
7 Purchas, D. B.: A non-guide to filter selection for liquids, *The Chemical Engineer*, No. 237, CH79-CH82 (1970).
8 Claussen, P. H.: *Treatment of effluents from the pulp industry by ultrafiltration and hyperfiltration for recovery of valuable by-products and pollution control*, Kem-Tek 4, Bella Centre, Copenhagen, 1977.
9 Kozeny, J.: Über kapillare Leitung des Wassers im Boden, *Berichte Wiener Akad.*, 136a, 271 (1927).
10 Ruston, A.: Effect of filter cloth structure on flow resistance, bleeding, blinding, and plant performance, *The Chemical Engineer*, No. 237, CE88–CE94 (1970).
11 Coulson, J. M. and J. F. Richardson: *Chemical Engineering*, Volume 2, Pergamon Press, Oxford, 2nd Edn., 1968.
12 Han, C. D. and H. J. Bixler: Washing of the liquid retained by granular solids, *A.I.Ch.E. J.*, **13**, 1058–1066 (1967).
13 Lemlich, R. et al.: *Adsorptive Bubble Separation Techniques*, Academic Press, New York and London, 1971.
14 Fuerstenau, M. C. et al.: *Flotation*, Volumes 1 and 2, American Institute of Mining, Metallurgical, and Petroleum Engineers, New York, 1976.
15 Lengler, P. and H. Hofmann: Eine neue Modellvorstellung zur Kinetik der Flotation, *Proc. Joint Meeting on Bubbles and Foames*, Sept. 1971, Nuernberg, Germany.
16 Perry, J. H.: *Chemical Engineers' Handbook*, McGraw-Hill, New York, 4th Edn., 1963.
17 Ruth, B. F.: Personal communication quoted by W. L. McCabe and J. C. Smith: *Unit Operations*, McGraw-Hill, New York, 3rd Edn., 1976.

CHAPTER 8
Atomization, Dispersion, Homogenization, Crushing, and Grinding

Atomization of a liquid in a gas

Three principles are used for atomizing a liquid in a gas: the liquid can be forced through a spray nozzle (Figure 8.1), it can pass over a rotating disc with or without vanes (Figure 8.2), or it is broken up by a two-fluid atomizer (Figure 8.3).

Spray nozzles are used for solid-free solutions of low or medium viscosity, as in atomizing solutions of detergent in spray dryers. Rotating discs are used for atomizing milk in spray dryers and for production of particularly small particles from pigment-containing liquids. Two-fluid atomization is used for yeast and pastes, including centrifuged pigments with viscosities up to 10,000 poise. An average diameter of 4 μm of the dried product is given from a spray dryer using 0.17 kg steam as atomizing fluid per kg paste[1]. For calculation of drop size refer to Marshall.[2]

Dispersion of a gas in a liquid

The simplest method of dispersing gas in a liquid is to introduce the gas through a sparger in the bottom of a vertical vessel of liquid. The sparger can be a perforated pipe, a perforated plate, or a porous septum.

Perforated plates and ring or cross style perforated pipe spargers usually have orifices from 3 to 12 mm in diameter. They are used without mechanical agitation to promote mass transfer, as in chlorination and biological sewage treatment. Fair[3] recommends orifice velocities less than 75 to 90 m/s.

Taking into account surface tension and buoyancy effects, slow bubbling gives a bubble diameter

$$D_b = \left(\frac{6\gamma D_0}{g\Delta\rho}\right)^{1/3} \qquad (8.1)$$

where D_0 is the orifice diameter, m, γ is the surface tension, N/m, and $\Delta\rho$ is the liquid density difference, $\rho_1 - \rho_g$, kg/m^3.

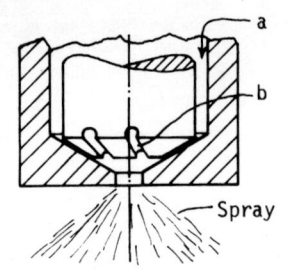

Figure 8.1 Spray nozzle: a, inlet; b, threadshaped grooves giving a rotating spray.

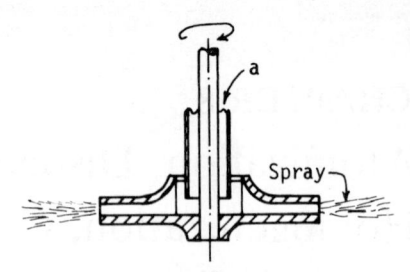

Figure 8.2 Rotating disc atomizer with vanes: a, inlet.

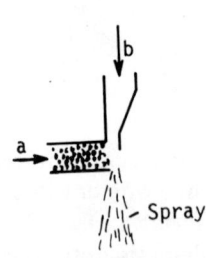

Figure 8.3 Two-fluid atomizer: a, inlet of liquid or paste; b, air or steam under pressure.

Fair[3] recommends use of equation (8.1) for low viscosity liquids and orifice velocities that correspond approximately to the equation

$$V < 1/(2000\, D_0^{1.1})\ \text{m/s}. \tag{8.2}$$

Higher orifice velocities give chain like bubble formation with a bubble diameter in low viscosity liquids,[3]

$$D_b = \left(\frac{72\rho_1}{\pi^2 g \Delta\rho}\right)^{1/5} Q_g^{0.4} \tag{8.3}$$

where Q_g is the volumetric flow rate through the orifice, m³/s.

Dispersion can be increased by an impeller. Experimental data are available for dispersed gas in tanks with volumetric gas fraction $\epsilon < 0.15$, and with agitation by a six-blade turbine impeller.[4] The average bubble diameter was determined to be

$$\bar{D}_b = 4.15 \frac{\gamma^{0.6}}{R^{0.4} \rho_1^{0.2}} \epsilon_g^{0.5} + 0.0009\ \text{m} \tag{8.4}$$

where R is the agitation intensity for the ungassed liquid, W/m³.

The volumetric gas fraction is estimated by the equation

$$\epsilon_g = \left(\frac{V_s \epsilon_g}{V_t}\right)^{0.5} + 0.0000543 \frac{R^{0.4} \rho_1^{0.2}}{\gamma^{0.6}} \left(\frac{V_s}{V_t}\right)^{0.5} \tag{8.5}$$

where V_s is the superficial gas velocity referred to the total cross-section of the vessel, m/s, and V_t is the bubble rise velocity (see Chapter 6), m/s.

Dispersion of a liquid in a liquid

In extraction columns the heavy liquid can be introduced through holes in the top, or the light liquid through holes in the bottom of the column (Figure 8.4). The perforated plate is oriented to give flow in the direction of the burrs from the punching.

An investigation of 14 liquid systems gave, within ±7.5%, drop sizes as shown in Figure 8.5.[5] The drops were of uniform size for liquid velocities less than 0.1 m/s through the holes. For velocities between 0.1 and 0.3 m/s, Figure 8.5 gives the diameter of the largest drops.

Temporary dispersions of relatively large drops for liquid–liquid extraction are also made in agitated vessels or in pipeline dispensers. The drop size varies with location in an agitated vessel, being smallest at the impeller and larger in regions with less turbulence, due to coalescence.[6]

The *surface volume mean diameter* or Sauter mean diameter, D_s, is a drop diameter that gives the same surface area per unit tank volume as the actual drops. Calderbank[7] found, for a baffled tank with a six-blade turbine with an impeller to tank diameter ratio of 1/3, the surface volume mean diameter

$$D_s = 2.24 \frac{\gamma^{0.6}}{R^{0.4} \rho_c^{0.2}} \epsilon_d^{0.5} \left(\frac{\mu_d}{\mu_c} \right)^{0.25} \tag{8.6}$$

where ϵ_d = fraction of dispersed liquid, $\epsilon_d < 0.2$
μ_c and μ_d = viscosity of continuous and of dispersed liquid, N s/m^2
ρ_c = density of continuous liquid, kg/m^3
R = agitation intensity, W/m^3
γ = surface tension, N/m.

Caution is needed in estimating the drop size, as data from different sources do not generally agree with each other.[8]

Homogenization

Homogenizers give a finely divided liquid dispersed in a continuous phase. Homogenized suspensions can be obtained from powerful agitators, colloid mills (Figure 8.6), and homogenizers as shown in Figure 8.7. To determine the need for stabilizer, and the drop size as a function of pressure in the homogenizer, it is recommended that tests be carried out in each individual case.

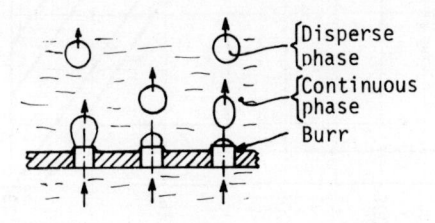

Figure 8.4 Dispersion of a light liquid in a heavy continuous liquid phase.

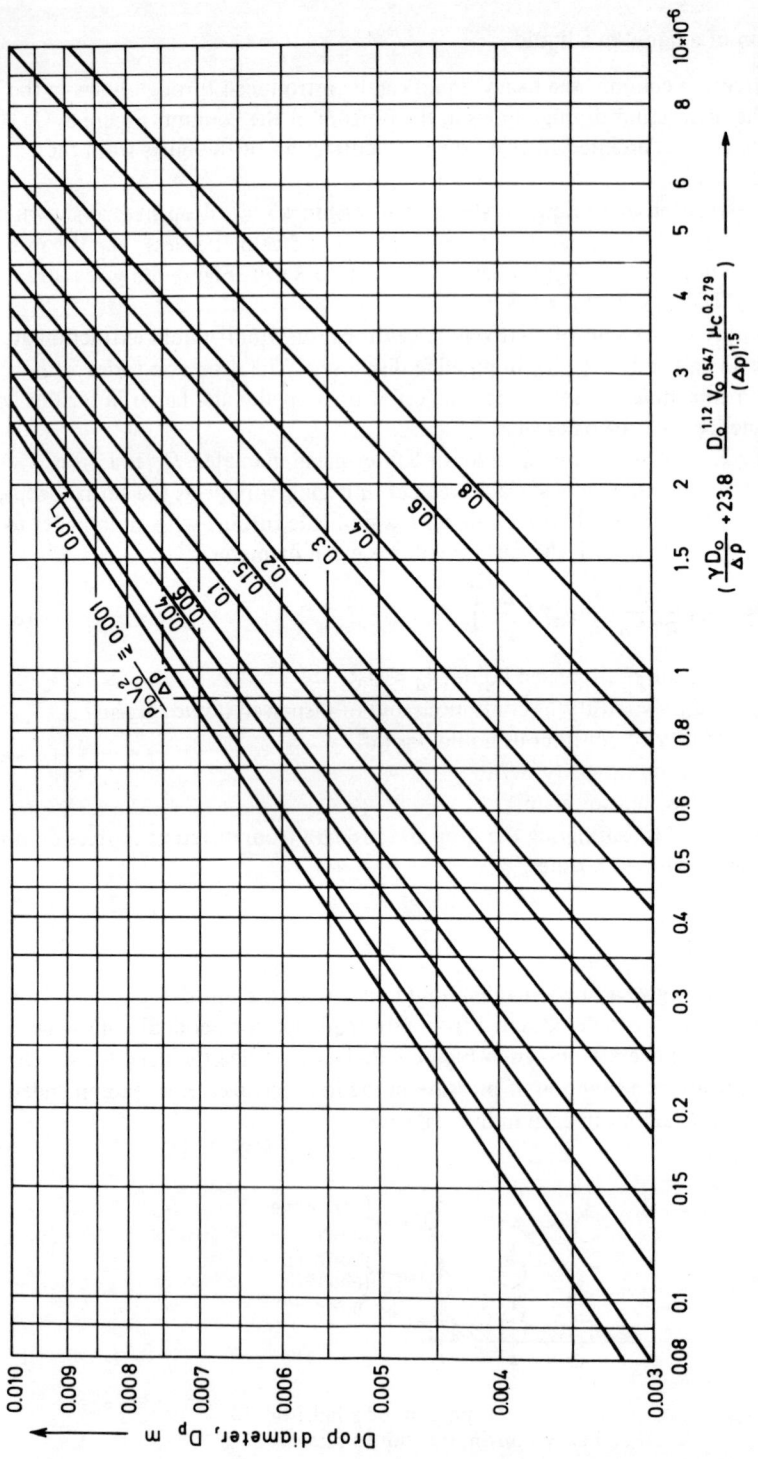

Figure 8.5 Drop diameter D_p in m plotted as a function of the following parameters: γ, surface tension, N/m; D_0, hole diameter, m; $\Delta\rho$ = density difference for the two liquids, kg/m^3; ρ_D, density of dispersed phase, kg/m^3; V_0, velocity through the holes, m/s; μ_c, viscosity of dense phase, Ns/m^2.

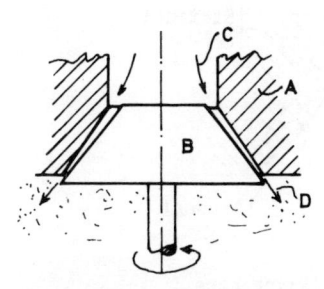

Figure 8.6 Colloid mill: A, stator; B, rotor, C, feed; D, homogenized liquid.

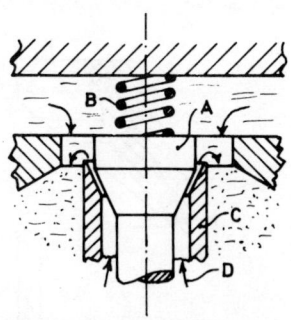

Figure 8.7 Homogenizer: A, precisely machined needle forced against the ring C by the spring B; D, feed forced through the annular space between A and C.

Crushing and grinding

Size reduction by means of crushers and grinders is often carried out in stages. An example is comminution of ores, where the first size reduction may be in a *jaw crusher* (Figure 8.8), the next reduction in a *gyratory crusher* (Figure 8.9), and the final reduction before flotation is usually by impact in a *ball* or *pebble mill* (Figure 8.10), operated dry or wet. Other types of size reducing equipment are *roller and pan grinders* (Figure 8.11), used for grinding of talc and in porcelain factories. *Buhrstone mills* (Figure 8.12) are used for grinding of special cereals and grains, and *roller mills* for paint grinding (Figure 8.13). *Pin mills* (Figure 8.14), are used as disintegrators for fibrous materials and for grinding cocoa.

Numerous other types of crushers and mills are used for different applications, such as the various *hammer mills* for grinding municipal wastes, chopping of seaweeds for cattle food, and in alginate production. Ultrafine grinders may be *fluid-energy* or *jet mills*.

Grinding to particle sizes in the range 1 μm to 7 μm is obtained by collisions between particles with different velocities. Rumpf[9] studied the influence of diameter, velocity, and length of the jet. Perry[8] gives fluid consumptions. The energy

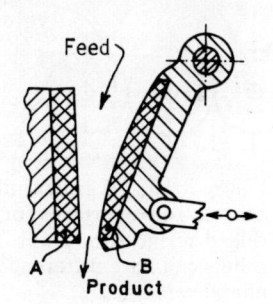

Figure 8.8 A jaw crusher: A, fixed jaw; B, swing jaw.

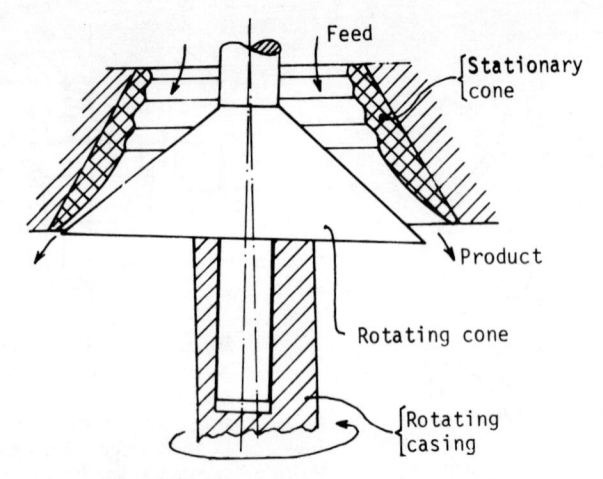

Figure 8.9 Gyratory crusher with the centre line of the lower cone at a slight angle. The centre line of the rotating casing is identical to the centre line of the stationary cone.

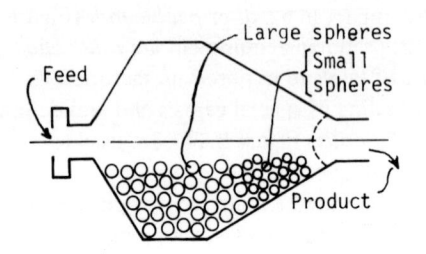

Figure 8.10 A ball mill.

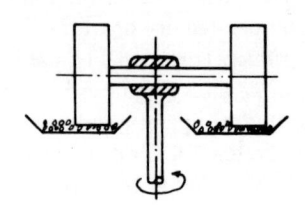

Figure 8.11 Roller and pan grinder.

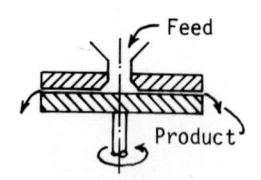

Figure 8.12 Buhrstone mill.

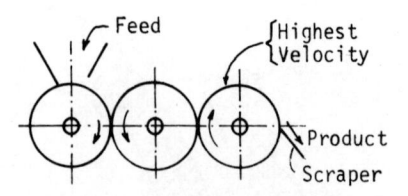

Figure 8.13 A roller mill: smooth roller surfaces for paint, ridged surfaces for grain. The rollers can have different peripheral velocities.

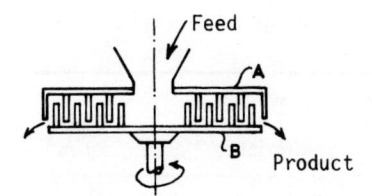

Figure 8.14 Pin mill for cocoa usually with horizontal shaft. A is a fixed and B a rotating pin wheel.

consumption ranges from 10 kWh to 1000 kWh per ton product. Types of products are pigments, dyes, chemicals, drugs, ceramic powders, waxes, and insecticides.

Wet grinding of pigments (Indian ink) was practised by the ancient Chinese. Today, wet grinding of organic pigments is used to obtain particle sizes less than 1 μm.[10]

Figure 8.15 gives cumulative particle sizes of some industrial products.[8]

Power is a major expense in crushing and grinding. It depends on the material, the particle size before crushing, and to a large extent on the final particle size. As an example, the power required in pulverizing slate in the cement industry is approximately[8]

$$P = 3e^{0.00635S} \tag{8.7}$$

where P is the power, kWh/ton, and S is the Blaine surface area of the product, m^2/kg (specific area determined by air permeability). Equation (8.7) is valid for Blaine surface areas between 300 and 600 m^2/kg. The power required in crushing can be estimated by Bond's half empirical equation[11],

$$P = W_i \left(\frac{1}{\sqrt{d_2}} - \frac{1}{\sqrt{d_1}} \right) \text{kWh/ton} \tag{8.8}$$

where W_i is the work index (average values given in Table 8.1) and d_1 and d_2 are particle sizes in mm before and after crushing. Here the definition of particle size is

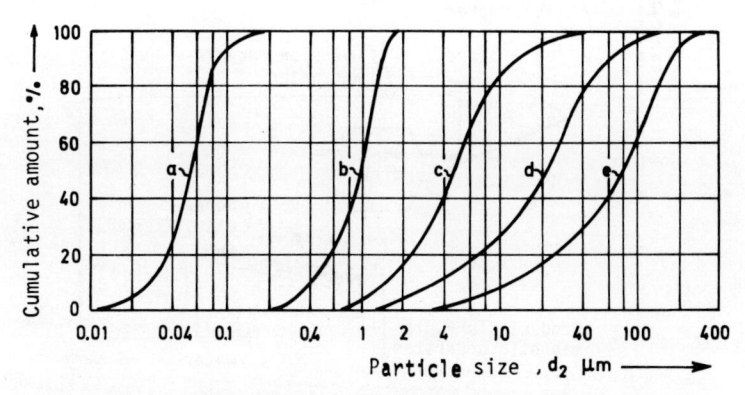

Figure 8.15 Typical particle size distribution of some industrial products: a, fine rubber fillers; b, pigments; c, mineral fillers; d, Portland cement; e, powdered coal.

Table 8.1 Average work indexes W_i in equation (8.8) for dry crushing and wet grinding. For dry grinding multiply by 4/3.[8]

Material	Density kg/m^3	Work index W_i	Material	Density kg/m^3	Work index W_i
Bauxite	2380	3.0	Iron ore		
Cement clinker	3090	4.3	haematite	3760	4.0
Cement raw material	2670	3.3	magnetite	3880	3.2
Clay	2230	2.2	Limestone	2690	3.7
Coal	1630	3.6	Manganese ore	3740	3.9
Coke, petroleum	1780	23.3	Nickel ore	3320	3.8
Copper ore	3020	4.2	Oil shale	1760	5.7
Dolomite	2820	3.6	Phosphate rock	2660	3.2
Feldspar	2590	3.7	Potash salt	2180	2.6
Ferrosilicon	4910	4.1	Pyrite ore	3480	2.8
Flint	2650	8.3	Quartz	2640	4.0
Fluorspar	2980	3.1	Sandstone	2680	3.6
Gabbro	2830	5.8	Shale	2580	5.2
Glass	2580	1.0	Silica	2710	4.3
Gneiss	2710	6.4	Silica sand	2650	5.2
Granite	2680	4.6	Silicon carbide	2730	8.3
Graphite	1750	14.2	Sinter	3000	2.8
Gravel	2700	8.0	Slag	2930	5.0
Ilmenite	4270	4.1	Slate	2480	4.4
			Titanium ore	4230	3.8

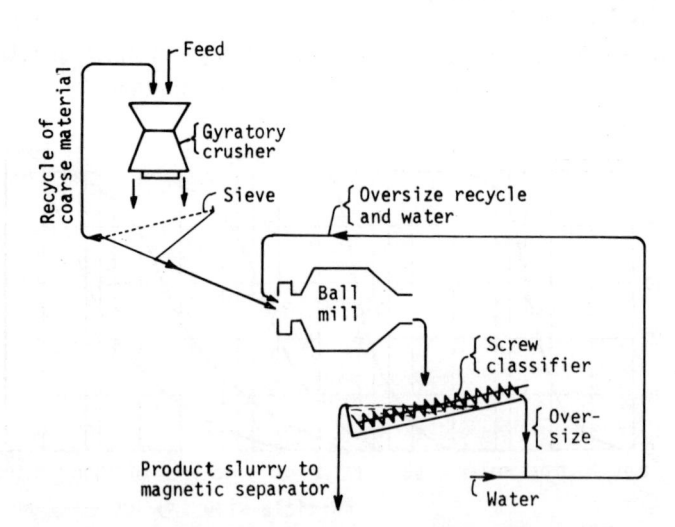

Figure 8.16 Flow sheet for closed circuit crushing and grinding of magnetic iron ore. The oversize recycle contains mixed grain.

Table 8.2   Aperture in mm for some mesh values.

Mesh	Tyler standard	British standard	IMM standard	Mesh	Tyler standard	British standard	IMM standard
5	4.00	3.35	2.54	120	–	0.124	0.107
10	1.68	1.68	1.27	150	0.105	0.104	0.084
14	1.19	1.20	–	170	0.088	0.089	–
20	0.841	–	0.635	200	0.074	0.076	0.064
28	0.595	–	–	240	–	0.066	–
30	–	0.500	0.422	250	0.063	–	–
44	–	0.353	–	270	0.053	–	–
48	0.297	–	–	300	–	0.053	–
60	0.250	0.251	0.211	325	0.044	–	–
65	0.210	–	–	350	–	0.046	–
100	0.149	0.152	0.127	400	0.037	–	–

that 80% passes through a sieve with openings d_1 and d_2 respectively, and about 65% passes through a sieve with half the opening.

The figures in Table 8.1 are average values and can only be used to estimate the order of magnitude of the power required.

In closed circuit operation a size classifier is used, and oversize particles are returned to the grinding machine. The closed circuit reduces both the power required and the fraction of particles that are ground unnecessarily fine. In Figure 8.16, the last step is wet grinding. Wet grinding reduces power consumption, eliminates dust, and makes classification easier.

Details about very small particles and their characteristic properties in surface films and in sintering are given in the book by Veale,[12] and about explosion hazards of organic materials in the book by Palmer.[13]

According to the proposed international standard (ISO-Recommendation), particle size determined in sieving should be given as mm or μm aperture of the sieves instead of 'mesh', which is the number of wires per inch. The wire diameters and the selected mesh values are somewhat different in the American Tyler standard, the British standard and the standard of the Institute of Mining and Metallurgy (IMM). Apertures for some selected values of these three standards are given in Table 8.2.

EXAMPLE 8.1. GAS DISPERSION IN AN AGITATED TANK

Air is dispersed in a cylindrical tank by a six blade turbine. The ratio of impeller diameter to tank diameter is 1/3. The tank diameter is 2.4 m and the liquid depth 2.6 m. The turbine runs at 2 rev./s. The surface tension is $\gamma = 0.07$ N/m.

(a)   Estimate the agitation intensity for the ungassed tank.

(b)   Estimate the average bubble diameter for a volumetric gas fraction $\epsilon_g = 0.06$.

(c)   Estimate the volumetric flow rate of air, assuming that the terminal velocity of air bubbles is as given by Figure 6.9.

(d)   Estimate the power consumption with volumetric gas fraction $\epsilon_g = 0.06$.

Solution

(a) Equation (5.1) and Figure 5.2 give

$$\frac{P}{\rho D_i^5 n^3} = 6$$

or $P = (6)(1000)(0.8^5)(2^3)$
$= 15{,}700$ W.

Hence

$$R = \frac{15{,}700}{2.6\,\dfrac{\pi}{4}\,2.4^2}$$

$= 1337$ W/m^3.

Note A high agitation intensity is necessary to obtain good gas dispersion.

(b) Equation (8.4) gives

$$\bar{D}_b = 4.15 \frac{0.07^{0.6}}{(1337^{0.4})(1000^{0.2})} 0.06^{0.5} + 0.0009$$

$= 0.0038$ m
$= 3.8$ mm.

(c) Figure 6.9 gives the terminal velocity $V_t = 0.25$ m/s. From equation (8.5),

$$0.06 = \left(\frac{0.06\,V_s}{0.25}\right)^{0.5} + 0.0000543\,\frac{(1337^{0.4})(1000^{0.2})}{0.07^{0.6}}\left(\frac{V_s}{0.25}\right)^{0.5}.$$

Superficial velocity,

$V_s = 0.013$ m/s

Volumetric rate of air flow,

$$Q = 0.013\,\frac{\pi}{4}\,2.4^2$$

$= 0.06$ m^3/s.

(d) Froude number (page 120),

$$\text{Fr} = \frac{n^2 D_i}{g} = \frac{(2^2)(0.8)}{9.82} = 0.33$$

and

$$\frac{Q}{nD_i^3} = \frac{0.06}{(2)(0.8^3)}$$

$= 0.059$,

i.e. in equation (5.5),

$P = [1 - (2.1)(0.059)]\,15.7$
$= 13.8$ kW.

EXAMPLE 8.2. LIQUID DISPERSION

Benzene is dispersed in water through punched holes in the bottom of an extraction column (Figure 8.4). Estimate the diameter of the holes in the distributor plate to give maximum drop diameter 4 mm with liquid velocity through the holes $V_0 = 0.3$ m/s. The column is operated at 20°C.

	Density kg/m^3	Viscosity cP	Surface tension dyn/cm	N/m
Water, 20°C	998	1.01	25.7	0.0257
Benzene, 20°C	879	0.65		

Solution

Figure 8.5 with drop diameter $D_p = 0.004$ m and the parameter

$$\rho_D V_0^2/\Delta\rho = (879)(0.3^2)/(998-879) = 0.67$$

gives

$$\frac{0.0257 D_0}{998-879} + 23.8 \frac{D_0^{1.12}(0.30^{0.547})(0.00101^{0.279})}{(998-879)^{1.5}} = (1.55)(10^{-6})$$

$D_0 = 0.0018$ m
   $= 1.8$ mm.

EXAMPLE 8.3. POWER CONSUMPTION IN CRUSHING OF ORE

75 ton/h magnetite ore with density 3850 kg/m^3 is crushed to a product size of approximately 60 mm in a jaw crusher, to less than 10 mm in a gyratory crusher, and finally ground in a ball mill to give particles of which 80% pass through a 100 mesh screen (0.15 mm aperture).
(a) Estimate the power consumption in the gyratory crusher.
(b) Estimate the power consumption of the ball mill for both wet and dry grinding.

Solution

(a) The value of d_2 in equation (8.8) is less than 10 mm. As a guess, 6 mm is assumed. This value, and work index $W_i = 3.2$ from Table 8.1, give the power consumption,

$$P = (75)(3.2)\left(\frac{1}{\sqrt{6}} - \frac{1}{\sqrt{60}}\right)$$
$= 67$ kW.

(b) Equation (8.8) for wet grinding gives,

$$P = (75)(3.2)\left(\frac{1}{\sqrt{0.15}} - \frac{1}{\sqrt{6}}\right)$$
$= 520$ kW,

and for dry grinding according to Table 8.1,

$P = (520)(4)/3$
$= 690 \text{ kW}$.

Problems

8.1. Water is dispersed in benzene through punched holes with diameter 2.0 mm in a distributor plate at the top of an extraction column operated at 20°C. Density, viscosity and surface tension are as in Example 8.2.

(a) Estimate the drop size for flow through the holes at 0.1 and 0.3 m/s.

(b) Estimate the area of the benzene water interface in the tank.
one hole.

8.2. 0.3 m³ benzene is temporarily dispersed in 1.7 m³ water in a baffled tank with diameter 1.8 m. The tank has a six blade turbine with diameter 0.6 m. The temperature is 20°C (physical data as in Example 8.2).

(a) Estimate the rotating speed of the turbine to give a mean drop diameter 1.5 mm.

(b) Estimate the area of the benzene water interface in the tank.

References

1. Cronan, C. S.: New easy way changes paste to powder, *Chem. Eng.*, **67**, 21 March, 83–84 (1960).
2. Marshall, W. R.: *Atomization and Spray Drying*, Chemical Engineering Progress Monograph Series, No. 2, 50 (1954).
3. Fair, J. R.: Design of gas-sparged reactors, *Chem. Eng.*, **66**, 3 July, 67–74 and 17 July, 207–214 (1967).
4. Calderbank, P. H., in V. W. Uhl and J. B. Gray (Eds.): *Mixing: Theory and Practice*, volume II, Academic Press, New York, 1967.
5. Hampworth, C. B. and R. E. Treybal: Drop formation in two-liquid-phase systems, *Ind. Eng. Chem.*, **42**, 1174–1181 (1950).
6. Schindler, H. D. and R. E. Treybal: Continuous-phase mass-transfer coefficients for liquid extraction in agitated vessels, *A.I.Ch.E.J.*, **14**, 790–798 (1968).
7. Calderbank, P. H.: The interfacial area in gas-liquid contacting with mechanical agitation, *Inst. of Chem. Eng.*, **36**, 443–463 (1958).
8. Perry, R. H. and C. H. Chilton: *Chemical Engineers' Handbook*, McGraw-Hill, New York, 5th Edn., 1973.
9. Rumpf, H.: Prinzipien der Prallzerkleinerung und ihre Anwendung bei der Strahlmahlung, *Chem.-Ing.-Techn.*, **32**, 129–135 (1960).
10. Giersiepen, G. and J. Schwedes: Erfahrungen mit Maschinen zur Naszerkleinerung in der chemischen Industrie, *Chem.-Ing.-Techn.*, **47**, 695–699 (1975).
11. Bond, F. C.: The third theory of comminution, *Trans. Am. Inst. Mining and Metallurgical Engrs.*, **193**, 484–494 (1952).
12. Veale, R. C.: *Fine Powders, Preparation, Properties and Uses*, Applied Science, London, 1972.
13. Palmer, K.N.: *Dust Explosions and Fires*, Chapman & Hall, London, 1973.

CHAPTER 9
Steady State Heat Transfer

The majority of chemical processes involve production or absorption of heat. In these processes, heat is transferred from a body of higher to a body of lower temperature. In heat transfer, three different mechanisms are involved.

Conduction in a solid or a fluid is transfer of vibrational energy from one molecule to another, and in fluids it is also the transfer of kinetic energy. In metals, free movement of electrons is also involved. This accounts for the high thermal conductivity of metals.

Radiation is transfer of energy by electromagnetic waves through vacuum or through a transmitting medium. The radiation that hits a body is partly absorbed, transmitted, or reflected. Only the absorbed fraction results in heat.

Convection is heat transfer by movement of macroscopic particles of the fluid with different temperatures. Hence, convection is limited to liquids and gases.

Heat conduction

Heat conduction through a solid or a fluid at rest can be described by Fourier's law

$$\dot{Q} = -kA \frac{dT}{dx} \quad (9.1)$$

where $\dot{Q}$ = rate of heat flow, W
k = thermal conductivity of the material, W/(K m) (characteristic values of k are given in Table 9.1)
A = area perpendicular to the heat flow, m^2
dT/dx = temperature gradient in direction of the heat flux, K/m.

For conduction through several layers, it is convenient to integrate equation (9.1) for each layer, solve for temperature difference and add up the temperature differences. Figure 9.1 shows three parallel layers with thickness $L_1, L_2,$ and L_3 with thermal conductivities $k_1, k_2,$ and k_3, respectively.

Table 9.1[2,3] Thermal conductivities, k in W/(K m), and specific heat capacities, c in kJ/(K kg), for gases $c_p = A + BT + CT^2$. T is in K and θ in °C[a]

Material	Temperature θ °C	Thermal conductivity k W/(K m)	Heat capacity c kJ/(K kg)
Metals			
Silver	20	420	0.23
Copper, commercial grade	20	350	0.63
Aluminium	20	200	0.77
Steel, cast iron, nickel	20	60	0.46
Acid resistant steel (18% Cr, 8% Ni, 1.5–2.5% Mo)	200	18	0.50
	600	22	
Nickel steel (40% Ni)	20	10	
Miscellaneous materials			
Rock	20	1.1–4.7	~0.8
Quartz sand	20	1.6	0.73
Refractory bricks	20	0.6–1.7	
Brick wall (building)	20	0.7–1.0	0.8
Insulating materials, plastic insulation, cork, glass, and mineral wool	20	0.029–0.06	0.8–2.1
Liquids			
Water	20	0.59	4.19
	150	0.69	4.27
	300	0.56	5.44
Acetic acid, 50%	20	0.34	3.12
100%	20	0.168	2.05
Lauric acid	100	0.172	1.8
Oleic acid	20	0.16	

Gases at pressure 1 bar = 10^5 Pa
H_2, $k = 0.173(1 + 0.003\,\theta)$, $A = 14.425$, $B = -0.000407$, $C = 0.985 \times 10^{-6}$
CO_2, $k = 0.0144(1 + 0.004\,\theta)$, $A = 0.603$, $B = 0.000965$, $C = -0.3249 \times 10^{-6}$
Air, $k = 0.0241(1 + 0.003\,\theta)$, $A = 0.922$, $B = 0.000254$, $C = -0.0384 \times 10^{-6}$

[a] In the absence of experimental data, refer to the estimation methods compiled by Reid, Prausnitz and Sherwood.[1]

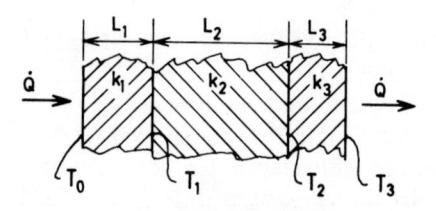

Figure 9.1 Heat conduction through parallel layers.

The solution of each of the layers in Figure 9.1 is,

$$\dot{Q} = k_1 A \frac{T_0 - T_1}{L_1} \quad \text{or} \quad T_0 - T_1 = \frac{\dot{Q}}{A}\frac{L_1}{k_1}$$

$$\dot{Q} = k_2 A \frac{T_1 - T_2}{L_2} \quad \text{or} \quad T_1 - T_2 = \frac{\dot{Q}}{A}\frac{L_2}{k_2}$$

$$\dot{Q} = k_3 A \frac{T_2 - T_3}{L_3} \quad \text{or} \quad T_2 - T_3 = \frac{\dot{Q}}{A}\frac{L_3}{k_3}$$

$$\text{sum} \quad T_0 - T_3 = \frac{\dot{Q}}{A} \Sigma \frac{L_x}{k_x}$$

$$\text{or} \quad \dot{Q} = A \frac{\Delta T}{\Sigma \frac{L_x}{k_x}} \tag{9.2}$$

where ΔT is the temperature difference between the outer surfaces, $\Delta T = T_0 - T_3$. The term $\Sigma(L_x/k_x)$ is known as the thermal resistance.

For a cylinder (Figure 9.2) equation (9.1) is in the form,

$$\dot{Q} = -k(2\pi r L) dT/dr \ .$$

or

$$\int_{r_1}^{r_2} \frac{dr}{r} = \frac{-2\pi L k}{\dot{Q}} \int_{T_1}^{T_2} dT$$

which gives

$$\dot{Q} = \frac{2\pi L k (T_1 - T_2)}{\ln(r_2/r_1)} \ . \tag{9.3}$$

For thin cylinders, heat conduction may be calculated as for a flat plate with area $2\pi r_m L$ where $r_m = (r_1 + r_2)/2$.

Heat conduction through concentric cylinders is calculated as shown for parallel

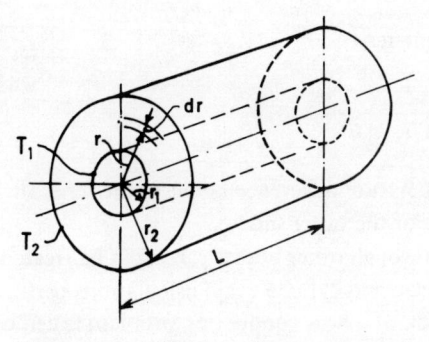

Figure 9.2 Heat conduction through a cylinder.

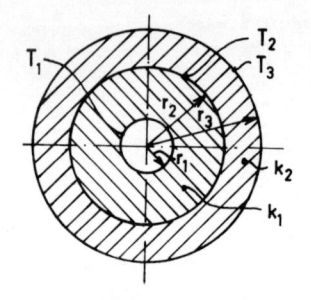

Figure 9.3 Heat conduction through concentric cylinders.

layers. Equation (9.3) is solved for the temperature difference for each cylinder and the equations are summed. With the symbols in Figure 9.3, this gives,

$$T_1 - T_2 = \frac{\dot{Q}}{2\pi L} \frac{\ln(r_2/r_1)}{k_1}$$

$$T_2 - T_3 = \frac{\dot{Q}}{2\pi L} \frac{\ln(r_3/r_2)}{k_2}$$

$$T_1 - T_3 = \frac{\dot{Q}}{2\pi L} \left[\frac{\ln(r_2/r_1)}{k_1} + \frac{\ln(r_3/r_2)}{k_2} \right]$$

or

$$\dot{Q} = \frac{2\pi L \Delta T}{\Sigma \ln(r_{x+1}/r_x)/k_x} \qquad (9.4)$$

where ΔT is the temperature difference between the inner surface of the inner cylinder and the outer surface of the outer cylinder in Figure 9.3, $T_1 - T_3$.

The same procedure is used for spheres. Integration of equation (9.1) for a hollow sphere with inner radius r_1 and outer radius r_2 gives,

$$\dot{Q} = \frac{4\pi k(T_1 - T_2)}{1/r_1 - 1/r_2} \qquad (9.5)$$

and for concentric spheres,

$$\dot{Q} = \frac{4\pi \Delta T}{\Sigma (1/r_x - 1/r_{x+1})/k_x} \qquad (9.6)$$

where ΔT is the temperature difference between the inner surface of the inner sphere and the outer surface of the outer sphere.

Heat conduction through other bodies can often be treated as if parts of them consist of slabs, spheres, and cylinders and calculations performed separately.

A graphical solution of a heat conduction problem is demonstrated in Figure 9.4. A grid of isotherms a and heat flux lines b is plotted in such a way that $\Delta x/\Delta y = 1.0$. The isotherms are perpendicular to the heat flux lines. With these conditions

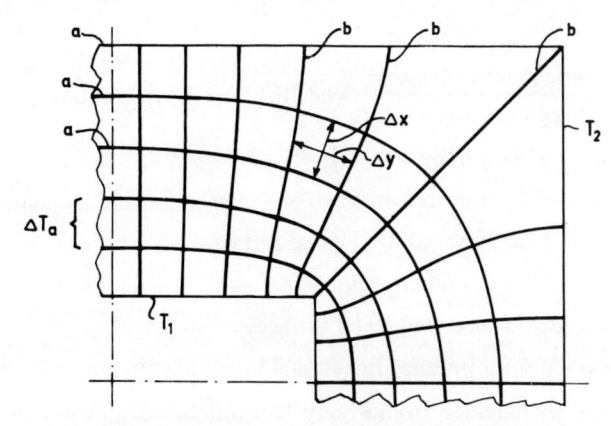

Figure 9.4 Graphical determination of heat conduction through an insulated rectangular duct (1/4 of the duct shown), with temperature T_1 at the inner and T_2 at the outer surface. Isotherms a and heat flux lines b are adjusted to be perpendicular to each other and to give $\Delta x/\Delta y = 1.0$.

fulfilled, the heat flux between any two heat flux lines is

$$\Delta \dot{Q} = k\Delta y L \frac{\Delta T_a}{\Delta x} = kL\Delta T_a \tag{9.7}$$

where L is the length of the duct and ΔT_a is the temperature difference between the two isotherms. With n isotherms

$$\Delta T_a = \frac{T_1 - T_2}{n - 1}. \tag{9.8}$$

Figure 9.4 shows 1/4 of a duct with 8.5 heat flux lines and $n = 6$ isotherms, giving a heat loss by conduction $\dot{Q} = (4)(8.5)k(T_1 - T_2)/5$.

Computer calculation is recommended if an adequate program is available. If not, graphical solution can often be carried out more quickly than the programming.

Radiation

Energy emission by radiation from a solid is given by the Stefan–Boltzmann law,

$$\dot{Q} = \epsilon C_b A T^4 \tag{9.9}$$

where

$\dot{Q}$ = emitted energy i W
C_b = radiation coefficient for a black body, $(5.7)(10^{-8})$ W/(m² K⁴)
A = surface area of the emitting body, m²
T = surface temperature of the emitting body, K

ϵ = emissivity (dimensionless)

$= \dfrac{\text{absorbed radiation energy}}{\text{received radiation energy}}$ when the same body is exposed to heat radiation

ϵ = 1.0 for a black body

= 0.85–0.95 for porcelain, glass, paper, paint (except aluminium paint)

= 0.6–0.8 for most oxidized metal surfaces

= 0.2–0.4 for aluminium paint

= 0.28–0.29 for molten metal surfaces

= 0.02–0.04 for polished metals and bright aluminium surfaces.

Heat transfer by radiation from a body 1 surrounded completely by a body 2 is derived using the following reasoning given by Ernst Schmidt.[2]

The Figures 9.5 and 9.6 show a sphere or a cylinder with surface area A_1 enclosed by an outer sphere or cylinder with surface area A_2. From every part of surface A_2, rays can be drawn in all directions (Figure 9.5). These rays can be grouped into an infinite number of sets of parallel rays, with one set as shown in Figure 9.6. The flux density is constant over the cross-section, and the fraction of radiation from A_2 that hits A_1 is

$$\phi = A_1/A_2.$$

The total radiation energy leaving a surface is the sum of emitted energy according to equation (9.9) and reflected radiation. Per unit surface area this sum is E_1 for A_1 and E_2 for A_2, i.e. $A_1 E_1$ leaves surface A_1 and $A_2 E_2$ leaves surface A_2, but of the latter only $\phi A_2 E_2 = A_1 E_2$ reaches surface A_1. Hence, the net heat transfer is

$$\dot{Q}_{1,2} = A_1 E_1 - A_1 E_2 = A_1 (E_1 - E_2)$$

where

$$E_1 = \epsilon_1 C_b T_1^4 + (1 - \epsilon_1) E_2.$$

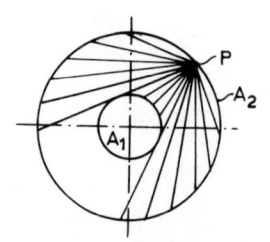

Figure 9.5 Heat exchange by radiation between concentric surfaces: rays from a point P on the outer surface.

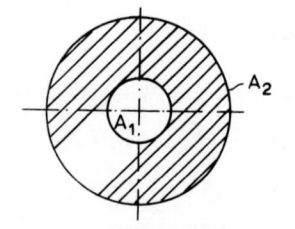

Figure 9.6 Parallel rays from an infinite number of points P (Figure 9.5).

The last term represents reflected radiation. At E_2, however, the incident radiation consists of two parts, one coming from A_1 and equal to ϕE_1, and the other coming from surface A_2 itself, equal to $(1-\phi)E_2$. We then have

$$E_2 = \epsilon_2 C_b T_2^4 + (1-\epsilon_2)\phi E_1 + (1-\epsilon_2)(1-\phi)E_2.$$

Combining the last three equations and $\phi = A_1/A_2$, we obtain the radiation heat transfer between the two surfaces:

$$\dot{Q}_{1,2} = \frac{C_b}{\dfrac{1}{\epsilon_1} + \dfrac{A_1}{A_2}\left(\dfrac{1}{\epsilon_2}-1\right)} A_1(T_1^4 - T_2^4) \tag{9.10}$$

or

$$\dot{Q}_{1,2} = C_{1,2} A_1 (T_1^4 - T_2^4) \tag{9.11}$$

where

$$C_{1,2} = \frac{C_b}{\dfrac{1}{\epsilon_1} + \dfrac{A_1}{A_2}\left(\dfrac{1}{\epsilon_2}-1\right)} \tag{9.12}$$

and it is termed the radiation exchange coefficient.

The equations are derived for spheres and long cylinders but they may give a fairly good estimate of heat exchange by radiation in other cases where one body is completely surrounded by another.

Graphical solutions may be convenient in cases where body 2 does not surround body 1. This solution is based on Lambert's cosine law, that the intensity of radiation is proportional to the cosine of the angle between the rays and a line perpendicular to the radiating surface. This gives the energy emitted between the four rays shown in Figure 9.7,

$$d\dot{Q}_1 = C \cos \alpha \, d\alpha \, d\beta.$$

The space angle $d\alpha \, d\beta$ cuts out the hatched surface $R^2 \, d\alpha \, d\beta$ on the half sphere Figure 9.7. The projection of this surface is

$$dA' = R^2 \cos \alpha \, d\alpha \, d\beta$$
$$d\dot{Q}_1 = C \, dA'/R^2.$$

Since the projection of the half sphere is πR^2, integration of this equation gives the total radiation from A_1,

$$\dot{Q}_1 = \pi C.$$

The last two equations give

$$d\dot{Q}_1 = \frac{dA'}{\pi R^2} \dot{Q}_1.$$

Integration of this between the rays that hit the edges of a surface A_2, gives the fraction of radiation from A_1 that hits A_2, $A'/(\pi R^2)$. As an approximation, the

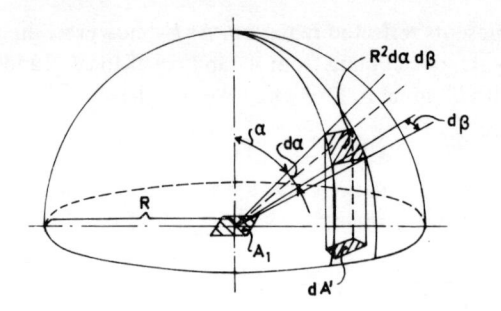

Figure 9.7 Radiation from A_1: α is the angle from the normal to A_1.

net radiation heat transfer calculated by equation (9.11) for a body 1 surrounded by a body 2 is multiplied by this factor,

$$Q_{1,2} = C_{1,2} \frac{A'}{\pi R^2} A_1 (T_1^4 - T_2^4) \qquad (9.13)$$

where $C_{1,2}$ is given in equation (9.12).

The approximation is best when one of the bodies almost surrounds the other or if the emissivities approach 1.0, as in boiler calculations where the surfaces usually have emissivities in the region of 0.8 to 0.9.

If the surface A_1 is too large to be approximated by a point in the centre, it may be divided into sections, and the radiation from each section estimated by equation (9.13).

Radiation from flames of oil, gas, or pulverized coals is radiation from a cloud of soot particles with diameters 0.02 to 0.14 μm. The emission from a flame does not follow the Stefan–Boltzmann law.

Radiation from non-luminous gases to surrounding surfaces is only significant at gas temperatures above 600°C. The radiation from diatomic gases such as N_2, O_2, and H_2 is negligible. Radiation from other gases such as SO_2, CO_2, HCl, H_2O, alcohols, and hydrocarbons is proportional to their partial pressure and the distance between the surrounding surfaces. For calculation methods refer to Kern[28] and Perry.[42]

Convection

The heat transferred between a surface and a moving gas or liquid is given by the equation

$$\dot{Q} = hA\Delta T \qquad (9.14)$$

where A is the surface area, m^2, ΔT is the temperature difference between the surface and the bulk of the fluid, °C or K, and h is the heat transfer coefficient due to convection, W/(m^2 K), which depends on properties of the fluid, geometry of the surface, and flow rate.

Natural or free convection results from density differences caused by temperature gradients in the fluid. This is in contrast to forced convection where the movement is produced by a pump, a stirrer, or some other outside means.

It is convenient to calculate heat transfer coefficients by means of dimensionless groups, that can be derived by dimensional analysis. The groups are

$\mathrm{Nu} = hD/k$ = Nusselt number

$\mathrm{Pr} = c_p \mu / k$ = Prandtl number

$\mathrm{Re} = \rho VD/\mu$ = Reynolds number

L/D = length ratio

μ/μ_w = viscosity ratio = $\dfrac{\text{fluid viscosity at bulk temperature}}{\text{fluid viscosity at wall temperature}}$

$\mathrm{Gr} = \rho^2 D^3 (\beta \Delta T) g / \mu^2$ = Grashof number

where β is the thermal coefficient of cubic expansion, K^{-1}, ΔT is the temperature difference between the surface and the bulk of the fluid, K, ρ is the fluid density, kg/m^3, k is its thermal conductivity, W/(m K), and c_p its specific heat at constant pressure, J/(kg K).

The general equation derived by dimensional analysis is

$$\mathrm{Nu} = C\, \mathrm{Re}^a \mathrm{Pr}^b\, \mathrm{Gr}^c \left(\frac{L}{D}\right)^d \left(\frac{\mu}{\mu_w}\right)^e. \tag{9.15}$$

Free convection

For conditions in which only natural convection occurs the velocity is dependent solely on buoyancy effects, represented by the Grashof number, and Reynolds number can be omitted. Measurements with vertical plates and cylinders show[3] that $b \approx c \approx 1/4$, and that the two last groups in equation (9.15) are insignificant. The equation simplifies to

$$\mathrm{Nu} = C(\mathrm{Pr}\, \mathrm{Gr})^{1/4} \tag{9.16}$$

where

$C = 0.47$ for horizontal and vertical pipes with the diameter as characteristic length

$C = 0.56$ for vertical plates and vertical cylinders with large diameter and the height as characteristic length and $(\mathrm{Gr}\, \mathrm{Pr}) < (2)(10^9)$

$C = 0.54$ for plates pointing upwards and 0.25 for plates pointing downwards, both with the width as characteristic length.

These values of C are valid for $10^3 < (\mathrm{Pr}\, \mathrm{Gr}) < 10^8$ for horizontal pipes. For horizontal pipes with $10^2 < (\mathrm{Pr}\, \mathrm{Gr}) < 10^3$

$$\mathrm{Nu} = (\mathrm{Pr}\, \mathrm{Gr})^{0.14}. \tag{9.17}$$

For vertical plates and $Gr < 10^9$, Kato et al.[4] recommend use of the equation

$$Nu = 0.683 \, Pr^{0.5} \left(\frac{Gr}{0.861 + Pr} \right)^{0.25} \pm 10\%. \tag{9.18}$$

For inclined surfaces with an angle α to the vertical less than $45°$, Tautz[5] found the same values as for vertical plates, while Rich[6] found a reduction proportional to $(\cos \alpha)^{1/4}$ for $\alpha < 40°$ and $10^6 < Gr < 10^9$.

For heat transfer by direct contact between liquid drops and a continuous liquid, refer to the literature.[7]

Forced convection

In forced convection the currents are set in motion by the action of a mechanical device such as a pump or agitator, and the dependence on Grashof number vanishes. The classical measurements by Sieder and Tate[8] gave $b \approx 1/3$ in equation (9.15) and $e \approx 0.14$, giving the equation,

$$Nu \, Pr^{-1/3} \left(\frac{\mu}{\mu_w} \right)^{-0.14} = f\!\left(Re, \frac{L}{D} \right) \tag{9.19}$$

where L is the length of pipe or duct, m, D is the hydraulic diameter, m, and μ_w is the viscosity of the fluid at the surface temperature of the pipe or duct, N s/m².

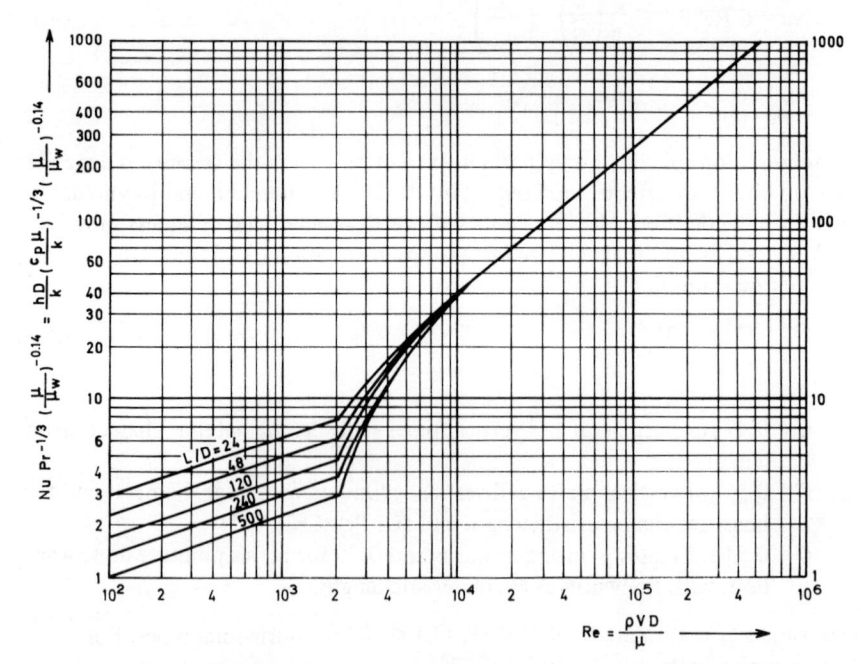

Figure 9.8 Heat transfer in straight pipes and ducts:[8] L, pipe-length; D, diameter; μ_w, viscosity at the surface temperature of the pipe. Usually $(\mu/\mu_w)^{-0.14} \approx 1.0$.

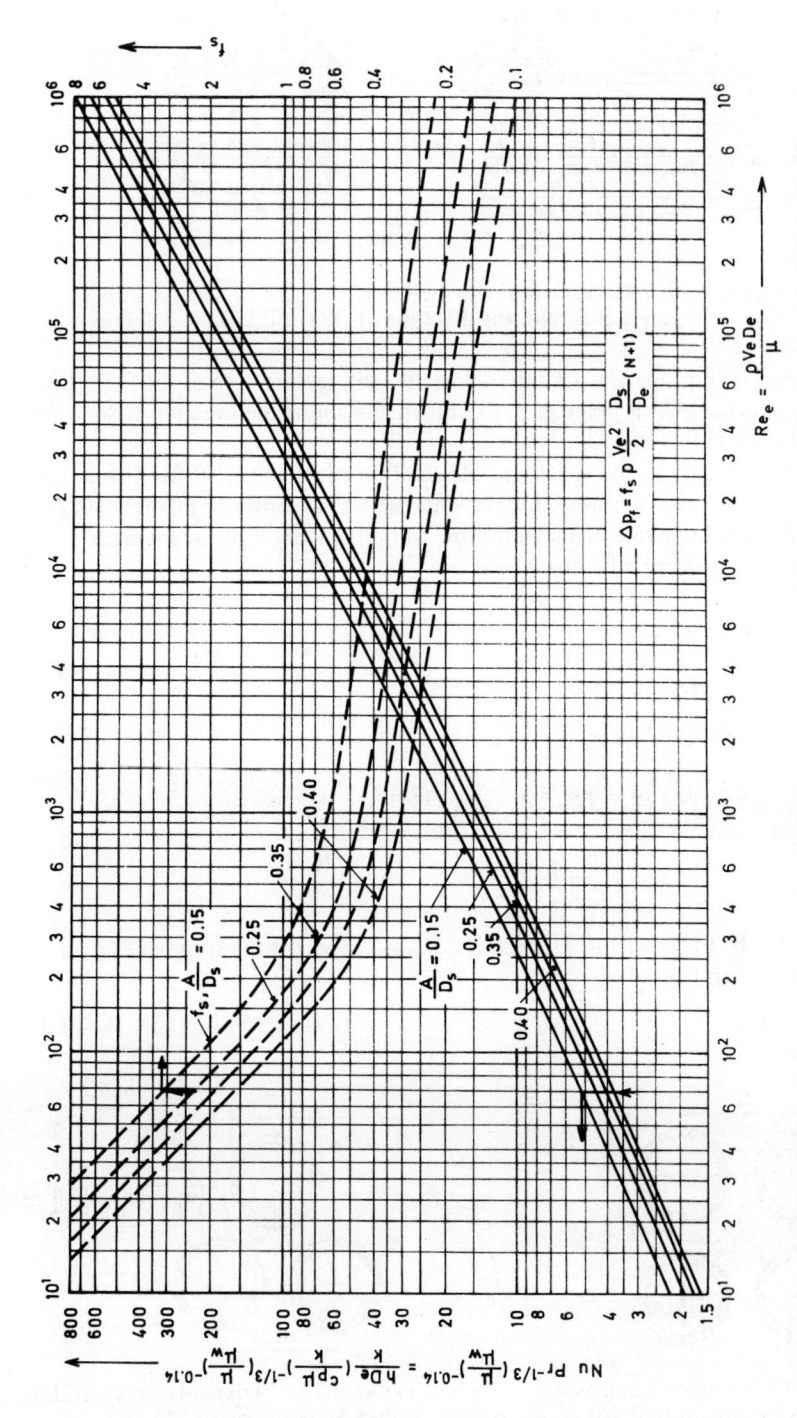

Figure 9.9 Heat transfer at the outside of the tubes in Figure 9.10 and friction factor f_s in equation (9.24) for baffle cuts $A/D_s = 0.15, 0.25, 0.35,$ and 0.40. $A/D_s = 0.25$ is usually used. D_e and V_e are defined by equations (9.20), (9.21), and (9.22).[9] Reproduced by permission of Wolverine Tube Division.

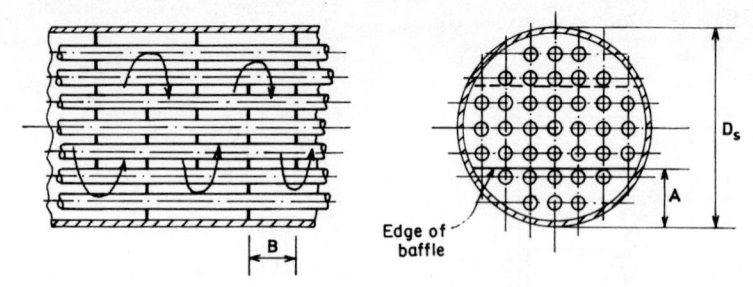

Figure 9.10 Fluid flow at the shell side of a heat exchanger with baffles. Figure 9.9 is valid for this type of flow if $0.2 D_s < B < D_s$.

All other fluid properties are referred to the bulk temperature of the fluid. For most cases, $(\mu/\mu_w)^{0.14}$ is practically 1.0. Figure 9.8 shows the function $f(\text{Re}, L/D)$ in equation (9.19).

Figure 9.9 illustrates the function $f(\text{Re}, L/D)$ in equation (9.19) for flow at the outside of the tubes in a heat exchanger with baffles[9] as shown in Figure 9.10. D_e is defined as the hydraulic diameter in flow without baffles. For square pitch (Figure 9.11), this hydraulic diameter is

$$D_e = \frac{4\left(P_t^2 - \frac{\pi}{4}D_0^2\right)}{\pi D_0} = \frac{4P_t^2}{\pi D_0} - D_0 \qquad (9.20)$$

and for triangular pitch (Figure 9.12),

$$D_e = \frac{4\left(\frac{1}{2}P_t\sqrt{P_t^2 - (P_t/2)^2} - \frac{1}{2}\frac{\pi}{4}D_0^2\right)}{\frac{1}{2}\pi D_0} = \frac{2\sqrt{3}P_t^2}{\pi D_0} - D_0. \qquad (9.21)$$

The velocity V_e to be used in Figure 9.9 is the velocity past the tube bank in, or closest to, the centre.

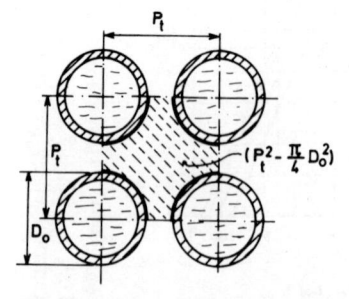

$$V_e = \dot{Q}/a_s \qquad (9.22)$$

Figure 9.11 Square pitch as described by equation (9.20).

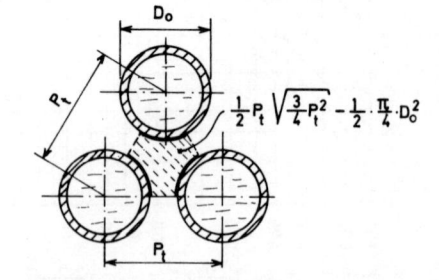

Figure 9.12 Triangular pitch as described by equation (9.21).

where $\dot{Q}$ is the fluid volume per unit time, m^3/s, and

$$a_s = \frac{D_s}{P_t}(P_t - D_0)B \tag{9.23}$$

with lengths as shown in Figures 9.10, 9.11, and 9.12.

The pressure drop is calculated by use of equation (9.24) with the friction factor f_s from Figure 9.9,

$$\Delta p_f = f_s \rho \frac{V_e^2}{2} \frac{D_s}{D_e}(N+1) \tag{9.24}$$

where N is the number of baffles.

Tube banks

For tube banks of smooth tubes (Figure 1.12 and 1.13), the Nusselt number is given by the Grimison equation,[10]

$$\text{Nu} = 0.32 f_A \text{Re}^{0.61} \text{Pr}^{0.31} \tag{9.25}$$

where $\text{Re} = \rho V_{max} D_0/\mu$ between 8000 and 40,000, D_0 is the outer tube diameter, and f_A is a factor depending on pitch, tube arrangement, and number of tubes. $f_A \approx 0.88$ for 4 tubes in the direction of flow and $P_T/D_0 \approx 2.0$, while $f_A \approx 1.0$ for more than 9 tubes in the direction of flow, for both staggered tubes and tubes in line. For other values of P_T/D_0 and for finned tubes, refer to the literature.[11]

Tube coils

Tube coils (Figure 9.13) have a heat transfer coefficient[12]

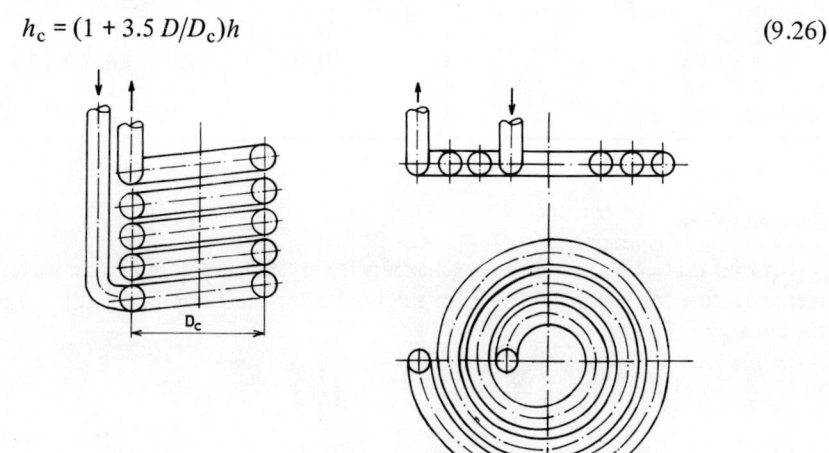

$$h_c = (1 + 3.5\, D/D_c)h \tag{9.26}$$

Figure 9.13 Tube coils: on the left a single helix and on the right a pancake coil.

where h is the heat transfer coefficient in a straight tube with the same inner diameter, D, and D_c is the diameter of the coil (Figure 9.13).

The heat transfer coefficient of spiral plate heat exchangers[13] (Figure 9.30) can be estimated by equation (9.26) with D as the hydraulic diameter of the channel.

Agitated vessels

For mechanically agitated vessels, equation (9.27) may be used,

$$\frac{h_j D_j}{k} = C \left(\frac{\rho n D_i^2}{\mu}\right)^{2/3} \left(\frac{c\mu}{k}\right)^{1/3} \left(\frac{\mu}{\mu_w}\right)^{0.14} \tag{9.27}$$

where h_j = heat transfer coefficient at the inner wall, W/(m²K)
  D_j = inner diameter of the vessel, m
  n = speed, revolutions per second
  D_i = agitator diameter, m
  k, ρ, c, μ, μ_w, see pages 344 and 345
  C is an empirical constant depending on the design.

For jacketed vessels the following values listed in Table 9.2 have been reported. For coils submerged in agitated vessels refer to the literature.[11,17,18]

Table 9.2 Some literature data for jacketed vessels

Agitator	Baffles	C	$Re = \frac{\rho n D_i^2}{\mu}$
Flat paddle[14]	None	0.36	$300 < Re < 5 \times 10^5$
Propeller,[15] pitch $h/D = 1.0$			
2 blades	4	0.47	
3 blades	4	0.50	$1700 < Re < 9.15 \times 10^5$
6 blades	4	0.58	
Six blade turbine[16]	4	0.73	$30 < Re < 4 \times 10^4$

Scraped tubes

Scraped tubes are used in heat exchangers for crystallization of certain waxes, for oleomargarine and shortening, and for sulphonization reactions. Skelland[19] suggests the equation

$$\frac{hD}{k} = 4.9 \left(\frac{\rho VD}{\mu}\right)^{0.57} \left(\frac{c\mu}{k}\right)^{0.47} \left(\frac{Dn}{V}\right)^{0.17} \left(\frac{D}{L}\right)^{0.37} \tag{9.28}$$

where D = inner tube diameter, m
  V = axial velocity of the liquid, m/s
  n = revolutions of the scraper, s^{-1}
  L = scraped length of the tube, m.

Granular beds

In granular beds of uniform spheres, the Nusselt number can be estimated by the equation,[20]

$$Nu = 1.03 \, e^{(1.1 - 2.75\epsilon)} Re^{0.6} \pm 8\% \tag{9.29}$$

where ϵ is the porosity = void fraction of the bed and $Re = \rho V_0 D/\mu$ with V_0 the superficial velocity referred to the cross section without spheres and D the sphere diameter.

Liquid films

The heat transfer coefficient h_f of a heated or cooled falling liquid film at a vertical wall (Figure 9.14) is measured to be[21]

$$\frac{h_f s}{k} = C \, Re^m \, Pr^{0.344} \left(\frac{\mu}{\mu_w}\right)^n \tag{9.30}$$

where

$$s = 0.3 \left(\frac{3\mu^2}{g\rho^2}\right)^{1/3} Re^{8/15}. \tag{9.31}$$

The hydraulic diameter of the falling film is $D_h = 4(Lx)/L = 4x$. With Γ kg liquid flowing per second per metre wall perpendicular to the direction of flow, the rate of flow is

$$V = \Gamma/\rho x \text{ m/s},$$

and Reynolds number,

$$Re = \frac{\rho(\Gamma/\rho x)4x}{\mu} = \frac{4\Gamma}{\mu}. \tag{9.32}$$

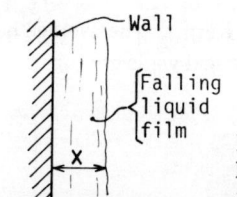

Figure 9.14 Falling liquid film at a vertical wall.

Table 9.3 The constants in equation (9.30) and limits of measurement

	C	m	n	Limits
Heating[22]	0.0614	8/15	0	$615 \, Pr^{-0.65} < Re < 400$
	0.00112	6/5	0	$400 < Re < 800$
	0.0066	14/15	0	$Re > 800$
Cooling[23]	0.00105	6/5	1.0	$Re < 800$
	0.00621	14/15	1.0	$Re < 800$

The constants in equation (9.30) are given in Table 9.3 for liquids not containing surface-active agents.

For heat transfer coefficients to the walls of packed tubes refer to VDI-Wärmeatlas,[11] for fluidized beds to Dahlhoff and Brachel[24] and Fritz,[25] and for other special cases to Perry.[42]

Condensation

Condensation takes place when a vapour comes into contact with a surface colder than the saturation temperature of the vapour. The resulting liquid, the condensate, will either produce a liquid film that flows down under the action of gravity, or it will form drops that grow until they are large enough to flow or drip down. The first mechanism is called film or filmwise condensation and the second dropwise condensation.

Dropwise condensation gives 4 to 8 times higher heat transfer coefficients than film condensation. However, dropwise condensation is mostly limited to water vapour. Special precautions may also have to be taken to ensure that the dropwise condensate remains.[26] In order not to underestimate the area necessary, almost all condenser calculations are based on the assumption of film condensation.

The following equations are based on Nusselt film theory,[27] i.e. the following assumptions are involved:

the condensate forms a film which falls with laminar flow due to gravity;

the heat delivered by the vapour is latent heat only;

the outer surface temperature of the condensate film is the saturation temperature of the vapour, i.e. thermodynamic equilibrium exists at the outer interface;

liquid density, heat capacity, viscosity, and thermal conductivity are constant across the condensate film.

Derivation of the equations is shown by Kern.[28] The average heat transfer coefficients in W/(m²K) for vertical tubes and vertical walls are,

$$h_{av} = 0.943 \left(\frac{k^3 \rho^2 \lambda g}{\mu L \Delta T_f} \right)^{1/4} \tag{9.33}$$

and for N horizontal tubes in a vertical row,

$$h_{av} = 0.725 \left(\frac{k^3 \rho^2 \lambda g}{\mu N^{2/3} D_0 \Delta T_f} \right)^{1/4} \tag{9.34}$$

where L = length of vertical tube, m
 D_0 = diameter of horizontal tube, m
 ΔT_f = temperature drop across condensate film, °C or K
 λ = heat of vaporization, J/kg

k, ρ, and μ (see pages 344 and 345) refer to the condensate. They are evaluated at the average film temperature, $T_f = T_{sa} - \Delta T_f/2$, where T_{sa} is the saturation temperature of the vapour.

In theory, the number N of tubes in a vertical row in the last equation should be to the first power. However, splashing of liquid as it drips over successive rows of tubes causes some increase in the average coefficient of heat transfer.

Condensation of superheated vapour can also be estimated by equations (9.33) and (9.34), if the surface temperature of the condenser is below the vapour saturation temperature. In this case, the heat of condensation in the equation, λ, is replaced by the difference between the enthalpy of the superheated vapour and that of the condensate at the saturation temperature of the vapour. The temperature difference in the equations is unchanged, equal to the temperature drop across the condensate film.

The presence of even small amounts of a non-condensable gas can seriously reduce the coefficient of heat transfer. A layer of the non-condensable gas will accumulate close to the cold surface, and the rate of condensation then depends on the diffusion rate of the vapour molecules through the layer of inert. In order to reduce the coefficient of heat transfer. A layer of the non-condensable gas will past the cold surfaces to sweep the inert gas to a part of the condenser where it can be removed.

Renker[29] suggested the half empirical equation

$$h = \frac{h_c}{1 + h_c/h_D} \qquad (9.35)$$

where h_c = heat transfer coefficient for pure vapour, W/(m²K)

$$h_D = 0.19 \frac{DP\lambda \, \text{Re}^{0.95} \, \text{Sc}^{0.92}}{(T-T_w) RT \, d} \left(\frac{p_v}{P-p_v}\right)^{0.38/\text{Sc}^{0.25}} \qquad (9.36)$$

where D = diffusivity of the gas vapour mixture, m²/s
P = total pressure of the mixture, Pa = N/m²
p_v = partial pressure of the vapour, Pa = N/m²
λ = heat of vaporization, J/kg
Re = Reynolds number, $\rho V d/\mu$
Sc = Schmidt number, $\mu/\rho D$
T = temperature of the gas vapour mixture, K
T_w = temperature of the condenser surface, K
R = the gas constant, 8314 J/(K kmol)
d = diameter of the condenser surface, m.

Boiling

In boiling liquids most of the heat is transferred to the liquid adjacent to the heat transfer surface. This gives a liquid temperature slightly above the saturation temperature, and most of the heat of vaporization is extracted from the liquid. Stroboscopic photographs reveal an increase in the volume of vapour bubbles from 100 to 4500% after the bubbles have left the heated surface,[30] i.e. most of the vaporization into the vapour bubbles takes place after the bubbles have started to rise.

Bubble formation is influenced by the finish of the surface and the wetting properties of the liquid. A high surface tension against the wall, b in Figure 9.15,

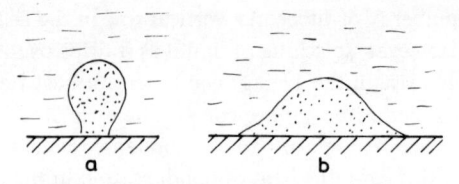

Figure 9.15 Bubble formation in boiling liquids with low (a) and high (b) surface tension against the wall.

gives vapour bubbles that have an insulating effect. To avoid this, a wetting agent may be added to certain liquids, such as magnesium in mercury boilers.

The vapour bubbles start at specific sites. A rough surface will have more sites where bubbles can start than a polished surface, and therefore have a higher coefficient of heat transfer. This is one reason for using evaporators with finned surfaces for some simple, clean liquids that do not tend to deposit scale, as the refrigerants R11 (trichlorofluoromethane) and R115 (chloropentafluoroethane). In most cases, however, smooth surfaces are preferred, to reduce the risk of scale formation and to make cleaning easier. In some cases, such as in evaporators for spent sulphite liquor, polished surfaces are used.

The heat transfer coefficient of a boiling liquid increases with increasing temperature difference up to a maximum value, h_{max}, at the critical temperature difference, $\Delta\theta_c$, and decreases again for higher temperature differences (Figure 9.16). The increase is due to the increased number of bubbles giving more interface between vapour and liquid, and increased agitation from the rising bubbles. The decrease at higher temperature differences is due to the insulating or blanketing effect of bubbles at the heat transfer surface. Figure 9.17 gives the order of magnitude of the critical

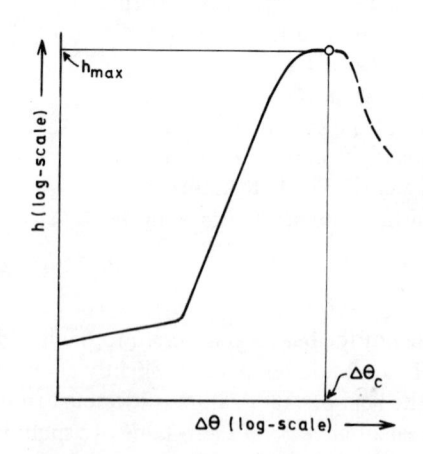

Figure 9.16 Heat transfer coefficient of a boiling liquid as a function of the temperature difference.

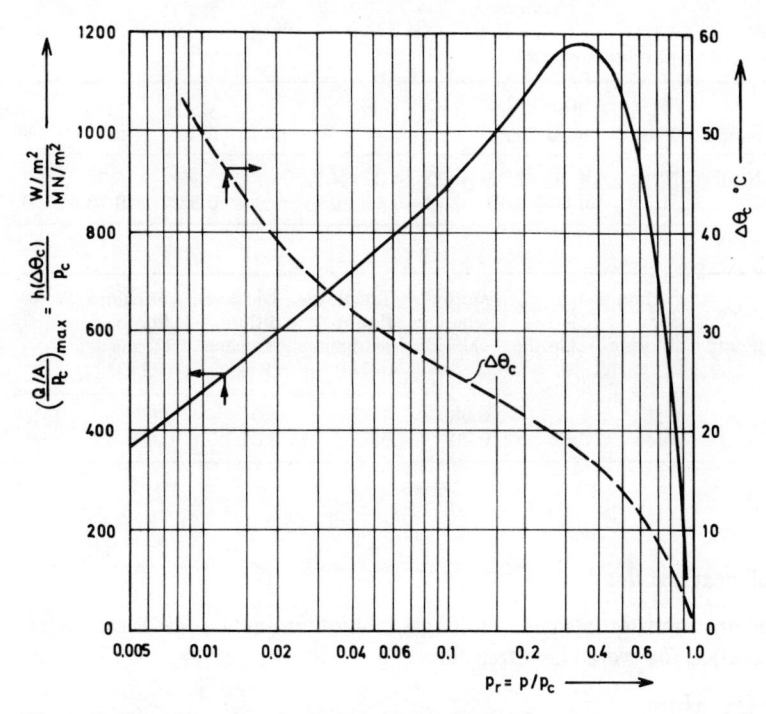

Figure 9.17 Heat flux by boiling with the critical temperature difference, $(\dot{Q}/A)_{max}$ in W/m² divided by the critical pressure p_c in MN/m², and the critical temperature difference, $\Delta\theta_c$, both plotted as functions of the reduced pressure, $p_r = p/p_c$.[31]

temperature difference, $\Delta\theta_c$, as a function of the ratio of the pressure to the critical pressure of the compound.

Industrial evaporators should be designed for temperature differences between the wall and the boiling liquid well below the critical temperature difference $\Delta\theta_c$. Kern[28] recommends that the heat flux, $q = h\Delta\theta$, be kept below 95,000 W/m² for dilute aqueous solutions and below 65,000 W/m² for organic liquids in forced convection and 40,000 W/m² in free convection.

At present, no simple, general equation is available for calculating the heat transfer coefficient for boiling liquids. Using the form suggested by Gorenflo,[32] the following equation gives the order of magnitude of the heat transfer coefficient for pure compounds at horizontal tubes,

$$h = C(1 + 15.7\ p/p_c)\dot{q}^{0.75} \quad \text{W}/(\text{°C m}^2) \tag{9.37}$$

where $\dot{q}$ is the heat flux, $h\Delta\theta$, W/m², p/p_c is the reduced vapour pressure, and C is an empirical constant depending on physical properties of the compound. Approximate values of C based on data from references 12 and 32 are given in Table 9.4. Data for a few binary mixtures are available in the literature.[11]

Table 9.4 Constant C in equation (9.37) for boiling liquids and critical pressure p_c for some pure compounds.

Compound	Water	Propane	Pentane	Benzene	Toluene	Acetone	Methanol	Ethanol	Sulphurdioxide
p_c MN/m²	22.1	4.56	3.37	4.92	4.22	4.72	7.95	6.38	7.88
C	1.3	0.80	0.50	0.52	0.41	0.71	0.83	0.70	0.97

Compound	Carbon tetra chloride	Methyl chloride	Methylene chloride	Trichlorofluoromethane (R11)	Dichlorodifluoromethane (R12)	Trichlorofluoroethane (R113)
p_c MN/m²	4.56	6.68	6.08	4.32	4.01	3.37
C	0.44	1.06	0.70	0.65	0.65	0.45

Overall heat transfer

The heat transfer from a liquid or gas to another liquid or gas is calculated with the equation for overall heat transfer,

$$Q = AU\Delta\theta \qquad (9.38)$$

where $\Delta\theta$ is the total temperature difference between one fluid and the other, $\Delta\theta = \theta_1 - \theta_2$, K, A is the area perpendicular to the heat flux, m², and U is the overall heat transfer coefficient, W/(m²K).

To determine the overall heat transfer coefficient, U, equation (9.14) for heat transfer and equation (9.1) for heat conduction for each surface and layer in Figure 9.18, are solved for the temperature differences that can be summarized as,

$$\theta_1 - \theta_a = (\dot{Q}/A)(1/h_1)$$
$$\theta_a - \theta_b = (\dot{Q}/A)(L_1/k_1)$$
$$\theta_b - \theta_c = (\dot{Q}/A)(L_2/k_2)$$
$$\theta_c - \theta_2 = (\dot{Q}/A)(1/h_2)$$

$$\theta_1 - \theta_2 = \frac{\dot{Q}}{A}\left(\frac{1}{h_1} + \frac{L_1}{k_1} + \frac{L_2}{k_2} + \frac{1}{h_2}\right).$$

Combination with equation (9.38) where $\Delta\theta = \theta_1 - \theta_2$ gives the overall resistance to heat transfer,

$$\frac{1}{U} = \Sigma\frac{1}{h} + \Sigma\frac{L}{k} \quad °C\,m^2/W \quad \text{or} \quad K\,m^2/W. \qquad (9.39)$$

This equation can be used to calculate the overall heat transfer coefficient, U W/(m²K).

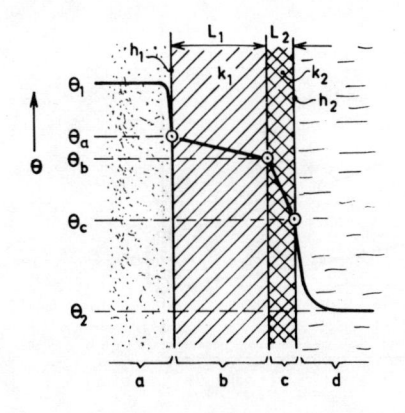

Figure 9.18 Heat transfer from condensing vapour a, through metal wall b, and deposit c, to cooling water d.

For heat transfer through a tube wall, the overall heat transfer coefficient can be referred to the outer, the average, or the inner tube surface. In general it is convenient to refer the coefficient to the surface with the lowest heat transfer coefficient. Referred to the inner diameter D_i, equation (9.38) can be rewritten,

$$\dot{Q} = A_i U_i \Delta\theta \tag{9.40}$$

and equation (9.39),

$$\frac{1}{U_i} = \frac{1}{h_i} + \frac{1}{h_o(D_o/D_i)} + \frac{D_o - D_i}{k(D_o + D_i)/D_i} \tag{9.41}$$

where k is the thermal conductivity of the wall and the suffixes i and o refer to the inner and the outer tube surfaces.

Temperature difference

In most heat exchangers one or both media vary in temperature as they pass through the unit, for example as shown in Figure 9.19.

The heat transferred is

$$\dot{Q} = AU\Delta\theta_m \tag{9.42}$$

where $\Delta\theta_m$ is the average or mean temperature difference. The ratio of the differentials (Figure 9.19) equals the ratio of the total changes in temperature,

$$\frac{d(\Delta\theta_1)}{d(\Delta\theta_2)} = \frac{\theta_1 - \theta_3}{\theta_2 - \theta_4}, \quad d(\Delta\theta_2) = \frac{\theta_2 - \theta_4}{\theta_1 - \theta_3} d(\Delta\theta_1).$$

The heat transferred through the area dA is,

$$d\dot{Q} = -w_1 c_1 d(\Delta\theta_1), \quad d(\Delta\theta_1) = -\frac{d\dot{Q}}{w_1 c_1}.$$

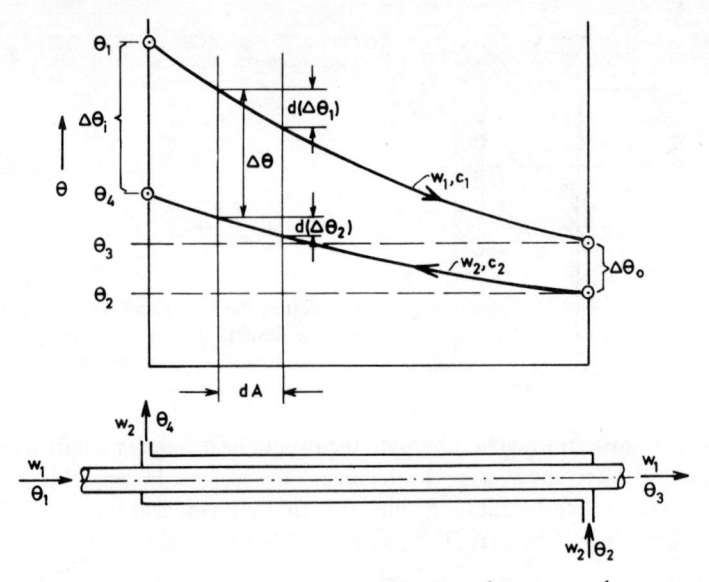

Figure 9.19 Double-pipe heat exchanger where w_1 and w_2 are amounts of liquid in kg per second and c_1 and c_2 the specific heat capacities in kJ/(kg K). The upper diagram shows the liquid temperature variation through the heat exchanger.

With these values for $d(\Delta\theta_1)$ and $d(\Delta\theta_2)$ inserted into the equation $d(\Delta\theta) = d(\Delta\theta_1) + d(\Delta\theta_2)$

$$d(\Delta\theta) = \frac{-\theta_1 + \theta_3 - \theta_2 + \theta_4}{w_1 c_1 (\theta_1 - \theta_3)} d\dot{Q} = \frac{\Delta\theta_o - \Delta\theta_i}{\dot{Q}} d\dot{Q}$$

$$d\dot{Q} = \frac{\dot{Q}}{\Delta\theta_o - \Delta\theta_i} d(\Delta\theta) = U \, dA \, \Delta\theta$$

$$\int_{\Delta\theta_i}^{\Delta\theta_o} \frac{d(\Delta\theta)}{\Delta\theta} = U \frac{\Delta\theta_o - \Delta\theta_i}{\dot{Q}} \int_0^A dA$$

$$\dot{Q} = AU \frac{\Delta\theta_i - \Delta\theta_o}{\ln(\Delta\theta_i / \Delta\theta_o)} .$$

This equation is identical to equation (9.42), i.e.

$$\Delta\theta_m = \frac{\Delta\theta_i - \Delta\theta_o}{\ln(\Delta\theta_i / \Delta\theta_o)} \tag{9.43}$$

which is the logarithmic mean temperature difference.

If $1 < \Delta\theta_i / \Delta\theta_o < 1.4$ the arithmetic mean temperature difference can be used giving an error less than 1%,

$$\Delta\theta_m = \frac{\Delta\theta_i + \Delta\theta_o}{2} . \tag{9.44}$$

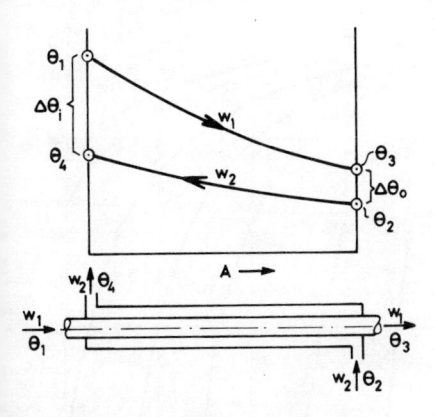

Figure 9.20 Counter-current heat exchanger: the upper diagram gives the temperatures.

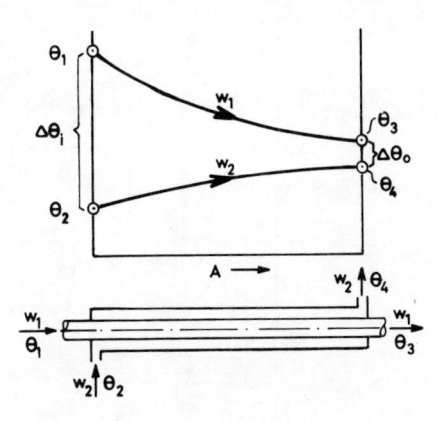

Figure 9.21 Co-current heat exchanger: the upper diagram gives the temperatures.

Equation (9.43) only applies when the streams have constant specific heat capacity, the overall heat transfer coefficient is constant over the whole heat transfer area, and either one stream has constant temperature (phase change) or it is counter-current or co-current flow (Figures 9.20 and 9.21).

Counter current flow gives the highest rate of heat transfer for a particular heat exchanger, and it should be preferred if there are no special reasons for other arrangements.

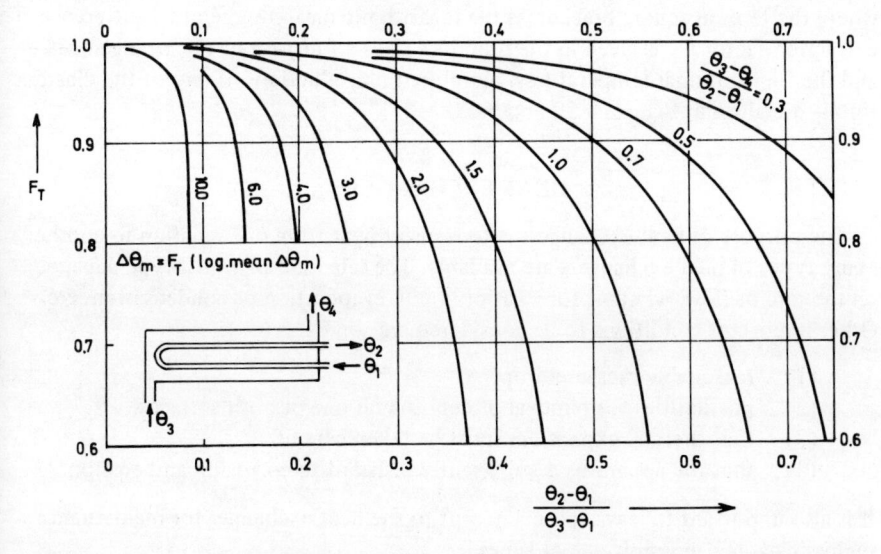

Figure 9.22 Correction factor F_T in equation (9.45) for the mean temperature difference for a heat exchanger with one outer pass and two or more inner passes. Reproduced by permission of Tubular Exchanger Manufacturers Association.

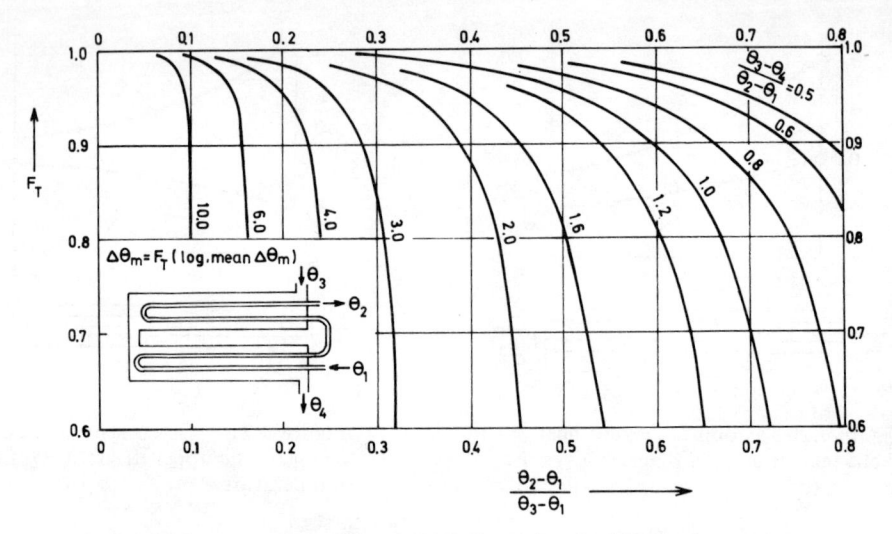

Figure 9.23 Correction factor F_T in equation (9.45) for the mean temperature difference for a heat exchanger with two outer passes and two or more inner passes. Reproduced by permission of Tubular Exchanger Manufacturers Association.

In mixed flow, the average or mean temperature difference is,

$$\Delta\theta_m = F_T \left[\frac{\Delta\theta_i - \Delta\theta_o}{\ln(\Delta\theta_i/\Delta\theta_o)} \right] \tag{9.45}$$

where the term in square brackets is the logarithmic mean temperature difference. The correction factor F_T is given in the literature[33] as a function of the arrangement and the four terminal temperatures. Simplified reproductions of two of the diagrams for F_T are given in Figures 9.22 and 9.23.

HEAT EXCHANGERS

The purpose of heat exchangers is to transfer heat from one medium to another. Many types of heat exchangers are available. The selection of which type depends on the rate of flow, whether for a gas or liquid, evaporation or condensation, etc. Other important conditions to be considered are

(1) reasonable pressure drop,
(2) possibilities for removal of deposits on one or both surfaces,
(3) that thermal expansions must be taken care of,
(4) that the apparatus is sufficiently resistant to corrosion and erosion.

It is also important to leave space adjacent to the heat exchanger for maintenance, such as cleaning and replacing of tubes.

Fouling or scaling on heat exchanger surfaces can be a serious problem in many chemical process industries. High liquid velocities ($V > 1$ m/s) reduce the fouling,

Table 9.5 Fouling resistances in heat exchangers, $(L/k)_f$ °C m^2/W. Reproduced by permission of Tubular Exchange Manufacturers Association.

Liquid, $V > 0.9$ m/s	
Distilled water and sea water, $\theta < 50°$C	0.0001
Brackish water	0.0002
River water (clean) and city water, $\Theta < 50°$C	0.0001
Water from cooling tower, $\Theta < 50°$C	0.0005
Cooling agents, Dowtherm, petrol, solvents, lubricating oils	0.0002
Light gas oil, diethanolamine solutions	0.0004
Vegetable oils	0.0005
Fuel oil	0.001
Gas and vapour	
Steam, oil-free	0.0001
Water vapour, exhaust from lubricated engines	0.0002
Dowtherm, solvents, stable organic vapours from distillation columns, natural gas	0.0002
Refrigerants from lubricated engines, compressed air, vapour from vacuum distillation and cracking	0.0004
Exhaust from diesel engines, manufactured gas	0.002

but it is often impossible to avoid scale formation on one of the surfaces. Increased thickness of the scale reduces the heat transfer, and periodic cleaning can be necessary. It is also recommended that a 'fouling factor' always be included in the design value of the overall heat transfer coefficient, U, or in the overall heat transfer resistance, $1/U$. Examples of resistances suggested by manufacturers of tubular heat exchangers are given in Table 9.5. For more complete data refer to the source of the data.[33]

Figures 9.24 to 9.35 and 9.38 to 9.40 show schematically the most common types of heat exchangers.

Double-pipe heat exchangers (Figure 9.24) are well suited for counter-current flow, and high velocities can easily be arranged for both fluids.

Cascade coolers (Figure 9.25) can be used as condensers and gas coolers. For corrosive fluids it is possible to use special materials for the tubes, such as graphite

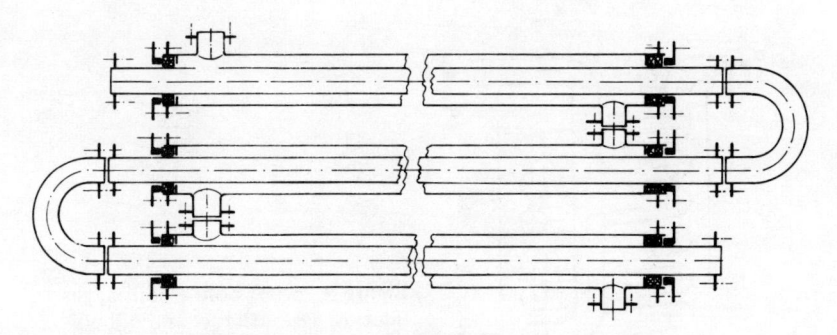

Figure 9.24 Double-pipe heat exchanger.

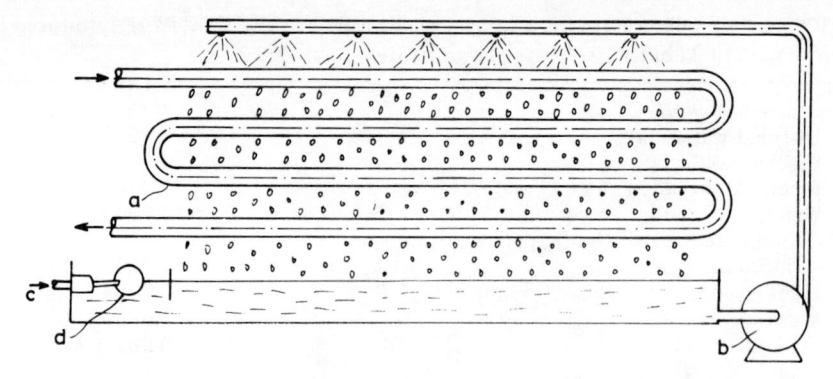

Figure 9.25 Cascade cooler a, with circulation pump b, for cooling water that drips over the tubes. The water distributor can be a trough. d is a float valve that opens for make-up water, c.

and glass. Cascade coolers with recirculation of cooling water are used extensively for smaller refrigeration plants in locations where cooling water is expensive. The water is cooled as some of it evaporates when it flows over and drips from the pipes. For heat transfer coefficients refer to the literature.[28]

Cooling coils (Figure 9.26) are mainly used for cooling of high-pressure gas when it is important to detect leaks which can be seen as gas bubbles in the cooling water.

Shell and tube heat exchangers have a wide variety of applications. They are used both for liquids and gases, and as evaporators and condensers. Two of the six designs described in TEMA's standards[33] are shown in Figures 9.27 and 9.28.

Plate heat exchangers consist of plates similar to a filter press. They have the advantages that they can be taken apart and cleaned on both sides, and that the number of plates and therefore the heat transfer area can be changed. The plates are rippled to increase the turbulence. At large liquid flow rates the exchanger can be operated with streams in parallel as indicated in Figure 9.29. Plate heat exchangers are used extensively in breweries and dairies. For details refer to Alfa Laval's handbook.[34]

Spiral plate heat exchangers can also be cleaned on both sides, as each cover gives

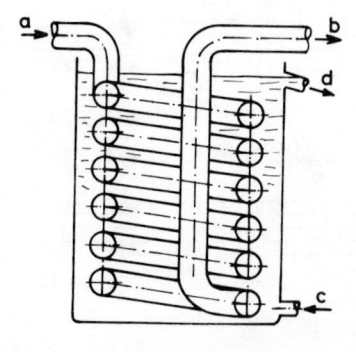

Figure 9.26 Cooling coil: a, gas inlet; b, gas outlet; c, and d, cooling water inlet and outlet.

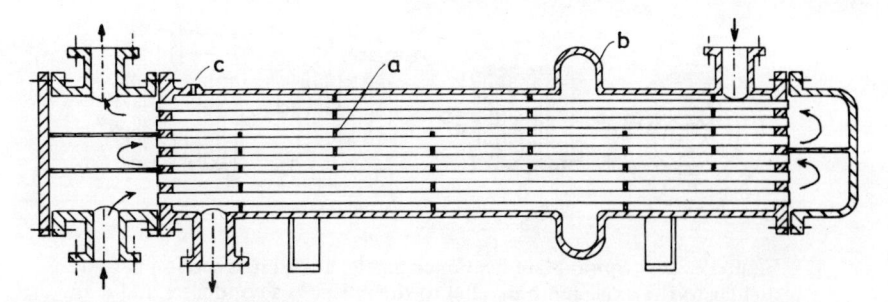

Figure 9.27 Shell and tube heat exchanger with channel type head to the left and bonnet type to the right. The channel type with removable cover makes it possible to clean the tubes without disconnecting the outside tube connections. a, baffles; b, bellow to take up thermal expansion; c, vent.

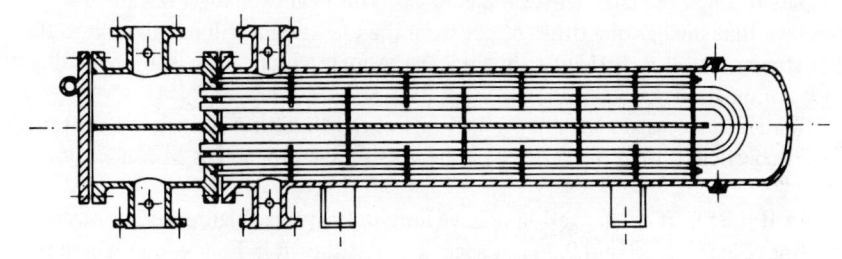

Figure 9.28 Shell and tube heat exchanger with hairpin tubes, suitable for counter-current flow.

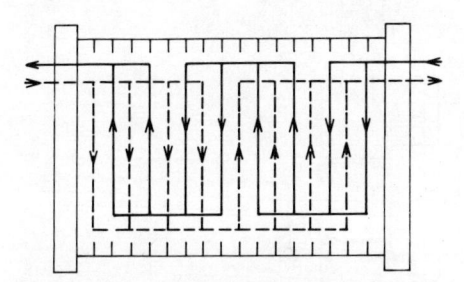

Figure 9.29 Diagram for a 4 x 2/2 x 4 plate heat exchanger, i.e. the cold liquid (solid lines) has 4 passes, each with 2 channels in parallel, and the hot liquid (broken lines) has 2 passes, each pass with 4 channels in parallel.

access to the channels for one fluid. Figure 9.30, a and b, shows the most common design for counter-current flow; c is a condenser, and d an evaporator.[13,34] It is a compact heat exchanger for use at moderate pressures. Stainless steel spiral plate heat exchangers are used extensively in the wood processing industry. Newer designs also include plastic-coated plates.

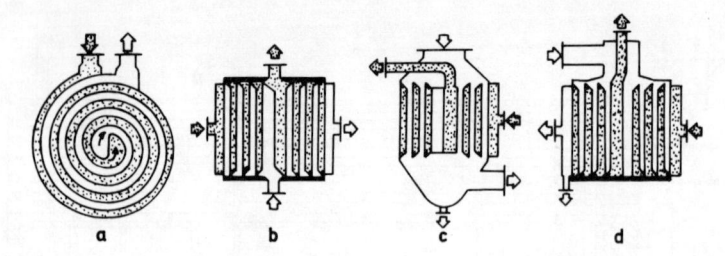

Figure 9.30 Spiral plate heat exchangers. a is a cross-section perpendicular to the axis, and b parallel to the axis. c is a condenser and d an evaporator.

Ljungstrom air preheater is a rotating, regenerative heat exchanger (Figure 9.31). The radial metal walls are heated while in contact with the hot gas and cooled when they pass through the channel for the cold gas. This heat exchanger has the disadvantage that smaller quantities of gas from the one stream will be mixed into the other stream, and it is difficult to avoid leaks. A major application is in preheating of air for combustion.

Extended surfaces are advantageous if one of the fluids has a much lower heat transfer coefficient than the other, as in heating of a gas by means of condensing steam.

Finned tube heat exchangers may have longitudinal fins (Figure 9.32) or transverse fins (Figures 9.33 and 9.34). A special case is shown in Figure 9.35 where the wall of a tank acts as longitudinal fins for tubes welded to the outside of the tank. This arrangement has the advantage of easy cleaning of the tank.

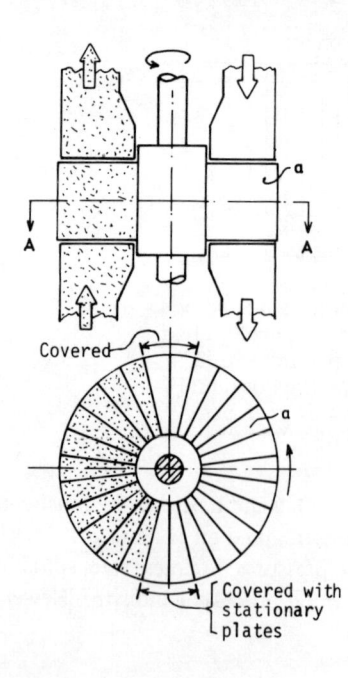

Figure 9.31 Ljungstrom heat exchanger: the segments a in the slowly rotating wheel are heated by hot gas to the left and cooled by cold gas to the right.

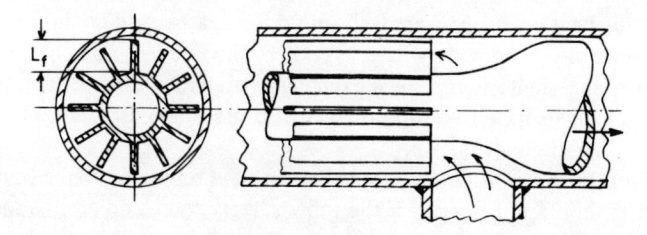

Figure 9.32 Double-pipe heat exchanger with longitudinal fins in the annular space.

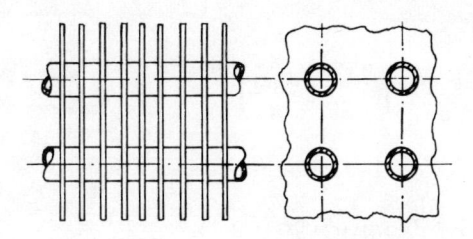

Figure 9.33 Part of a finned tube element with continuous fins.

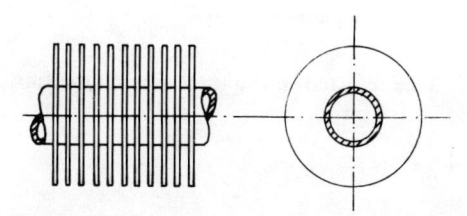

Figure 9.34 Finned tube with transverse fins.

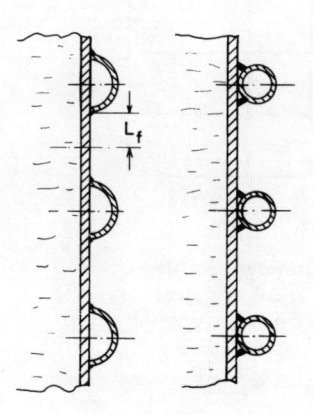

Figure 9.35 Tubes welded to the tank wall: half tubes on the left and whole tubes on the right.

Finned tube heat exchangers are well suited for heating and cooling air. In recent years, they have also been widely used as coolers and condensers in oil refineries where water cooled shell and tube heat exchangers used to dominate the market. These air cooled exchangers have two to four rows of tubes in the direction of the air flow and use forced air circulation.

Aluminium fins spiral-wound or forced on to steel tubes are used extensively. Such elements should not be used if the tube is likely to be cooled to a temperature that makes the fins loosen (Figure 9.36).[35]

The heat transfer coefficient to the longitudinal fins in Figure 9.32 is calculated as for flow in tubes with the hydraulic diameter of the cross section between the fins. The heat transfer coefficient for flow vertical to finned tube banks is given by the equation[36]

$$\frac{hD_o}{k} = C\left(\frac{\rho V_{max} D_o}{\mu}\right)^{0.625} \left(\frac{A_{tot}}{A_o}\right)^{-0.375} Pr^{1/3} \qquad (9.46)$$

where

$C = 0.45$ for staggered tubes
$C = 0.30$ for tubes in line
D_o = outer diameter of the tubes, m
V_{max} = maximum velocity through the tube bank, m/s
A_{tot} = total outside surface area (including fins)
A_o = outer surface area without fins.

The heat transfer resistance referred to the total area on the finned side is,

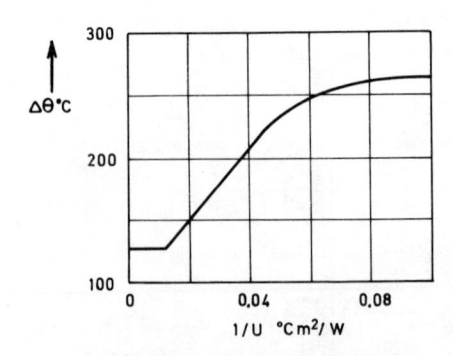

$$\frac{1}{U} = \frac{1}{h_a} + \frac{A_{tot}}{A_i}\left(\frac{1}{h_i} + \frac{r_o - r_i}{k_t}\right) \qquad (9.47)$$

Figure 9.36 Maximum permissible temperature difference, $\Delta\theta$, for steel tubes with aluminium fins as a function of the heat transfer resistance, $1/U$.[35]

where

A_i = inner surface area of the tube, m^2
h_i = heat transfer coefficient on the inner side, W/(m^2 K)
$r_o - r_i$ = wall thickness of the tube, m
k_t = thermal conductivity of the tube, W/(m K)
h_a = artificial heat transfer coefficient which takes into account the temperature drop through the fins, W/(m^2 K).

The fin efficiency η_f is the ratio of the mean temperature difference between the fins and the surrounding fluid and the same temperature difference at the base of the fins,

$$\eta_f = \frac{(T_{fin} - T_{ambient})_{mean}}{T_{base\ fin} - T_{ambient}}.$$

The corresponding value of the artificial heat transfer coefficient in equation (9.46) is given by

$$h_a A_{tot} = h(A_{tot} - A_f) + h\eta_f A_f$$

or

$$h_a = h[1 - (1 - \eta_f)A_f/A_{tot}] \qquad (9.48)$$

where A_f is the fin surface area and A_{tot} is the surface area of fins and of the bare tubes between the fins.

Derivation of equations for fin efficiency is given in most books on heat transfer, and only the results are given here:

$$\eta_f = \frac{\tanh \varphi}{\varphi} = \frac{e^\varphi - e^{-\varphi}}{e^\varphi + e^{-\varphi}} \frac{1}{\varphi} \qquad (9.49)$$

where

$$\varphi = L_f \sqrt{2h_f/k_f s_f} \qquad (9.50)$$

for fins of constant cross section perpendicular to the heat flux through the fin, as in the longitudinal fins of Figure 9.32, where L_f is the height of the fin, m, and

$$\varphi \approx [(B_f/D_o)^a - b]D_o\sqrt{h_f(2k_f s_f)} \qquad (9.51)$$

for fins perpendicular to tubes (Figure 9.33 and 9.34), where

h_f = heat transfer coefficient at the fin side, W/(m^{2}K)
s_f = thickness of fin, m
k_f = thermal conductivity of fin, W/(m K)
D_o = outer tube diameter, m
B_f and the constants a and b are given in Figure 9.37.

The equations are derived for fins with the same heat transfer coefficient on both sides of the fins. However, equation (9.49) is also valid when there are different

a = 1.17
b = 1.0

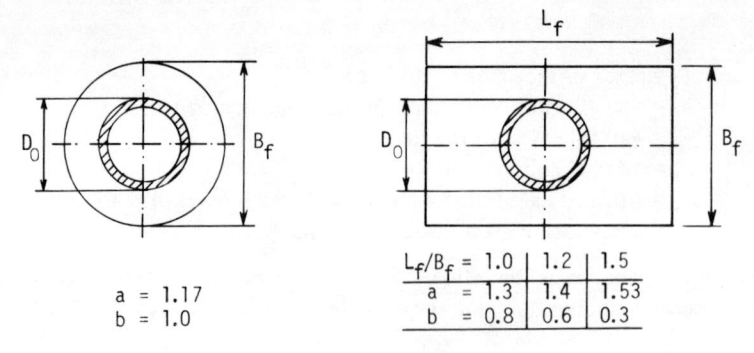

L_f/B_f =	1.0	1.2	1.5
a =	1.3	1.4	1.53
b =	0.8	0.6	0.3

Figure 9.37 Constants a and b in equation (9.51) for circular fins (left) with $B_f/D_0 < 3.5$, and for rectangular fins (right) with $0 < B_f/D_0 < 4$.

heat transfer coefficients on the two sides if h_f in equations (9.50) and (9.51) is replaced by the arithmetic average of the heat transfer coefficients on the two sides, $(h_{f_1} + h_{f_2})/2$. For a tank wall (Figure 9.37) the heat transfer from the outside will usually be small, and h_f in equation (9.50) can be replace by $h_{f_1}/2$ where h_{f_1} is the heat transfer coefficient of the liquid in the tank. Corrections for varying thickness of the fins and equations for needle-shaped fins are given in VDI-Wärmeatlas.[11]

Direct contact between heat exchanging fluids is used occasionally if one fluid is a gas or vapour and the other a liquid, or if the two fluids are liquids which are immiscible.

The barometric condenser (Figure 4.51) is an example of an apparatus allowing direct contact between vapour and liquid. One design requirement for direct contact condensers is that the temperature difference between the condensing vapour and the water leaving the condenser should be approximately 3°C,[28,37] but data for heat and mass transfer in direct contact condensers are scarce.[38]

Other examples of direct contact heat and mass transfer are spray dryers (Figure 6.50) and concentration towers for sulphuric acid in which the acid flows counter-current to hot flue gases. Direct liquid–liquid contact[7] is also used in a process for sea water desalination.

Secondary heat transfer media are used to give a simpler design, to reduce any possible danger of local overheating, or to reduce the harmful effects of possible leaks. Heat transfer oils can be used in the temperature range from −10 to 320°C, at temperatures above 300°C under a slight nitrogen overpressure.[39] Inhibited ethylene glycol-containing liquids are recommended in the temperature ranges from −54 to 177°C and 20 to 400°C, and heat transfer salts from 155 to 540°C.

For heating, condensing water vapour is a convenient secondary heat transfer medium in the temperature range from 100 to about 200°C, and in some cases for higher temperatures as well. In the range from 200 to 260°C evaporating and condensing liquids based on *o*-chlorobenzene can be used, and from 200 to 350°C a mixture of 26.5% biphenyl and 73.5% diphenyl oxide with melting point 10–12°C and normal boiling point 258°C [Trade marks 'Dowtherm' (The Dow Chem. Co.),

Table 9.6   Data for a 26.5% biphenyl + 73.5% diphenyl oxide mixture.

Temperature, °C		200	250	300	350	400
Vapour pressure,	bar	0.24	0.87	2.64	5.44	10.43
Heat of vaporization,	kJ/kg	322	304	279	252	219
Density, liquid,	kg/m^3	910	861	809	750	681
Density, vapour,	kg/m^3	1.03	3.40	8.85	19.6	39.4

'Thermex' (ICI Ltd.), 'Santotherm 66' (Monsanto)]. Some characteristic data are given in Table 9.6.

For cooling, water is a convenient secondary heat transfer medium in the temperature range from 0 to 100°C, and aqueous solutions of sodium chloride, calcium chloride, ethylene glycol, propylene glycol, ethanol, and methanol can be used for temperatures below 0°C.

Evaporation

The object of evaporation is usually to remove water by vaporization from an aqueous solution containing a non-volatile solute that either becomes a more concentrated solution, or crystals form in suspension or at the walls. In evaporation some special problems may arise. Density and viscosity increase as the concentration increases, and crystals must be removed before the tubes become clogged. It may also be necessary to keep supersaturation below that level at which crystals form at and stick to surfaces, for example, on a propeller used for forced circulation over the heat transfer surfaces.

Scales depositing on the heating surfaces are another problem, for example calcium sulphate precipitating during evaporation of spent sulphite liquor. One possible solution is an arrangement of valves that allow transfer, at intervals, of the evaporating liquid and the condensing steam to opposite sides of the heat transfer tubes. In this way, the condensate may wash off scales formed in the previous interval.

Foaming can give serious loss of liquid with exhaust vapour. Organic substances in particular can give trouble, and it may be necessary to add surface active agents to break down stable foams.

Entrainment of drops from the bubbles when they burst is another problem in evaporation, and some kind of demister is often required. Cyclones will often do the job, but knitted wire mesh demisters (chapter 6) have a higher efficiency as long as they do not become fouled by solids in the liquor.

Heat sensitive materials can be evaporated under vacuum or in film evaporators with short residence times. Figure 9.38 shows a film evaporator used for concentration of gelatine, antibiotics, fruit juice, etc. The upper part of the rotating vanes, h, force entrained material to the wall where it flows down, while the lower part of the vanes, i, maintains a thin liquid film. Heat transfer coefficients for this type of film evaporator are given by Hauschild.[40]

Figures 9.39 and 9.40 show two common types of evaporators. The callandria

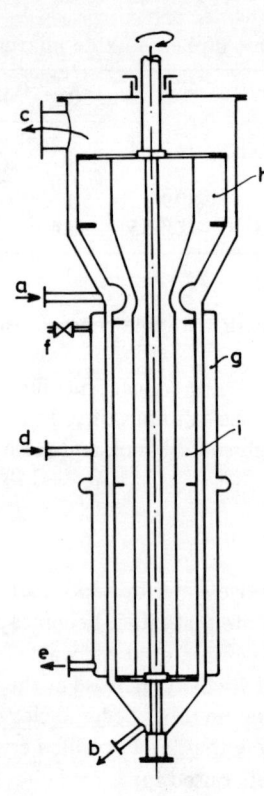

Figure 9.38 Luwa film evaporator: a, feed; b, concentrate; c, exhaust vapour; d, steam; e, condensate; f, vent; g, steam jacket; h and i, rotating vanes.

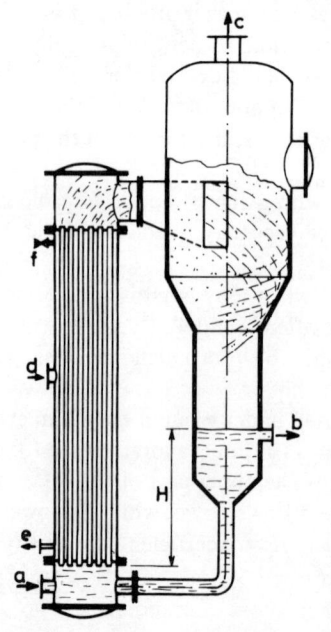

Figure 9.39 Recirculating long-tube vertical evaporator (Kestner evaporator). The symbols are the same as in Figure 9.38.

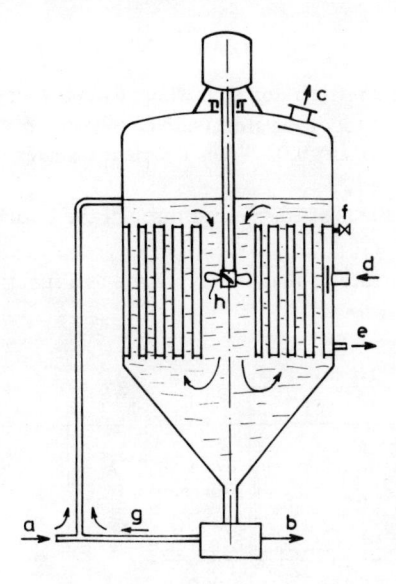

Figure 9.40   Propeller callandria evaporator: a, solution; b, coarse crystals; c, exhaust vapour; d, steam; e, condensate; f, vent; g, recycled fine crystals; h, propeller.

evaporator in Figure 9.40 is a crystallizer in which fine crystals are returned with the stream g to provide the seeds necessary to avoid crystallization on the propeller, h.

For calculation of the heat transfer area in an evaporator, the increase in boiling temperature due to the solute and due to the static head must be taken into account.

The static head corresponds to the weight of liquid above the heat transfer surface. Vertical evaporator tubes will contain a mixture of liquid and vapour, and the highest rate of heat transfer is obtained with a liquid level H in Figure 9.39, considerably less than the height of the tubes. Figure 9.41 shows experimental results where the highest overall heat transfer coefficient was obtained with a static head corresponding to 1/4 of the height of the tubes.[41]

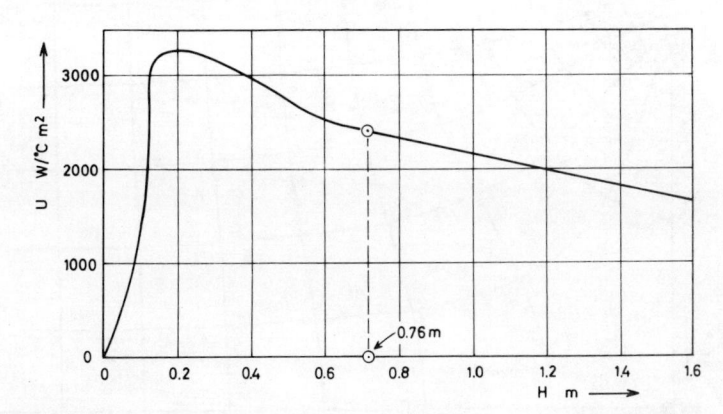

Figure 9.41   Overall heat transfer coefficient, U, in vertical 0.76 m long tubes plotted as a function of the liquid level, H in Figure 9.41 in the downcomer.

EXAMPLE 9.1. GRAPHICAL SOLUTION OF HEAT CONDUCTION

Figure 9.42 shows the insulated hull of a refrigerated ship where θ_1 is the air temperature in the hull and θ_0 the temperature of the steel (side of the ship and ribs). Thermal conductivity of the insulation, $k = 0.04$ W/(m°C). Heat transfer coefficient to the air, $h_1 = 6$ W/(m²°C).

(a) Estimate the percentage increase in the heat transferred through the insulation, caused by the ribs.

(b) Estimate the heat transfer per m² surface of the hull, expressed as a function of the temperature difference, $\theta_0 - \theta_1$.

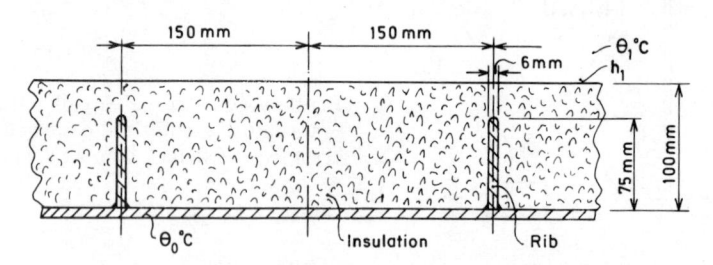

Figure 9.42 Insulated ship's hull with ribs in the insulation.

Solution

(a) The calculations are made for a section from the centre of a rib to the middle line between two ribs, and the heat transfer resistance at the surface, $1/h_1$, is replaced

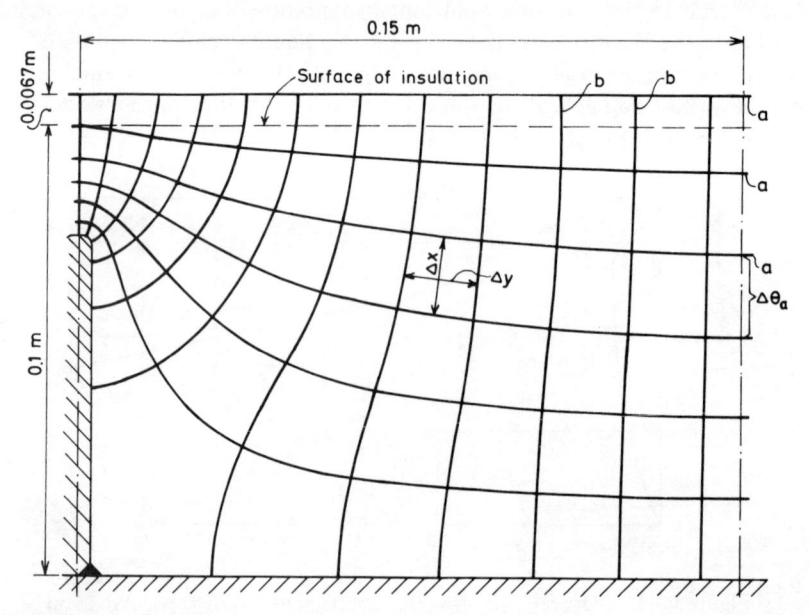

Figure 9.43 The insulation of Figure 9.42 with the superficial layer $s_1 = 0.0067$ m and isotherms a and flux lines b.

by that of a superficial layer of insulation with thickness,

$$s_1 = k/h_1 = 0.04/6$$
$$= 0.0067 \text{ m}.$$

This thickness is plotted in Figure 9.43. In the same figure flux lines and isotherms are drawn, such that $\Delta x/\Delta y = 1.0$ (Flux lines are always perpendicular to isotherms.)

Figure 9.43 contains 7 isotherms, a (6 temperature intervals), and about 11.4 intervals between the flux lines, b. With 6 temperature intervals and without ribs, the distance between the flux lines would have been the same as for the isotherms, $0.1067/6 = 0.0178$ metre, and the number of intervals between flux lines, $0.15/0.0178 = 8.4$. With $\Delta x/\Delta y =$ constant, the heat transferred is proportional to the number of intervals between the flux lines, i.e. the increase in heat transfer due to the ribs is,

$$100(11.4-8.4)/8.4 = 36\%$$

(b) One m² surface with width 0.15 m has a length $L = 1/0.15 = 6.67$ m. The heat transfer between two flux lines, equation (9.7), is

$$\Delta \dot{q} = (0.04)(6.67)(\theta_0 - \theta_1)/6 = 0.0445(\theta_0 - \theta_1)$$

With 11.4 intervals between flux lines, the heat transfer per m² is,

$$\dot{q} = (11.4)(0.0445)(\theta_0 - \theta_1)$$
$$= 0.51(\theta_0 - \theta_1) \text{ W/m}^2.$$

EXAMPLE 9.2. GRAPHICAL DETERMINATION OF THE SURFACE TEMPERATURE IN A CORNER

A brick building has 0.3 m thick walls with thermal conductivity $k = 0.7$ W/(m°C). The heat transfer coefficient at the inside is $h_i = 5.5$ W/(m²°C), and at the outside $h_0 = 17$ W/(m²°C). The air temperature at the outside is 0°C, and in the room +16°C. Estimate maximum relative humidity in the room without condensation in the corner.

Note The diagram in Figure 9.44 is a general diagram for corners, including heat transfer resistance on the two sides.

Solution

Heat transfer resistance at the inside is equivalent to

$$k/h_i = 0.7/5.5 = 0.127 \text{ m wall}.$$

Heat transfer resistance at the outside is equivalent to

$$k/h_0 = 0.7/17 = 0.041 \text{ m wall}.$$

Thickness of the brick wall, 0.300 m wall.

Total heat transfer resistances are equivalent to 0.468 m wall.

A thickness of 0.468 m corresponds to the distance a in Figure 9.44, i.e. the inner surface is $(0.127/0.468)a = 0.27a$ from the inner isotherm. This gives the broken line through A as the inner surface of the wall in Figure 9.44. A finer grid of flux lines and isotherms in the corner gives the temperature at point A approximately 3.3 intervals from the inner isotherm, i.e. the temperature in the corner is

$$\theta_A = 16 - (3.3)(16)/6 = 7.2°C.$$

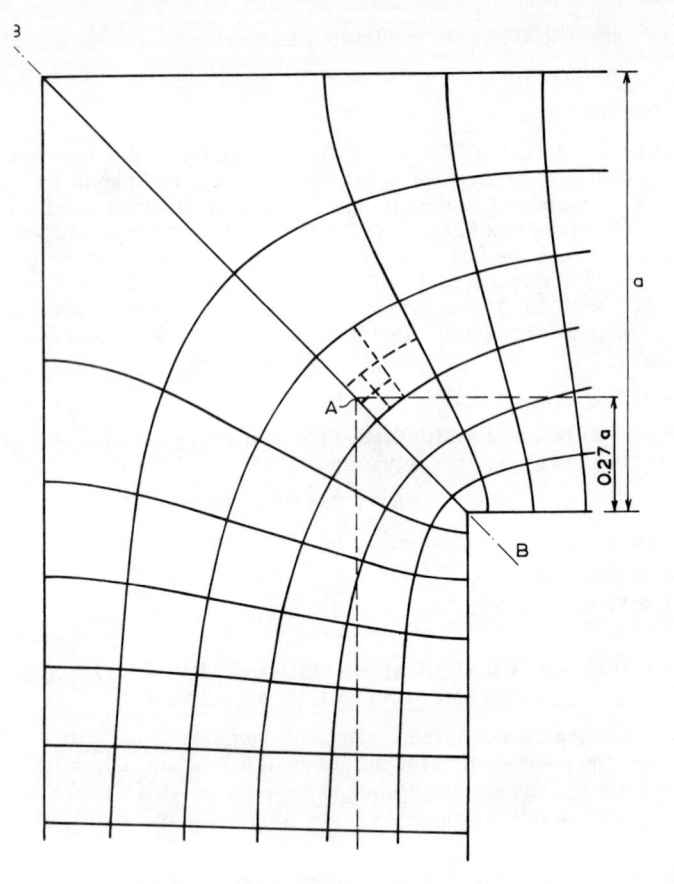

Figure 9.44 Isotherms and heat flux lines in a corner.

Steam tables give the saturation pressure of water

at 7.2°C, 1.02 kN/m².

at 16°C, 1.82 kN/m².

i.e. condensation occurs in the corner if the relative humidity exceeds

$\varphi = (100)(1.02)/1.82$
  $= 56\%$.

EXAMPLE 9.3. RADIATION FROM A BODY SURROUNDED BY ANOTHER BODY

Figure 9.45 shows a plate radiator with surface area 1.1 m² on each side.
Estimate the heat transferred by radiation from the radiator when the surface temperature of the radiator is 100°C or 373 K and the temperature of walls, ceiling and floor in the room are 18°C or 291 K. The emissivity of all surfaces is estimated to be $\epsilon = 0.85$.

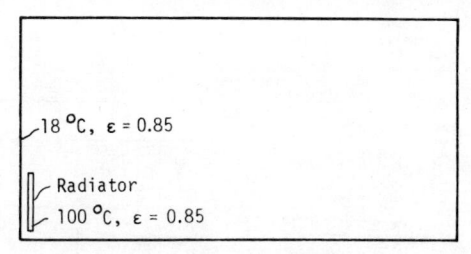

Figure 9.45 Room containing a radiator.

Solution

In equation (9.12) $A_1/A_2 = 1.0$ for the left and 0 for the right hand side of the radiator. This gives for the left side,

$$C_{1,2} = \frac{(5.7)(10^{-8})}{\dfrac{1}{0.85} + \left(\dfrac{1}{0.85} - 1\right)} = (4.2)(10^{-8}) \frac{W}{m^2 K^4}$$

and for the right side,

$$C_{1,2} = (0.85)(5.7)(10^{-8}) = (4.85)(10^{-8}) \frac{W}{m^2 K^4}.$$

Equation (9.11) gives

$$\dot{Q}_{1,2} = C_{1,2}(1.1)(373^4 - 291^4) = (1.34)(10^{10})C_{1,2}.$$

From the left side,

$$\dot{Q}_{1,2} = (1.34)(10^{10})(4.2)(10^{-8}) = 560 \text{ W}$$

and from the right side,

$$\dot{Q}_{1,2} = (1.34)(10^{10})(4.85)(10^{-8}) = \underline{650 \text{ W}}$$

Total heat transfer by radiation,   1210 W.

Note There will also be heat transfer by convection. In addition the wall near the radiator will be warmer than the other walls.

EXAMPLE 9.4. GRAPHICAL DETERMINATION OF HEAT EXCHANGE BY RADIATION

Figure 9.46 shows a steam boiler and a grate with glowing coals. The area of the grate is $A_1 = 0.55$ m^2. The surface temperature of the coals is estimated to be 1350°C or 1623 K and of the boiler 180°C or 453 K. The emissivity of the coal layer is $\epsilon_1 = 0.98$ and of the boiler $\epsilon_2 = 0.9$. Estimate the heat transferred to the boiler by radiation from the grate.

Solution

In Figure 9.46 straight lines are drawn from the centre of the grate to the outer edges of the 'visible' part of the boiler surface. Projections of the interceptions between these lines and the half sphere over the grate give the four points transferred to the lower drawing. These four points are sufficient for an estimation of the whole projection of interceptions of limiting rays with the half sphere. It is the hatched area, A' in Figure 9.46, measured to be 32% of the whole circle.

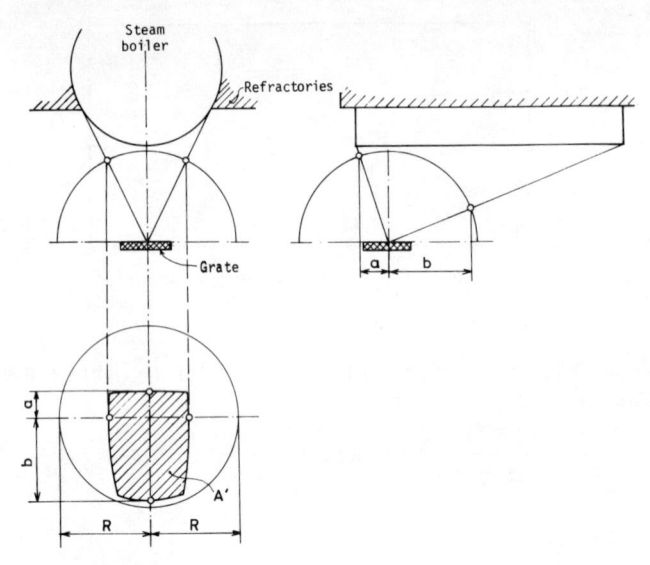

Figure 9.46 Radiation from a grate to the bottom of a steam boiler.

From the figure it can be seen that the area of the boiler 'seen' from the grate is approximately 1/10 the area of the grate, i.e. $A_1/A_2 \approx 0.1$. If A_2 had surrounded A_1, the term

$$\frac{A_1}{A_2}\left(\frac{1}{\epsilon} - 1\right)$$

in equation (9.12) would be 0.01. In this case the influence of radiation from the boiler to the grate is even smaller, and $C_{1,2}$ in equation (9.12) is well approximated by

$$C_{1,2} \approx \epsilon_1 C_s = (0.98)(5.7)(10^{-8}) = (5.586)(10^{-8}).$$

Equation (9.13) gives the net heat transfer by radiation from the grate to the boiler,

$$\dot{Q}_{1,2} = (0.32)(5.586)(10^{-8})(0.55)(1623^4 - 453^4)$$
$$= 67,800 \text{ W}.$$

Note Heat will also be transferred to the boiler by radiation from the refractory bricks, by radiation from the hot gases, and by convection of the gases.

EXAMPLE 9.5. HEAT TRANSFER BY RADIATION AND CONVECTION. ERROR IN TEMPERATURE MEASUREMENT

Figure 9.47 shows a thermometer A_t located between the boiler tubes in a steam boiler. All radiation from the thermometer will hit boiler tubes. The surface temperature of the tubes is estimated to be about 230°C and the heat transfer coefficient by convection $h = 30$ W/(m^{2}K). The emissivity of both thermometer and tubes $\epsilon \approx 0.9$.

(*a*) Estimate the flue gas temperature when the thermometer reads 411°C.

(*b*) Estimate the measurement error if the thermometer is placed inside a cylindrical metal shield polished at the outer side and with dimensions as given in Figure

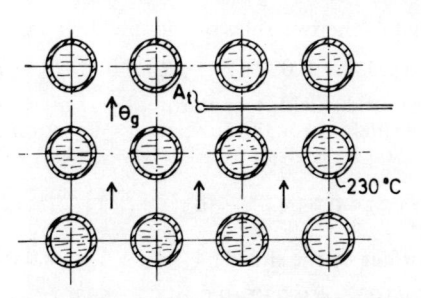

Figure 9.47 Thermometer A_t described in Example 9.5 between boiler tubes in a steam boiler.

9.48. The thermometer reads 575°C. The heat transfer coefficient by convection is assumed to be $h = 30$ W/(m²K) both for thermometer and shield. 5% of the radiation from the thermometer bulb pass through the openings of the shield.

Solution

(a) The thermometer bulb with surface area A_t will receive heat by convection from the flue gas and give off heat by radiation to the boiler tubes. With flue gas temperature θ_g the heat received by convection is [equation (9.14)],

$$\dot{Q} = A_t h \Delta\theta = A_t(30)(\theta_g - 411). \tag{a}$$

With A_1/A_2 in equation (9.12) equal to 0 ($A_2 \gg A_t$), equations (9.11) and (9.12) give the heat given off,

$$\dot{Q} = (0.9)(5.7)(10^{-8})A_t[(273 + 411)^4 - (273 + 230)^4] = 7945\, A_t. \tag{b}$$

At steady state,

$$A_t(30)(\theta_g - 411) = 7945\, A_t,$$
$$\theta_g = 676°C.$$

i.e. measurement error $676 - 411 = 265°C$.

(b) The temperature of the shield is estimated making the following simplifying assumptions:
(α) radiation from the thermometer to the shield can be neglected in calculation of the shield temperature;
(β) radiation from the inside of the shield through the openings with 30 mm diameter is treated as radiation from a black body, i.e. $C_{1,2}$ for the openings is $(5.7)(10^{-8})$ W/(m²K⁴).

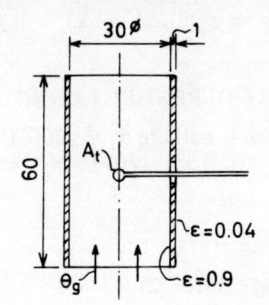

Figure 9.48 Radiation shield: all dimensions are in mm.

The heat transferred to the two sides of the shield by convection is,

$$\dot{Q} = (30)(0.06)\pi(0.030 + 0.032)(T_g - T_s) = 0.35(T_g - T_s) \tag{c}$$

where T_g and T_s are gas and shield temperature, K.

Radiation from the shield through the openings, equation (9.11) with $C_{1,2} = C_b$ and $T_2 = 273 + 230 = 503$ K, is given by

$$\dot{Q}_1 = (5.7)(10^{-8})(2)\frac{\pi}{4}0.03^2(T_s^4 - 503^4) = (8)(10^{-11})T_s^4 - 5.16$$

and from the outer surface of the shield ($A_1/A_2 \approx 0$, $\epsilon_1 = 0.04$), by

$$\dot{Q}_2 = (0.04)(5.7)(10^{-8})\pi(0.032)(0.06)(T_s^4 - 503^4)$$
$$= (1.4)(10^{-11})T_s^4 - 0.88.$$

Heat from the shield by radiation,

$$\dot{Q} = \dot{Q}_1 + \dot{Q}_2 = (9.4)(10^{-11})T_s^4 - 6.04. \tag{d}$$

Equations (c) and (d) give

$$T_g - T_s = (2.7)(10^{-10})T_s^4 - 17.3. \tag{e}$$

The same procedure is used for the thermometer bulb:

$$\dot{Q} = 30(T_g - 848)A_t. \tag{f}$$

95% of the radiation hits the shield,

$$\dot{Q}_1 = (0.95)(0.9)(5.7)(10^{-8})(848^4 - T_s^4)A_t$$
$$= [25,200 - (4.87)(10^{-8})T_s^4]A_t$$

and 5% hits the boiler tubes,

$$\dot{Q}_2 = (0.05)(0.9)(5.7)(10^{-8})(848^4 - 503^4)A_t = 1162\, A_t$$
$$\dot{Q} = \dot{Q}_1 + \dot{Q}_2 = [26,360 - (4.87)(10^{-8})T_s^4]A_t. \tag{g}$$

From equations (f), (g), and (e),

$$T_s = 1744 - (1.89)(10^{-9})T_s^4,$$
$$T_s = 833 \text{ K}$$

and from equation (e),

$$T_g = 833 + (2.7)(10^{-10})(833^4) - 17.3 = 946 \text{ K}$$
$$\theta_g = 946 - 273$$
$$= 673°C,$$

i.e., measurement error $673 - 575 = 98°C$.

Note The error is reduced to 1/3 of the value without shield, but is still considerable. In addition the polished surface can easily be oxidized, which will give a higher emissivity.

EXAMPLE 9.6. TEMPERATURE-DEPENDENT CONDUCTIVITY

The surface temperature of a 0.24 m thick wall are $\theta_1 = 1000°C$ and $\theta_2 = 200°C$. The thermal conductivity of the wall is $k = 0.21\theta^{0.3}$ W/(m°C).

(a) Calculate the heat flux through the wall.

(b) Estimate the percentage error using an average thermal conductivity

$$k = 0.21\,(1000^{0.3} + 200^{0.3})/2 = 1.349 \text{ W/(m°C)}.$$

Solution

(a) From equation (9.1),

$$\dot{q} = \frac{\dot{Q}}{A} = -0.21\theta^{0.3}\frac{d\theta}{dx}$$

$$\dot{q}\int_0^{0.24} dx = -0.21\int_{1000}^{200} \theta^{0.3}\,d\theta$$

$$\dot{q} = -\frac{0.21}{(0.24)(1.3)}(200^{1.3} - 1000^{1.3})$$

$$= 4687\ \text{W/m}^2.$$

(b) $\dot{q} = -1.349(200 - 1000)/0.24$

$= 4497\ \text{W/m}^2,$

∴   error, $(100)(4687 - 4497)/4687 = 4\%.$

EXAMPLE 9.7. TEMPERATURE CHANGE IN A HEAT EXCHANGER DUE TO CHANGED FLOW RATE

A shell-and-tube heat exchanger has saturated steam condensing at 160°C at the shell side, and a vegetable oil at the tube side where the oil is heated from 20°C to 106°C. At the average temperature the density of the oil is 920 kg/m³ and the viscosity 7.6 cP. The thermal conductivity of the oil is unknown.

The heat exchanger has 22 tubes of Cr–Ni–Mo-steel, each tube 3.8 m long, inner diameter 19 mm and outer diameter 22 mm, and thermal conductivity $k = 17\ \text{W/(m°C)}$. The following data are obtained under the present operating conditions:

heat transferred,	$\dot{Q} = 123\ \text{kW},$
average velocity in the tubes,	$V = 0.75\ \text{m/s},$
estimated heat transfer coefficient for condensing steam (including fouling),	$h_0 = 6800\ \text{W/(m}^2\text{°C)}.$

Estimate the exit oil temperature with the same heat exchanger used with doubled flow rate.

Solution

Equation (9.43) gives the logarithmic mean temperature difference

$$\Delta\theta_m = \frac{\Delta\theta_i - \Delta\theta_o}{\ln(\Delta\theta_i/\Delta\theta_o)} = \frac{(160-20)-(160-106)}{\ln[(160-20)/(160-106)]}$$

$$= 90°\text{C}.$$

Overall heat transfer coefficient [equation (9.42)], referred to the inner tube surface,

$$U_i = \frac{\dot{Q}}{A_i\Delta\theta_m} = \frac{123{,}000}{(22)\pi(0.019)(3.8)(90)} = 274\ \text{W/(m}^2\text{°C)}.$$

Table 9.5 gives the fouling resistance for vegetable oil, $(L/k)_f = 0.0005\ °\text{C m}^2/\text{W}.$

From equation (9.41),

$$\frac{1}{U_i} = \frac{1}{274} = \frac{1}{h_i} + 0.0005 + \frac{1}{(6800)(22)/19} + \frac{0.022-0.019}{(17)(0.022+0.019)/0.019}$$

$$= \frac{1}{h_i} + 0.000709,$$

$$h_i = 340 \text{ W/(m}^2{}^\circ\text{C)}.$$

Re = $(920)(0.75)(0.019)/0.0076$ = 1725, and L/D = 3.8/0.019 = 200, gives (Figure 9.8) Nu Pr$^{-1/3}$ ≈ 3.9. Twice the flow rate gives Re = (2)(1725) = 3450, and Nu Pr$^{-1/3}$ ≈ 11, or

$$h_2 = (340)(11)/3.9 = 960 \text{ W/(m}^2{}^\circ\text{C)}$$
$$1/U_2 = 1/960 + 0.000709,$$
$$U_2 = 571 \text{ W/(m}^2{}^\circ\text{C)}.$$

The new logarithmic mean temperature difference is

$$\Delta\theta_{m2} = \frac{(160-20)-(160-\theta_o)}{\ln(160-20)-\ln(160-\theta_o)} = \frac{\theta_o - 20}{4.9416 - \ln(160-\theta_o)} \tag{a}$$

where θ_o is the exit oil temperature.

$$\dot{Q}_2 = (2w)c(\theta_o - 20) \tag{b}$$

where $2w$ is the mass flow rate, kg/s, and c is the specific heat capacity, J/(kg°C). Heat transfer, with velocity V = 0.75 m/s, is given by

$$\dot{Q}_1 = 123{,}000 = wc(106-20),$$
$$wc = 1430 \text{ J/(s °C)}.$$

From equation (b),

$$\dot{Q}_2 = (2)(1430)(\theta_o - 20) = 2860(\theta_o - 20).$$

Equation (9.38), with $\Delta\theta_m$ from equation (a), gives

$$2860(\theta_o - 20) = (571)(22)\pi(0.019)(3.8)\frac{\theta_o - 20}{4.9416 - \ln(160-\theta_o)}$$

$$\theta_o = 180°\text{C}.$$

Note This is a special case, because the increased flow rate changes the flow from laminar to turbulent. Usually an increased flow rate will reduce the exit temperature of the heated fluid.

EXAMPLE 9.8. HEAT EXCHANGER FOR GASES WITH CONSTANT HEAT CAPACITY

Figure 9.49 shows a heat exchanger for recovery of heat from the gases in a methanol reactor (H_2, CO, CH_3OH, and inert). The following specifications are given for the gas mixture.

		Gas to reactor*	Gas from reactor
Flow rate,	kmol/h	1150	1289
	kg/h	12,200	14,250
Pressure,	MN/m²	26.5	25.5
Temperature,			
gas to heat exchanger,	°C	30	380
gas from heat exchanger,	°C	330	140
Density,			
gas to heat exchanger,	kg/m³	103	49
gas from heat exchanger,	kg/m³	53	76
Viscosity,	cP	0.0167	0.0194
Thermal conductivity,	W/(m°C)	0.124	0.145
$Pr^{-1/3}$		1.39	1.28
Heat transferred,	kW	2890	

*Only the fraction that passes through the heat exchanger.

Use of the 1-1 shell and the tube heat exchanger specified below is being considered with reactor gas on the tube side and counter-current flow of gas from

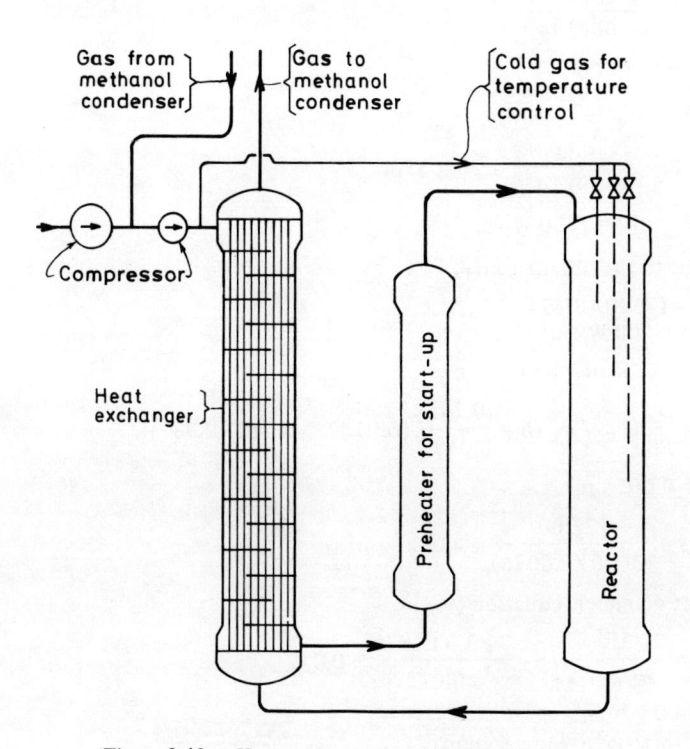

Figure 9.49   Heat exchanger in a methanol factory.

the methanol condenser on the shell side:

tubes, 98% Cu and 1.5% Mn, thermal conductivity	$k \approx 350$ W/(m°C)
131 tubes in triangular pitch, pitch	$P_t = 18.7$ mm
inner and outer tube diameter,	8 and 12 mm
tube length,	6 m
shell diameter,	$D_s = 245$ mm
baffle spacing,	$B = 87$ mm
baffle cut 20%, i.e. A/D_s (Figure 9.10)	$= 0.2$.

The gas is well purified. Assume that a heat exchange area 20% in excess of what is needed with clean surfaces is sufficient. A pressure drop less than 0.5 MN/m² is required for both streams.

Check whether or not the heat exchanger will fulfil the requirements.

Solution

1. Overall heat transfer coefficient referred to the outside tube area: tube side,

$$\rho V = \frac{14{,}250}{(3600)(131)\frac{\pi}{4}\,0.008^2}$$

$$= 600 \text{ kg/(m}^2\text{s)}.$$

$$\text{Re} = \frac{(600)(0.008)}{0.0000194}$$

$$= 247{,}400$$

and Figure 9.8 give

$$\text{Nu Pr}^{-1/3} = \frac{0.008 h_i}{0.145}\,1.28 = 530,$$

$$h_i = 7500 \text{ W/(m}^2\text{°C)}$$

or referred to the outer surface,

$$h_1 = (7500)(8/12)$$
$$= 5000 \text{ W/(m}^2\text{°C)}.$$

Shell side area, equation (9.23),

$$a_s = \frac{D_s}{P_t}(P_t - D_o)B = \frac{0.245}{0.0187}(0.0187 - 0.012)(0.087)$$

$$= 0.0076 \text{ m}^2.$$

$$\rho V_e = \frac{12{,}200}{(3600)(0.0076)} = 446 \text{ kg/(m}^2\text{s)}.$$

Equivalent diameter, equation (9.21),

$$D_e = \frac{2\sqrt{3}P_t^2}{\pi D_o} - D_o = \frac{2\sqrt{3}\,0.0187^2}{\pi 0.012} - 0.012$$

$$= 0.020 \text{ m}.$$

$$\text{Re}_e = \frac{\rho V_e D_e}{\mu} = \frac{(446)(0.020)}{0.0000167}$$

$$= 534{,}000.$$

Figure 9.9, with $A/D_s = 0.2$, gives
$$\text{Nu Pr}^{-1/3} = \frac{h_0 0.020}{0.124} 1.39 \approx 510,$$
$$h_0 = 2275 \text{ W/(m}^2\,^\circ\text{C)}.$$

Equation (9.41) for the outer surface becomes
$$\frac{1}{U_0} = \frac{1}{2275} + \frac{1}{(7500)(8/12)} + \frac{0.012 - 0.008}{(350)(0.008 + 0.012)/0.012} = 0.000646.$$

Overall heat transfer coefficient,
$$U_0 = 1550 \text{ W/(m}^2\,^\circ\text{C)}.$$

2. Logarithmic mean temperature difference:
$$\Delta\theta_i = 140 - 30 = 110\,^\circ\text{C},$$
$$\Delta\theta_0 = 380 - 330 = 50\,^\circ\text{C}.$$
$$\Delta\theta_m = (110 - 50)/\ln(110/50) = 76\,^\circ\text{C}.$$

3. Necessary heat exchange area A_0, clean surface, equation (9.40),
$$(2890)(10^3) = A_0(1550)(76),$$
$$A_0 = 24{,}5 \text{ m}^2,$$

i.e. excess area
$$\frac{(131)\pi(0.012)(6) - 24.5}{24.5} \, 100 = 21\%.$$

4. Pressure drop at the tube side: mass rate of flow per unit area,
$$\frac{14{,}250}{(3600)(131)\frac{\pi}{4}0.008^2} = 600 \text{ kg/(m}^2\text{ s)}.$$

Entrance loss (Table 1.2),
$$(0.5)(49)\frac{(600/49)^2}{2} = 1840 \text{ N/m}^2$$

and exit loss (Table 1.2),
$$(1.0)(76)\frac{(600/76)^2}{2} = 2370 \text{ N/m}^2.$$

$$\text{Re} = \frac{(600)(0.008)}{0.0000194} = 247{,}400.$$

From Table 1.1, $\epsilon/D = 0.0015/8 = 0.00019$, and from Figure 1.8, friction factor $f = 0.0042$.

Friction loss with average density $(49 + 76)/2 = 62.5$ kg/m^3 are given by equation (1.8),
$$\Delta p_f = (4)(0.0042)(62.5)\frac{6}{0.008}\frac{(600/62.5)^2}{2}$$
$$= 36{,}290 \text{ N/m}^2.$$

Total pressure drop,
$$1840 + 2370 + 36{,}290 = 40{,}500 \text{ N/m}^2$$
$$= 0.04 \text{ MN/m}^2.$$

5. Pressure drop at the shell side can also be determined. Figure 9.9 with $Re_e = 534{,}000$ and $A/D_s = 0.2$ gives the friction factor $f_s = 0.22$. Equation (9.24) with average density $(103 + 53)/2 = 78$ kg/m^3, number of baffles $N = 68$, and average velocity $V_{av} = 446/78 = 5.7$ m/s, gives

$$\Delta p_f = (0.22)(78)\frac{5.7^2}{2}\frac{0.245}{0.020}(68 + 1) = 236{,}000 \text{ N/m}^2$$
$$= 0.24 \text{ MN/m}^2,$$

i.e. the pressure drops are within the specified limit.

EXAMPLE 9.9. TEMPERATURE DIFFERENCE IN HEAT EXCHANGER FOR FLUID WITH VARIABLE HEAT CAPACITY

Figure 9.50 shows the heat exchanger I for methane in Example 4.9. The points 3, 4, 5, and 6 refer to the same points in Figure 4.39. Data from Example 4.9 are given in the table.

Point (Figures 4.39 and 9.50):		3	4	5	6
Flow rate, kg per kg in stream 5		2.11	2.11	1.0	1.0
Pressure,	MN/m^2	0.1	0.1	14	14
Temperature	°C	−164	−25	15	−158
Enthalpy,	kJ/kg	181	515	453	−250

Calculate the mean temperature difference in a counter-current heat exchanger, assuming constant overall heat transfer coefficient U.

Figure 9.50 Heat exchanger I in Figure 4.39.

Solution

With the symbols of Figure 9.51 the heat transferred through the area ΔA per kg methane is

$$\Delta \dot{Q} = h_{b1} - h_{b2} = 2.11(h_{a4} - h_{a3}) = U \Delta A \Delta \theta$$

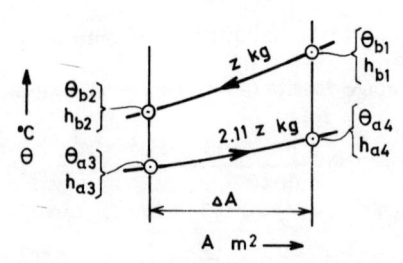

Figure 9.51 Gas temperatures across an area ΔA in Figure 9.50.

or

$$h_{a4} = h_{a3} + (h_{b1} - h_{b2})/2.11 \qquad (a)$$

and

$$\Delta AU = (h_{b1} - h_{b2})/\Delta\theta \qquad (b)$$

where

$$\Delta\theta = \frac{(\theta_{b1} - \theta_{a4}) - (\theta_{b2} - \theta_{a3})}{\ln[(\theta_{b1} - \theta_{a4})/(\theta_{b2} - \theta_{a3})]}. \qquad (c)$$

Equation (c) is valid for small temperature intervals, $\theta_{b1} - \theta_{b2}$, where the heat capacity is almost constant. The following table is calculated column to column for five such temperature intervals.

interval		ΔA_I	ΔA_{II}	ΔA_{III}	ΔA_{IV}	ΔA_V	ΔA_{VI}	Notes
b1	°C	−129*	−100	−70	−40	−15	15	chosen
b2	kJ/kg	−250†	−150.8	−48	77	228	338	= h_{b1} in preceding column
b2	°C	−158†	−129	−100	−70	−40	−15	= θ_{b1} in preceding column
a3	kJ/kg	181†	228	276.7	335.9	407.5	459.6	= h_{a4} in preceding column
a3	°C	−164†	−164	−136	−109	−78	−51	= θ_{a4} in preceding column
b1	kJ/kg	−150.8*	−48	77	228	338	453	diagram, 14 MN/m², $\theta = \theta_{b1}$
a4	kJ/kg	228*	276.7	335.9	407.5	459.6	514.1	from equation (a)
a4	°C	−164	−136	−109	−78	−51	−27	diagram, 0.1 MN/m², $h = h_{a4}$
$\Delta\theta$	°C	16.4	35.5	37.5	38.5	37.0	38.9	equation (c)
ΔAU	W/°C	6.03	2.90	3.34	3.92	2.97	2.95	equation (b)

*In the first interval h_{a4} is chosen as 228 kJ/kg. This corresponds to point 8 in Figure 4.41 where all
 liquid is evaporated. In the same interval h_{b1} is calculated by equation (a), and the corresponding value
 of θ_{b1} is read from the enthalpy diagram, page 350.
†From the preceding table.

The total heat transferred is

$$\dot{Q} = 453 - (-250) = 703 = (\Sigma \Delta AU)\Delta\theta_m$$

where $\Sigma \Delta AU = AU$ is the sum of the last line in the table and equals 22.11 W/°C or

$$\Delta\theta_m = 703/22.11$$

$$= 31.8°C.$$

Comment The temperatures θ_a and θ_b are given in Figure 9.52 as a function of the product AU. The corresponding areas ΔA are valid for constant overall heat transfer coefficient U. The logarithmic mean temperature difference is

$$\Delta\theta_{lgm} = \frac{(15 + 25) - (164 - 158)}{\ln[(15 + 25)/(164 - 158)]} = 17.9°C$$

which gives an error of

$$(100)(31.8 - 17.9)/31.8 = 44\%.$$

EXAMPLE 9.10. FINNED TUBE HEAT EXCHANGER

Calculate the overall heat transfer coefficient referred to the total outer surface area of a finned tube heat exchanger (Figure 1.28), with tube-side heat transfer coefficient including fouling, $h_i = 5500$ W/(m²°C), and air with pressure 1.0 bar and

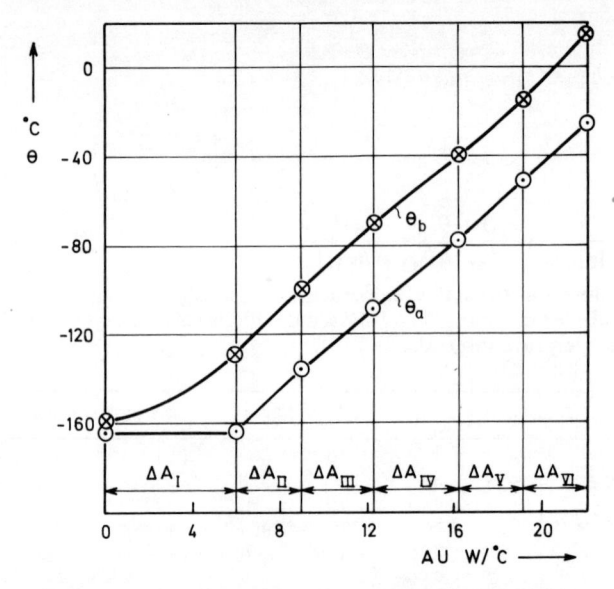

Figure 9.52 Temperatures of methane through the heat exchanger, with points as calculated in the table, page 279.

temperature 20°C on the outside. The air velocity referred to the total cross section is 9.5 m/s.

	Heat capacity	Thermal conductivity	Dynamic viscosity
Air	1.0 kJ/(kg°C)	0.0257 W/(m°C)	0.0182 cP

	Inner diameter	Outer diamter	Pitch	Thermal conductivity
Copper tubes	21 mm	25 mm	55 mm	350 W/(m°C)

	Number per m	Outer diameter	Thickness	Thermal conductivity
Aluminium fins	360	50 mm	0.5 mm	180 W/(m°C)

Solution

1. Outside heat transfer coefficient, from equation (9.46).

$\rho = Mp/RT = (29)(10^5)/(8314)(293) = 1.19$ kg/m^3.

Free cross section of area perpendicular to the air flow,

$[0.055 - 0.025 - (360)(0.0005)(0.05 - 0.025)]/0.055$
$= 0.543$ m^2/m^2 total cross section

∴ $V_{max} = 9.5/0.543$
$= 17.5$ m/s.

$$\frac{A_{tot}}{A_o} = \frac{0.025\pi[1-(0.0005)(360)] + 2\frac{\pi}{4}(0.05^2 - 0.025^2)(360)}{0.025\pi}$$

$$= \frac{1.124}{0.0785} = 14.3.$$

$$\Pr = \frac{(1000)(0.0000182)}{0.0257}$$

$$= 0.71.$$

From equation (9.46),

$$\frac{0.025h}{0.0257} = 0.45 \left[\frac{(1.19)(17.5)(0.025)}{0.0000182}\right]^{0.625} 14.3^{-0.375} 0.71^{1/3}$$

$$h = 46 \text{ W}/(\text{m}^2{}^\circ\text{C})$$

2. Fin efficiency, η_f: from equation (9.51) with $B_f/D_0 = 50/25 = 2.0$ and a and b from Figure 9.37,

$$\varphi = (2^{1.17} - 1)(0.025)\sqrt{46/[(2)(180)(0.0005)]} = 0.5$$

and from equation (9.49),

$$\eta_f = \frac{\tanh 0.5}{0.5} = 0.92.$$

3. Overall heat transfer coefficient: the heat transfer resistances referred to the total outer area are for the tube side

$$\frac{1}{5500} \frac{1.124}{0.02\pi} = 0.0031 \text{ m}^2{}^\circ\text{C/W}$$

and for conduction through the tube,

$$\frac{0.002}{350} \frac{1.124}{0.023\pi} = 0.00009 \text{ m}^2{}^\circ\text{C/W}.$$

Heat transfer at the outside,

$$\dot{Q} = h\left(\frac{A_b}{A_{tot}} + \frac{\eta_f A_f}{A_{tot}}\right) A_{tot} \Delta\theta_o \tag{a}$$

where

A_b = outer tube area between the fins, per m tube
    = $0.025\pi[1 - (0.0005)(360)] = 0.0644 \text{ m}^2$

and

A_f = area of the fins, per m tube
    = $2\frac{\pi}{4}(0.05^2 - 0.025^2)(360) = 1.060 \text{ m}^2.$

Equation (a) with $h = 46 \text{ W}/(\text{m}^2{}^\circ\text{C})$ corresponds to a heat transfer resistance,

$$\frac{1}{(46)[0.0644/1.124 + (0.92)(1.06)/1.124]} = 0.0235 \text{ m}^2{}^\circ\text{C/W}$$

$$\therefore 1/U = 0.0031 + 0.00009 + 0.0235$$

$$= 0.0267 \text{ m}^2{}^\circ\text{C/W}$$

$$U = 37.5 \text{ W}/(\text{m}^2{}^\circ\text{C}).$$

EXAMPLE 9.11. TANK WALL AS FIN, FREE CONVECTION

Figure 9.53 shows the top of a horizontal storage tank for light beer. Cooling is obtained by circulation of a refrigerant through the two half tubes welded to the outer side of the tank; the tank wall acts as a fin. The tank wall under the half tubes is kept at a constant temperature, 0°C.

Estimate the heat transfer per m length of one of the tubes as a function of the temperature of the beer in the temperature range 5 to 20°C.

Light beer has a maximum density at a temperature of 1.5°C. For free convection heat transfer calculations, the physical properties of the beer are estimated as equal to those for water at a temperature 2.5°C higher than that of the beer. The tank is made of $s = 6$ mm thick aluminium plates with thermal conductivity, $k_f = 200$ W/(m °C).

Solution

1. The heat transfer coefficient, h: Grashof number and Prandtl number with average values for water in the temperature region in question,

$$Gr = \frac{1000^2 D^3 (\beta \Delta \theta) 9.82}{0.0014^2} = (5)(10^{12})(\beta \Delta \theta) D^3$$

$$Pr = \frac{(4190)(0.0014)}{0.58} = 10.1.$$

Equation (9.16) with $C = 0.56$, height H as characteristic length D, the correction coefficient from Rich, $(\cos 45°)^{1/4} = 0.917$, and thermal conductivity of the beer 0.58 W/(m °C), gives

$$h = (0.58)(0.917)(0.56)[(10.1)(5)(10^{12})(\beta \Delta \theta)/H]^{1/4}$$

$$= 794(\beta \Delta \theta/H)^{1/4}. \tag{a}$$

Equation (9.18) with the correction coefficient from Rich, gives

$$h = 950(\beta \Delta \theta/H)^{1/4}. \tag{b}$$

The estimates are based on the following assumptions:

(i) The effective height H in the equation for heat transfer in free convection is $H \approx 0.6$ m.
(ii) The average tank wall temperature over the height H is assumed to be mid way between the lowest wall temperature (0° C) and the temperature of the beer (θ_b °C), giving a mean temperature difference $\Delta \theta_m = \theta_b/2$ in equations (a) and (b) over the height $H = 0.6$ m.

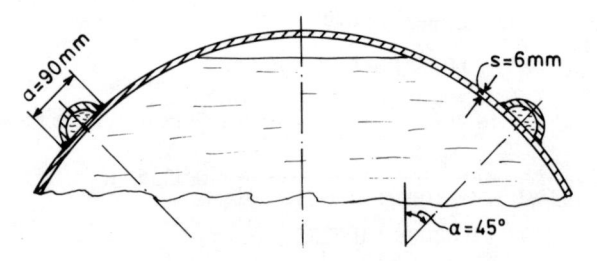

Figure 9.53 Horizontal, cylindrical storage tank with half tubes for cooling at the outer side.

(iii) The thermal coefficient of cubic expansion, β, is assumed to be the same as for water at a temperature $(\theta_b/2 + 2.5)°C$.
(iv) An average value of h from equations (a) and (b) is used,

$$h = \frac{794 + 950}{2} \left[\frac{\beta(\theta_b - 0)}{(2)(0.6)}\right]^{1/4} = 833(\beta\theta_b)^{1/4}. \qquad (c)$$

The first two assumptions will be checked later.

Table a

Beer temperature, θ_b, °C	5	10	15	20	Notes
$\beta \times 10^6$ $(10^6 \, °C)^{-1}$	15.3	51.6	84.5	115	
h W/(m² °C)	78	126	157	182	equation (c)

2. Heat transfer through the 'fins': height of 'fins', $L_f = (H - 0.09)/2$ m. Equation (9.50) with heat transfer from one side of the fin only (h_f substituted by $h_f/2$) gives

$$\varphi = L_f \sqrt{h_f/[(200)(0.006)]} = 0.91 L_f \sqrt{h_f} \qquad (d)$$

and fin efficiency, equation (9.49),

$$\eta_f = \frac{\tanh \varphi}{\varphi}. \qquad (e)$$

The heat transfer per metre length of the fin is

$$\dot{Q}_f = h_f L_f \eta_f (\theta_b - 0) = 1.1 \, h_f^{1/2} \theta_b \tanh \varphi. \qquad (f)$$

The assumed height, $L_f = (0.6 - 0.09)/2 = 0.255$ m, gives $\varphi = 0.232 \sqrt{h_f}$, or with h_f from 78 to 182 (Table a), $\tanh \varphi$ between 0.97 and 0.996. The assumed height takes care of 97% to 99.6% of the heat transferred through the fins.

3. Heat transfer per metre tube: heat transfer over the distance a (Figure 8.53),

$$\dot{Q}_a = 0.09 h \theta_b. \qquad (g)$$

Heat transfer through the 'fins' (to both sides of the half tube), equation (f) with $\tanh \varphi = 1.0$,

$$\dot{Q}_f = (2)(1.1) h_f^{1/2} \theta_b. \qquad (h)$$

Table b

Beer temperature, θ_b °C	5	10	15	20	Notes
h W/(m² °C)	78	126	157	182	from Table a
$\dot{Q}_a$ W/m tube	35	113	212	328	equation (g)
$\dot{Q}_f$ W/m tube	97	247	413	594	equation (h)
$\dot{Q}$ W/m tube	132	360	625	922	$\dot{Q} = \dot{Q}_a + \dot{Q}_b$

4. To check of the assumption $\Delta\theta_m \approx \theta_b/2$:

$$\Delta\theta_m = \frac{0.09 + (0.6 - 0.09)\eta_f}{0.6} \Delta\theta_a \qquad (i)$$

where $\Delta\theta_a$ is the temperature difference at the base of the 'fins', in this example equal to θ_b.

Equation (i) with η_f from the equations (d) and (e) and $L_f = 0.255$ gives

for $\theta_b = 5°C$, $\Delta\theta_m = 0.55\theta_b$
for $\theta_b = 20°C$, $\Delta\theta_m = 0.42\theta_b$.

These values are close enough to the value $\Delta\theta = \theta_b/2$ used in equation (c); the last line in Table b gives an estimate of the heat transfer as a function of the temperature of the beer.

EXAMPLE 9.12. HEAT PIPES FOR COUNTER-CURRENT HEAT TRANSFER

Figure 9.54 shows an arrangement where standard finned tubes arranged as heat pipes are used to approach counter-current heat transfer between two gas streams. The tubes in the lower duct with hot gas contain boiling ammonia, and the ammonia vapour condenses in the finned tubes in the upper duct with the cold gas stream. The tubes in Figure 9.54 are arranged in $n = 3$ sections. The diagram in the upper part of the figure shows the gas and ammonia temperatures.

Calculate the ratio between the mean temperature difference for the system with n sections and with an infinite number of sections (pure counter-current flow), $\Delta\theta_n/\Delta\theta_\infty$.

The following data are the same for both ducts:

1. the heat transfer area, A_1;
2. the overall heat transfer coefficient between the ammonia and the gas, U_1;
3. the product of gas mass flow rate and heat capacity of the gas, wc_p.

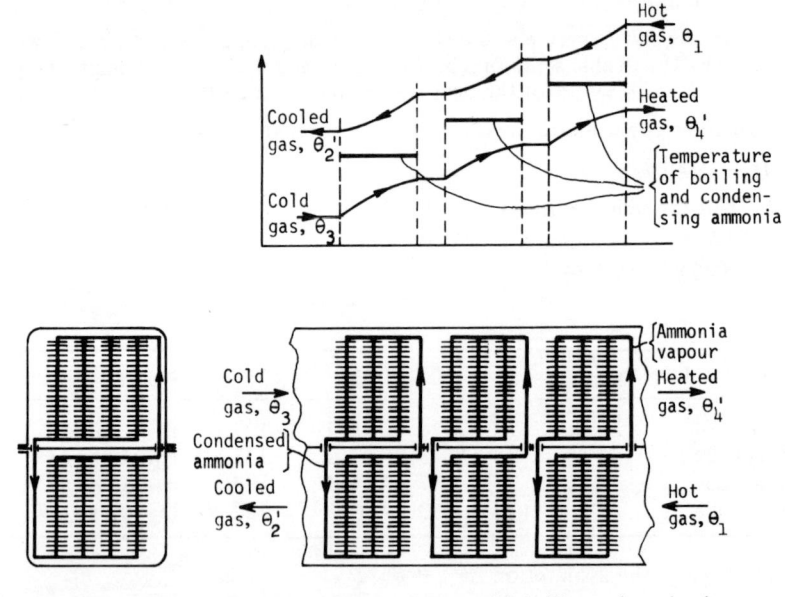

Figure 9.54 Finned tube gas–gas heat exchanger with boiling and condensing ammonia as heat transfer medium in the tubes: lower left, cross-section of the ducts; lower right, longitudinal section through the ducts; upper diagram, gas and ammonia temperatures.

As a consequence of statements 1 and 2, the heat transferred from the hot to the cold gas is $\dot{Q} = U_1 A_1 (\Delta\theta/2) = UA_1 \Delta\theta$, where $U = U_1/2$.

Solution

1. Figure 9.55 gives the gas temperatures with an infinite number of sections, $n = \infty$. The corresponding temperature difference is

$$\Delta\theta_\infty = \theta_1 - \theta_4$$

and the amount of heat transfer with $U = U_1/2$,

$$\dot{Q}_\infty = UA_1(\theta_1 - \theta_4) = wc_p(\theta_4 - \theta_3)$$

Rearranging to eliminate θ_4, gives

$$\dot{Q}_\infty = ab/(a+1) \tag{a}$$

where

$$a = UA_1/(wc_p) \tag{b}$$

and

$$b = wc_p(\theta_1 - \theta_3) \tag{c}$$

2. Figure 9.56 gives the gas and the ammonia temperatures with only one section ($n = 1$).

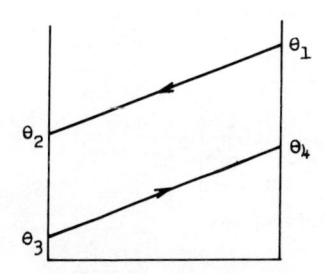

Figure 9.55 Counter-current heat transfer with infinite number of sections, $n = \infty$.

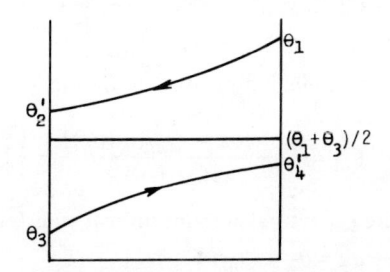

Figure 9.56 Heat transfer with one section, $n = 1$, and heat transfer medium with temperature $\theta_a = (\theta_1 + \theta_3)/2$.

With the same product $U_1 A_1$ in both ducts, the temperature of the ammonia in the tubes will be

$$\theta_a = (\theta_1 + \theta_3)/2.$$

The inlet and outlet temperature differences of the hot gas are

$$\Delta\theta_i = \theta_1 - \frac{\theta_1 + \theta_3}{2} = \frac{\theta_1 - \theta_3}{2}$$

and

$$\Delta\theta_o = \theta_2' - \frac{\theta_1 + \theta_3}{2} = \theta_1 - \frac{\dot{Q}_1}{wc_p} - \frac{\theta_1 + \theta_3}{2} = \frac{\theta_1 - \theta_3}{2} - \frac{\dot{Q}_1}{wc_p}.$$

The logarithmic mean temperature difference is

$$\Delta\theta_1 = \frac{(\theta_1 - \theta_3) - \left(\theta_1 - \theta_3 - \frac{2\dot{Q}_1}{wc_p}\right)}{2\ln\frac{\theta_1 - \theta_3}{\theta_1 - \theta_3 - 2\dot{Q}_1/(wc_p)}} = \frac{\dot{Q}_1}{wc_p \ln\frac{b}{b - 2\dot{Q}_1}}$$

$$\dot{Q}_1 = U_1 A_1 \Delta\theta_1 = 2\frac{UA_1}{wc_p}\frac{\dot{Q}_1}{\ln\frac{b}{b - 2\dot{Q}_1}} = 2a\frac{\dot{Q}_1}{\ln\frac{b}{b - 2\dot{Q}_1}}$$

$$\dot{Q}_1 \ln\frac{b}{b - 2\dot{Q}_1} = 2a\dot{Q}_1,$$

$$\frac{b}{b - 2\dot{Q}_1} = e^{2a}$$

$$\dot{Q}_1 = \frac{b}{2}(1 - e^{-2a}) = wc_p(\theta_1 - \theta'_2). \tag{d}$$

3. With n sections the change in gas temperature in each section is
$(\theta_1 - \theta'_2)/n = (\theta'_4 - \theta_3)/n$.

The ammonia temperature in the first section is

$$\theta_{a1} = \frac{\theta_1 + [\theta'_4 - \dot{Q}_n/(nwc_p)]}{2} = \frac{1}{2}\left[\theta_1 + \theta_3 \frac{(n-1)\dot{Q}_n}{nwc_p}\right]$$

where $\dot{Q}_n$ = total amount of heat transferred with n sections

$\theta'_4 = \theta_3 + \dot{Q}_n/(wc_p)$.

The same procedure as for one section gives the amount of heat transferred,

$$\dot{Q}_n = \frac{nb(e^{2a/n} - 1)}{(n+1)e^{2a/n} - (n-1)}. \tag{e}$$

The ratio of transferred heat for system with n sections to transferred heat for system with ∞ sections, equals the ratio of the mean temperature differences, and equations (a) and (e) give,

$$\frac{\Delta\theta_n}{\Delta\theta_\infty} = \frac{n(a+1)(e^{2a/n} - 1)}{a[(n+1)e^{2a/n} - (n-1)]}. \tag{f}$$

In Figure 9.57 numerical values of equation (f) are plotted as a function of

$$a = \left[\frac{\theta_1 - \theta_2}{\theta_1 - \theta_4}\right]_{n = \infty} \tag{g}$$

where θ_2 and θ_4 are the temperatures obtained with an infinite number of sections.

$a = 1.0$ in Figure 9.57 corresponds to cold gas heated to the same temperature as the leaving hot gas in a heat exchanger with pure counter-current flow ($n = \infty$).

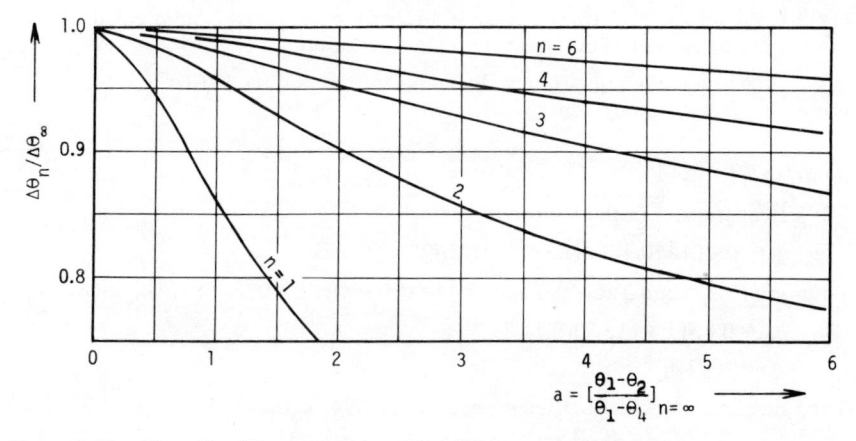

Figure 9.57 The ratio of the mean temperature differences, $\Delta\theta_n/\Delta\theta_\infty$, plotted as a function of the temperature ratio $[(\theta_1 - \theta_2)/(\theta_1 - \theta_4)]_{n=\infty}$: n is the number of sections.

Even in this case, only three sections are required to give a mean temperature difference 98% of the mean temperature difference in pure counter-current flow.

EXAMPLE 9.13. EVAPORATOR FOR METHANOL

The feed to the shell side of the evaporator Figure 9.58 is liquid methanol, with pressure varying between 1.6 and 2.1 bar and temperature from $-18°C$ in winter to $20°C$ in summer. A differential pressure cell DP and a controller C maintain a constant pressure of 1.25 bar in the insulated vapour line after the valve V. The feed line A is connected to a storage tank at a higher level. Excess vapour forces liquid back from the evaporator to the storage tank and reduces the liquid level in the evaporator. Heat is supplied by condensing steam in the tubes, steam quality $x = 0.98$, pressure 8.5 bar, saturation temperature $173°C$, and heat of condensation 1997 kJ/kg. Viscosity and thermal conductivity of the condensate, $\mu \approx 0.22$ cP and $k \approx 0.7$ W/(m°C).

Data for methanol:

Pressure	Saturation temperature °C	Heat of vaporization, kJ/kg	Heat capacity of liquid, kJ/(kg °C)
$p = 1.6$ bar	77	470	2.6
$p = 2.1$ bar	85		

Tubes, inner diameter 34 mm, outer diameter 42 mm, and thermal conductivity $k = 58$ W/(m °C).

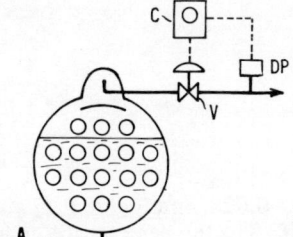

Figure 9.58 Methanol evaporator: A, liquid feed; V, control valve.

(a) Estimate the heat transfer area if 50% of the tube area is submerged in boiling methanol and 1900 kg is evaporated per hour.

(b) Check that the temperature difference of the boiling methanol is below the critical temperature difference.

Solution

(a) Heat transfer in the winter

$$\dot{Q} = 1900[470 + 2.6(85 + 18)]/3600 = 390 \text{ kW}.$$

Equation (9.37) and Table 9.4 give the heat transfer coefficient at the shell side,

$$h_o = 0.83[1 + (15.7)(2.1)/79.5]\dot{q}_o^{0.75}$$
$$= 1.17\dot{q}_o^{0.75}.$$

Tube side heat transfer coefficient, equation (9.34) with $N = 1$,

$$h_i = 0.725 \left[\frac{(0.7^3)(1000^2)(1,997,000)(9.82)}{(0.00022)(0.034)\Delta\theta_i} \right]^{1/4}$$
$$= 22{,}300(\Delta\theta_i)^{-1/4}.$$

Table 9.5 indicates that the fouling resistance is 0.0001 °C m²/W on both sides. This gives an overall heat transfer resistance referred to the outer surface,

$$\frac{1}{U_o} = \frac{1}{1.17\dot{q}_o^{0.75}} + 0.0001 + \frac{42}{38}\frac{0.042 - 0.034}{(2)(58)} + \frac{42}{34}\left[\frac{(\Delta\theta_i)^{1/4}}{22{,}300} + 0.0001\right]$$
$$= 0.000300 + 0.855\dot{q}_o^{-0.75} + 0.000055(\Delta\theta_i)^{0.25}.$$

With steam temperature 173°C and methanol temperature 85°C the temperature difference $\Delta\theta = 173 - 85 = 88$°C. The heat flux referred to the outer surface is

$$\dot{q}_o = U_o\Delta\theta = \frac{88}{0.0003 + 0.855\dot{q}_o^{-3/4} + 0.000055(\Delta\theta_i)^{1/4}}. \qquad (a)$$

In addition $\dot{q}_i = h_i\Delta\theta_i$ and $\dot{q}_o = (34/42)\dot{q}_i$, i.e.

$$\dot{q}_o = \frac{34}{42}\frac{22{,}300}{(\Delta\theta_i)^{1/4}}\Delta\theta_i = 18{,}050(\Delta\theta_i)^{3/4}. \qquad (b)$$

Equations (a) and (b) can be solved by trial and error or by iteration, giving

$$\Delta\theta_i = 18.9\text{°C}$$

and

$$\dot{q}_o = 164{,}000 \text{ W/m}^2$$

or submerged heat transfer area,

$$A_s = \dot{Q}/\dot{q}_o = 390{,}000/164{,}000$$
$$= 2.4 \text{ m}^2.$$

Total outer tube area,

$$A = (2)(2.4) = 4.8 \text{ m}^2.$$

(b) The reduced pressure is $p_r = 2.1/79.5 = 0.026$, and Figure 9.17 gives the critical temperature difference, $\Delta\theta_c = 36.5$°C. The temperature difference at the outer tube surface is

$$\Delta\theta_o = \dot{q}_o/h_o = 164{,}000/[(1.17)(164{,}000^{3/4})] = 17.2°C.$$

This is well below the critical temperature difference.

EXAMPLE 9.14. TEMPERATURE DIFFERENCE IN A CONDENSER FOR A VAPOUR MIXTURE

Figure 9.59 shows the top of a distillation column for separation of a hydrocarbon mixture under a pressure of 20.7 bar. C is a condenser, S a safety valve, R the reflux, and D the distillate. The vapour to the condenser is assumed to be in equilibrium with the liquid in the top tray in the column. This liquid contains 8 mole% ethane and 92 mole% propane. Equilibrium data are given in Figure 9.60.

Figure 9.59   Top of a distillation column with condenser C.

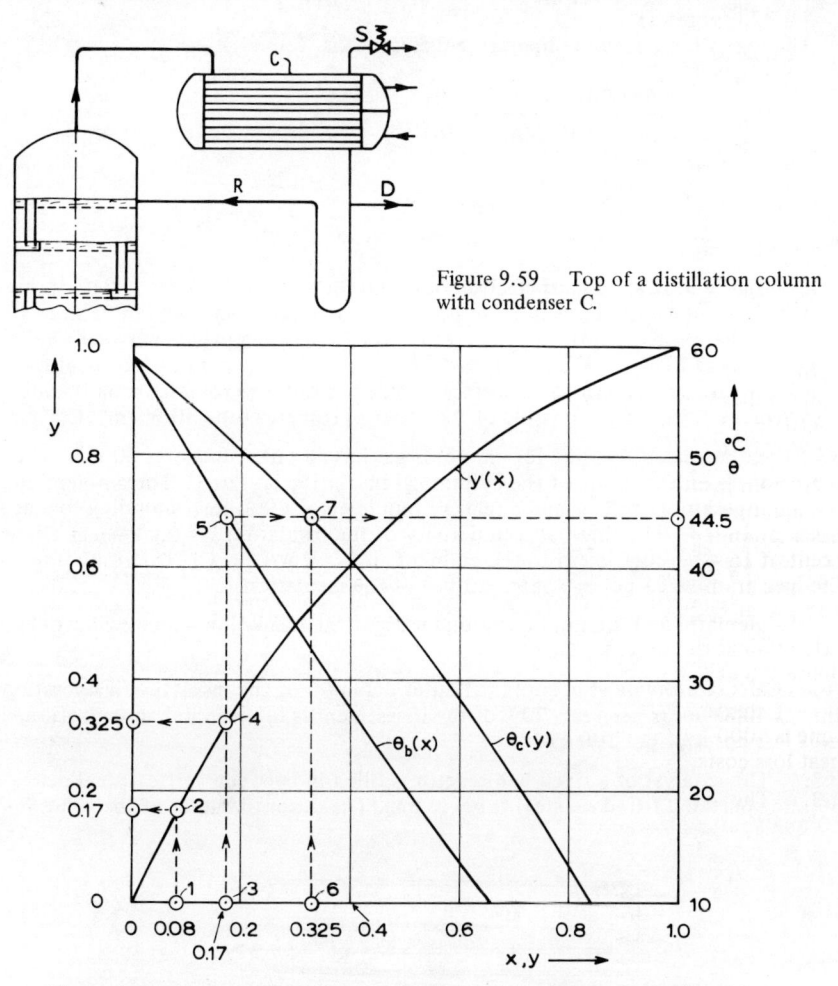

Figure 9.60   Vapour–liquid equilibrium data for a mixture of ethane and propane under a pressure of 20.7 bar. y is the mole fraction of ethane in the vapour phase, and x that in the liquid. θ_b is the boiling point of the liquid and θ_d is the dew point of the vapour [*Chem. Eng. Data,* 7, 232 (1962)].

Calculate the logarithmic mean temperature difference between the condensing vapour and the cooling water with inlet temperature 20°C and outlet temperature 30°C.

Solution

0.08 mole fraction ethane in the liquid in the top tray gives a vapour with 0.17 mole fraction ethane (point 2 in Figure 9.60) and 0.83 mole fraction propane. The vapour close to the cooling surface in the condenser will be enriched in ethane until steady state is obtained, giving condensate with the same composition as the vapour. Condensate with mole fraction ethane $x = 0.17$ (point 3 in Figure 9.60) is in equilibrium with vapour containing a mole fraction ethane $y = 0.325$ (point 4). This corresponds to a boiling point of 44.5°C for the condensate film ($x = 0.17$, point 5) or the same dew point for the vapour adjacent to the condensate film ($y = 0.325$, point 7).

The logarithmic mean temperature difference is

$$\Delta \theta_m = \frac{(44.5 - 20) - (44.5 - 30)}{\ln[(44.5 - 20)/(44.5 - 30)]}$$
$$= 19°C.$$

Problems

9.1. Figure 9.61 shows a special suction pyrometer with an extra radiation shield, designed to reduce the measurement error due to radiation at high temperatures. Estimate the error when this pyrometer reads 610°C when placed between the boiler tubes in Figure 9.47, and all surfaces have emissivity $\epsilon = 0.9$. The heat transfer coefficient by convection to all surfaces inside the outer pyrometer tube is about 50 W/(m²°C), and to the outside of the outer pyrometer tube 30 W/(m²°C).

9.2. An insulated pipeline for ammonia gas has an outer diameter 60 mm. The temperature on the inside of the cylindrical insulation is −20°C. The ambient air has a temperature of 20°C and a relative humidity of 80%, corresponding to a dew point of 16.4°C. The thermal conductivity of the insulation $k = 0.04$ W/(m°C), and the heat transfer coefficient to the ambient air $h = 6$ W/(m²°C). The insulation costs £65 per m³ plus £4 per m² outer surface of the insulation.

(a) Calculate the minimum outer diameter of the insulation which will avoid condensation at the outer surface.

(b) Calculate the most economical outer diameter of the insulation if operation time is 4000 hours per year, 20% of the investment is interest and amortization, and heat loss costs £1 per 100 kWh.

9.3. The catalyst of a fixed bed reactor is filled in between vertical steel plates 50 mm apart and fitted on steel tubes in line. The outer diameter of the tubes is

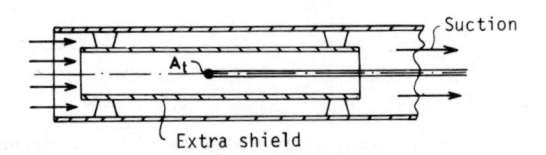

Figure 9.61 Suction pyrometer with extra radiation shield.

38 mm and their pitch (centre distance) is 133 mm in both directions. The plates are $s = 4$ mm thick with thermal conductivity $k = 52$ W/(m°C).

(a) Calculate the fin efficiency for the plates with heat transfer coefficients $h = 10, 15,$ and 30 W/(m²°C).

(b) Calculate the heat transferred per m³ reactor volume if there is a temperature difference of 40°C between the outer surface temperature of the tube and the gas in the catalyst bed; heat transfer coefficient $h = 15$ W/(m²°C).

9.4. Figure 9.62 is a unit for producing fresh water from sea water on a ship. You are given the following data:
Heat transferred in the evaporator, $\dot{Q} = 180$ kW, and pressure of condensing vapour, $p = 0.07$ bar.

	Length	Inner diameter	Outer diameter	Thermal conductivity
Evaporator tubes	0.7 m	27 mm	32 mm	35 W/(m°C)

The sea water contains 3.4 weight% salts. The boiling point increases by about 0.2°C per weight% salt and the heat capacity decreases by about 0.033 kJ/(kg°C) per weight% salt.

	Density	Viscosity	Thermal conductivity
Sea water	1030 kg/m³	1.34 cP	0.6 W/(m°C)

Point (Figure 9.62)	A	B	C and E
Water temperature, °C	64	58	10

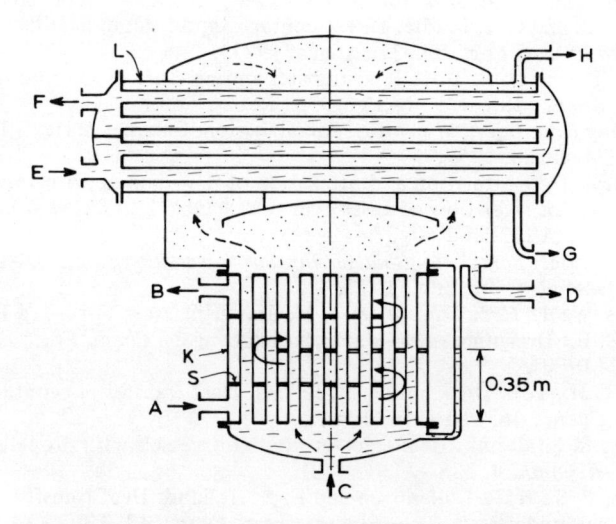

Figure 9.62 A marine fresh water generator. A is inlet and B the outlet for hot water from the diesel engine. C is the sea water inlet and D the outlet for excess sea water. Cooling water (sea water) comes in at E and out at F. G is the fresh water outlet and H is the connection to the steam ejector, K is a vertical evaporator and L a horizontal condenser.

(a) Estimate the arithmetic mean temperature difference in the evaporator when 25% of the water entering at C evaporates, and the water is assumed to be already boiling in the bottom part of the tubes where the liquid head is 0.35 m.

(b) Estimate the number of tubes needed in the evaporator if the baffles are spaced to give a heat transfer coefficient $h_o = 1500$ W/(m$^2\,^\circ$C) on the outside.

(c) Find the production of fresh water in tons per 24 hours.

9.5. Estimate number and length of tubes in the condenser in problem 9.4, using the same tubes as in the evaporator, with tube side pressure drop 0.25 bar and temperature difference 4°C between outlet and inlet. Condensate viscosity 1.0 cP and thermal conductivity 0.6 W/(m°C). The heat load is 150 kJ/s.

References

1. Reid, R. C., J. M. Prausnitz and Th.K. Sherwood: *The Properties of Gases and Liquids*, McGraw-Hill, New York, 3rd Edn., 1977.
2. Schmidt, E.: *Thermodynamics. Principles and Applications to Engineering*, Dover, New York, 1966.
3. Coulson, J. M. and J. F. Richardson: *Chemical Engineering*, Volume 1, Pergamon, Oxford, 2nd Edn., 1964.
4. Kato, H., N. Nishiwaki and M. Hirata: On the turbulent heat transfer by free convection from a vertical plate, *Int. J. Heat and Mass Transfer*, 11, 1117–1125 (1968).
5. Schiller, L.: Messung von thermischen Konvektionsströmungen und Wärmeübergang an einer quadratischen Platte, *Wärme- und Kältetechnik*, 43, 6–12 (1941).
6. Rich, R. B.: An investigation of heat transfer from an inclined flat plate in free convection, *Trans. Am. Soc. Mech. Engrs.*, 75, 489–499 (1953).
7. Virdee, G. S. and C. J. Liddle: Direct contact, liquid/liquid heat transfer, *Chem. and Process Eng.*, 29–31 (August 1971).
8. Sieder, E. N. and G. E. Tate: Heat transfer and pressure drop of liquids in tubes, *Ind. Eng. Chem.*, 28, 1429–1435 (1936).
9. *Engineering Data Book*, Wolverine Tube Division, Calumet & Hecla Inc., Southfield Rd., Michigan, 1959.
10. Grimison, E. D.: Correlation and utilization of new data on flow resistance and heat transfer for cross flow of gases over tube banks, *Trans. Am. Soc. Mech. Eng.*, **59**, 383 (1937) and **60**, 381 (1938).
11. *VDI-Wärmeatlas. Berechnungsblätter für den Wärmeübergang*, VDI-Verlag GmbH, Düsseldorf, 3rd Edn., 1977.
12. McAdams, W. H.: *Heat Transmission*, McGraw-Hill, New York, 3rd Edn., 1963.
13. Minton, P. E.: Designing spiral-plate heat exchangers, *Chem. Eng.*, 77 (No. 10), 103–112 (1970).
14. Chilton, T. H., T. B. Drew and R. H. Jebens: Heat transfer in agitated vessels, *Ind. Eng. Chem.*, 36, 510–516 (1944).
15. Strek, F. and S. Masink: Heat transfer in mixing vessels with propeller agitators, *Verfahrenstechnik*, 4, 238–241 (1970).
16. Chapman, F. S., H. R. Dallenbach and F. A. Holland: Heat transfer in baffled jacketed, agitated vessels, *Trans. Instn. Chem. Engrs.*, 42, T398–T406 (1964).
17. Uhl, V. W. and J. B. Gray: *Mixing*, Volume 1, Academic Press, New York, 1966.
18. Holland, F. A. and F. C. Chapman: *Liquid Mixing and Processing in Stirred Tanks*, Reinhold, New York, 1966.
19. Skelland, A. H. P.: Correlation of scraped-film heat transfer in the votator, *Chem. Eng. Sci.*, 7, 166–175 (1958).

20. Lydersen, A. L.: *Untersuchungen über Wärmeübergang und Druckabfall in Kugelstapeln beim Durchblasen von Luft*, PhD Thesis, NTH, Trondheim, 1950.
21. Brauer, H.: *Strömung und Wärmeübergang bei Rieselfilmen*, VDI-Forschungsheft 457, VDI-Verlag, Düsseldorf, 1956.
22. Wilke, W.: *Wärmeübergang an Rieselfilme*, VDI-Forschungsheft 490, VDI-Verlag, Düsseldorf, 1956.
23. Horn, R. K.: *Messungen zum Wärme- und Stoffübergang am Rieselfilm*, PhD Thesis, Karlsruhe, 1970.
24. Dahlhoff, B. and H.v. Brachel: Wärmeübergang an horizontalen in einem Fluidatbett angeordneten Rohrbündeln, *Chem.-Ing.-Techn.*, 40, 372–376 (1968).
25. Fritz, W.: Wärmeübergang zwischen Wärmeaustauschflächen und Fliessbetten verchiedener geometrischen Formen, *Chem.-Ing.-Techn.*, 41, 435–442 (1969).
26. Butcher, D. W. and C. W. Honour: Tetrafluoroethylene coatings on condenser tubes, *Int. J. Heat and Mass Transfer*, 9, 835 (1966).
27. Nusselt, N.: Die Oberflächenkondensation des Wasserdampfes, *Zeitschr. VDI*, 60, 541–546 and 569–575 (1916).
28. Kern, D. Q.: *Process Heat Transfer*, McGraw-Hill, New York, 1950.
29. Renker, W.: Der Wärmeübergang in Anwesenheit von nichtkondensierender Gase, *Chem.-Ing.-Techn.*, 7, 451–461 (1955).
30. Jakob, M. and W. Fritz: Versuche über den Verdampfungsvorgang, *Forsch. a.d. Gebiete des Ing.-Wesens*, 2, 434–447 (1931).
31. Chilli, M. T. and C. F. Bonilla: Heat transfer to liquids boiling under pressure, *Trans. A.I.Ch.E.*, 41, No. 6, 755 (1945).
32. Gorenflo, D.: Zur Druckabhängigkeit des Wärmeübergangs an siedende Kältemittel bei freier Konvektion, *Chem.-Ing.-Techn.*, 40, 757–762 (1968).
33. *Standards of Tubular Exchanger Manufacturers Association*, Tubular Manufacturers Ass. Inc., New York, 6th Edn., 1978.
34. Alfa Laval: *Thermal Handbook*, Aeroconsulter AB, Västerås, Sweden, 1969.
35. Gardner, K. A.: *Rational Design Temperature Limits for Interference-Fit Finned Tubing*, Symposium on Air-Cooled Heat Exchangers, Cleveland, Ohio, 10th Aug. 1964, American Society of Mechanical Engineers, New York.
36. Schmidt, Th.E.: Der Wärmeübergang an Rippenrohre und die Berechnung von Rohrbündel-Wärmeaustauschern, *Kältetechnik*, 15, 98–102 and 370–378 (1963).
37. How, H.: How to design barometric condensers, *Chem. Eng.*, 63 (2), 174–182 (1956).
38. Kopp, J. H.: *Über den Wärme- und Stoffaustauch bei Mischkondensatoren*, PhD Thesis, ETH, Zürich, 1956.
39. B.P.: *Heat Transfer Oils*, British Petroleum, East Grinstead, 1965.
40. Hauschild, W.: Leistung von Dünnschichtverdampfern mit zwangsläufig ausgebildeten Filmen, *Chem.-Ing.-Techn.*, 25, 573–574 (1953).
41. Badger, W. L. and W. L. McCabe: *Elements of Chemical Engineering*, McGraw-Hill, New York, 2nd Edn., 1936.
42. Perry, R. H. and C. H. Chilton: *Chemical Engineers' Handbook*, McGraw-Hill, New York, 5th Edn., 1973.

CHAPTER 10
Unsteady State Heat Transfer

In unsteady state heat transfer the temperature is a function of both space and time. The basic equations from steady state heat transfer are applied, but time has to be included as a new variable.

An analytical solution is recommended for simple cases, such as some problems with phase changes and simple boundary conditions.

A graphical solution is frequently the least time consuming method. Also, this method is a suitable basis for computer programming.

Computer calculation is recommended if an adequate program is available, or if the calculations have to be repeated several times.

With phase change

Unsteady state heat transfer with phase change occurs in freezing, melting, sublimation, or condensation of vapour to solid. Approximate calculations are carried out with the simplifying assumption that only the latent heat of phase change need be taken into account. As a second approximation sensible heat can be added to the latent heat as shown in Example 10.1.

Figure 10.1 shows a liquid being frozen from below. Heat conduction through the frozen layer is given by the equation

$$dQ/dt = kA \frac{\theta_1 - \theta_0}{x}. \tag{10.1}$$

(It is convenient to write dQ/dt instead of $\dot{Q}$.) The latent heat removed in freezing the layer dx is

$$dQ = \lambda \rho_0 A\, dx \tag{10.2}$$

where λ is the latent heat of freezing, ρ_0 is the density of the frozen material, and A is the area perpendicular to the heat flux.

Combination of equations (10.1) and (10.2) and separation of the variables yields

$$\int_{x_1}^{x_2} x\, dx = \frac{k}{\lambda \rho_0} (\theta_1 - \theta_0) \int_0^t dt.$$

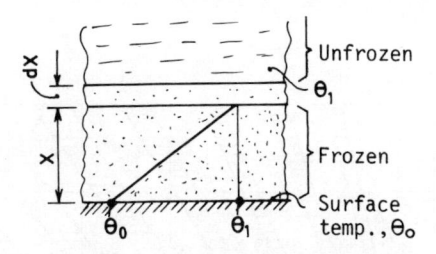

Figure 10.1 Liquid freezing from below: a thickness dx freezes in time dt.

Integration gives the freezing time from thickness x_1 to thickness x_2,

$$t = \frac{\lambda \rho_0}{k(\theta_1 - \theta_0)} \left(\frac{x_2^2}{2} - \frac{x_1^2}{2} \right) . \quad (10.3)$$

The corresponding equation for a cylinder is derived in Example 10.3. For freezing from more than one surface, reference should be made to the literature.[1]

Animal meat does not have a fixed freezing point, but freezes over a temperature range as shown for cod fillets in Figure 10.2. Figure 10.3 gives the time required for freezing of the same fillets when wrapped and placed between refrigerated plates as shown in Figure 10.4. Figures 10.2 and 10.3 are based on empirical measurements.[2]

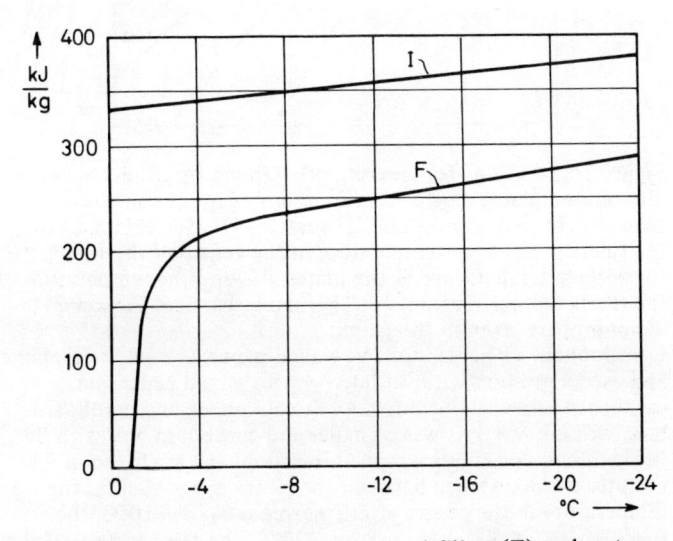

Figure 10.2 Heat removed from cod fillets (F) and water (I), at temperature 0°C before freezing, plotted against the final temperature.

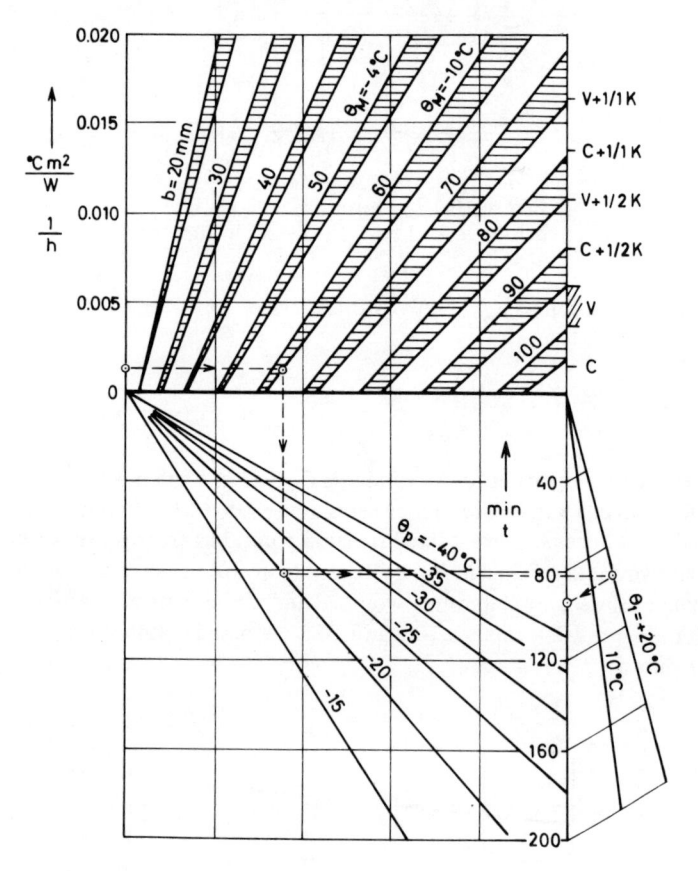

Figure 10.3   Time for freezing of wrapped fish fillets between refrigerated plates plotted as a function of the thermal resistance between the plates and the fillet surface, $1/h$, the thickness of the fillets b, the final temperature in the centre of the fillets, θ_M, the surface temperature of the plates, θ_p, and the temperature of the fillets before freezing, θ_1. Thermal resistances for some wrappings are given in the graph:
C, cellophane or plastic foil; V, waxed paper, C + ½K, cellophane and cardboard box without lid; V + ½K, waxed paper and cardboard box without lid; C + ¼K, cellophane and cardboard box with lid; V + ¼K, waxed paper and cardboard box with lid. Broken line: cod fillets wrapped in cellophane are fozen in 93 minutes if the distance between the plates is b = 60 mm, the temperature in the centre after freezing is $\theta_M = -10°C$, the temperature of the plates is $\theta_p = -20°C$, the temperature of the fillets before freezing +20°C, and the thermal resistance of the wrapping $1/h$ = 0.00125 °C m²/W.

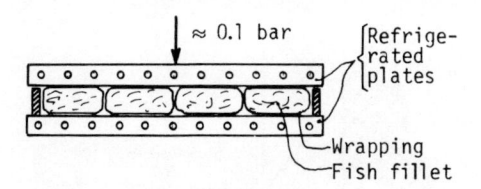

Figure 10.4 Freezing of wrapped fish fillets between refrigerated plates.

Without phase change

Figure 10.5 shows a thin slab with area A and thickness dx and oriented perpendicular to the heat flux. The amount of heat entering the slab and the amount of heat leaving the slab are shown in Figure 10.5. The difference,

$$(dQ/dx)_x - (dQ/dx)_{x+dx} = kA(\delta^2\theta/\delta^2 x)\,dx$$

is accumulated in the slab per unit time and equals $A\,dx c_p(\delta\theta/\delta t)$, i.e.

$$\frac{\delta\theta}{\delta t} = \alpha \frac{\delta^2\theta}{\delta x^2} \tag{10.4}$$

where α is the thermal diffusivity, $\alpha = k/\rho c_p$ m^2/s.

The second order partial differential equation (10.4) is solved by means of Laplace's theorem,

$$\theta = f(t)g(x) \tag{10.5}$$

where the functions $f(t)$ and $g(x)$ are determined by the boundary conditions. With an asymmetrical temperature profile, it may be simpler to deal with the related symmetrical profile as indicated in Figure 10.6. The use of this method is demonstrated in Example 10.5.

Equation (10.4) is solved in the literature for many cases with constant starting temperature.[3] Some of the solutions are given in diagrams originally introduced by

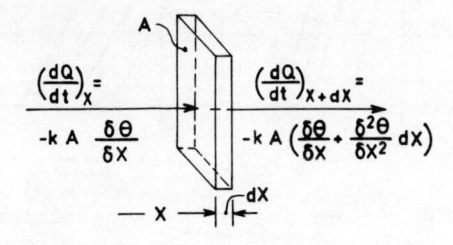

Figure 10.5 Heat flux into and out of a slab with area A perpendicular to heat flux.

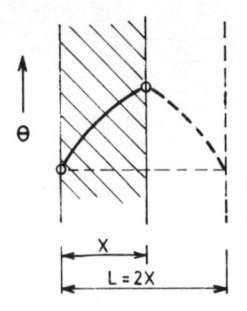

Figure 10.6 A symmetrical temperature profile over x treated as a symmetrical profile over $L = 2x$.

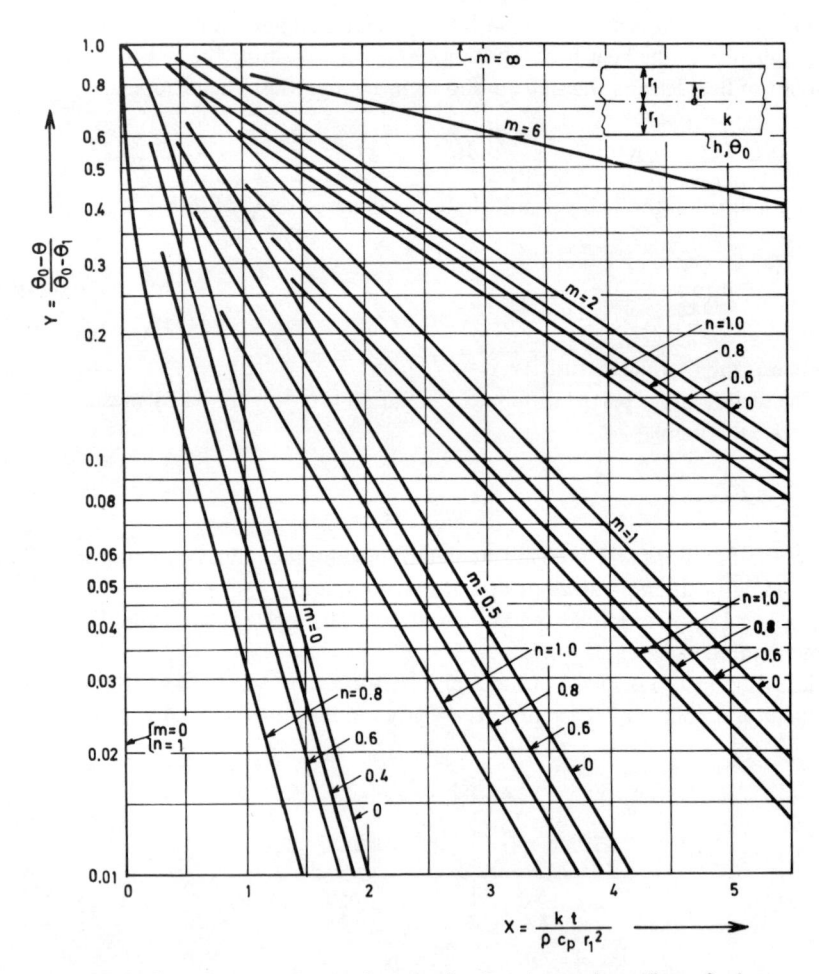

Figure 10.7 Gurney–Lurie diagram for heating and cooling of a large slab: $m = k/hr_1$ and $n = r/r_1$

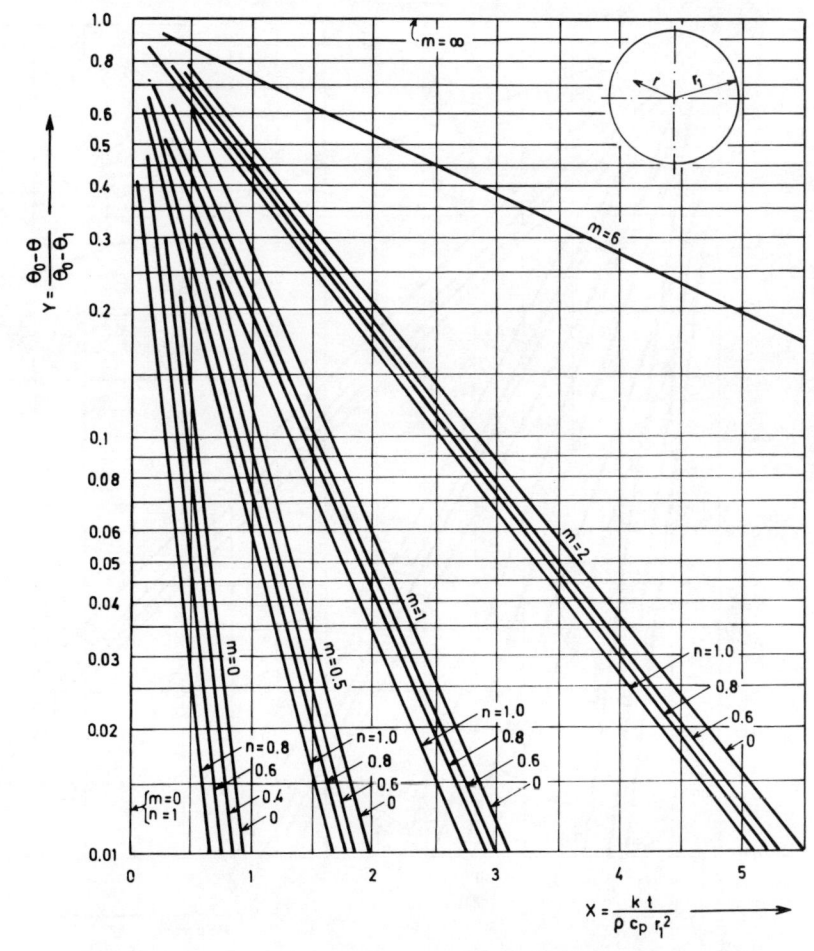

Figure 10.8 Gurney–Lurie diagram for heating and cooling of an infinitely long cylinder: $m = k/hr_1$ and $n = r/r_1$.

Gurney and Lurie.[4] Examples are given in Figures 10.7, 10.8, and 10.9, where

θ_0 = the temperature of the surroundings, °C or K

θ_1 = initial temperature of the body, °C or K

θ = temperature at distance r from the centre at time t, °C or K

k = thermal conductivity of the body, W/(m K)

h = heat transfer coefficient at the surface, W/(m² K)

ρc_p = specific heat capacity per unit volume, J/(m³ K) = Ws/(m³ K)

r_1 = distance from the centre to the surface, m.

A diagram showing the temperature in the centre of various bodies with constant surface temperature is given in Figure 10.10.[5]

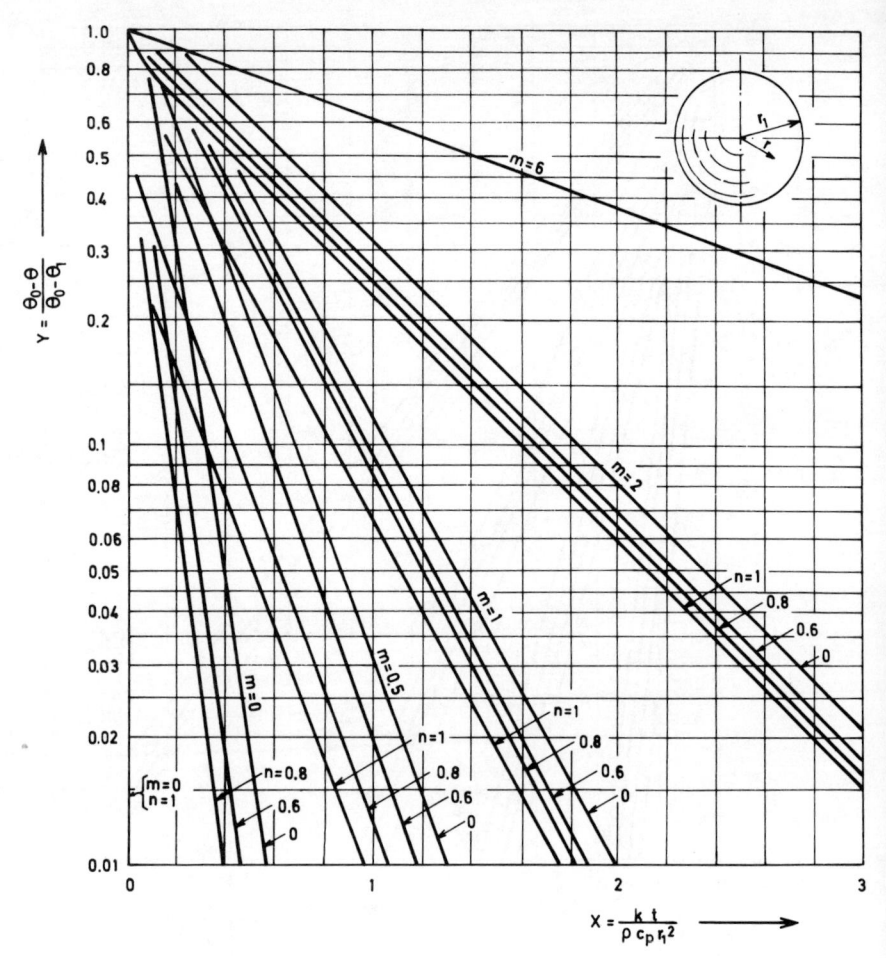

Figure 10.9 Gurney–Lurie diagram for heating and cooling of a sphere: $m = k/hr_1$ and $n = r/r_1$.

Schmidt's graphical method

In the graphical method of Ernst Schmidt the differential equation is replaced by a finite difference equation.[6,7] For a slab, the differentials in equation (10.4) are replaced by

$$\frac{\delta\theta}{\delta t} \approx \frac{\theta_{x,(t+\Delta t)} - \theta_{x,t}}{\Delta t}$$

$$\frac{\delta^2\theta}{\delta x^2} \approx \frac{(\theta_{(x+\Delta x),t} - \theta_{x,t})/\Delta x - (\theta_{x,t} - \theta_{(x-\Delta x),t})/\Delta x}{\Delta x}$$

$$= \frac{\theta_{(x+\Delta x),t} - 2\theta_{x,t} + \theta_{(x-\Delta x),t}}{(\Delta x)^2}.$$

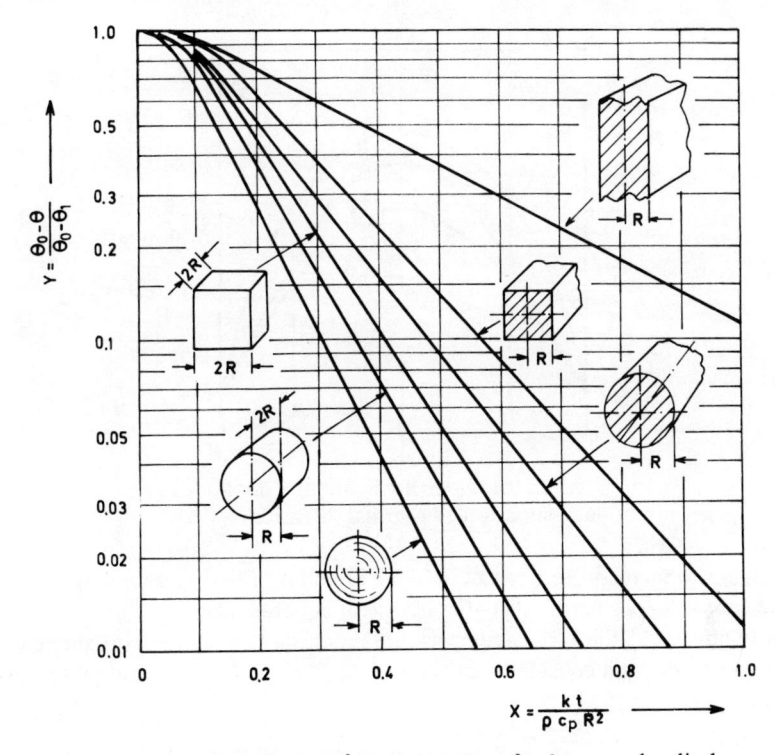

Figure 10.10 Temperature θ in the centre of spheres and cylinders with length equal to the diameter, cubes, square rods and cylinders with infinite length, and infinite slabs, as a function of the dimensionless group $X = kt/\rho c_p R^2$ when the surface temperature θ_0 is constant.

These differences inserted in equation (10.4), give

$$\theta_{x,(t+\Delta t)} - \theta_{x,t} = \alpha \frac{\Delta t}{(\Delta x)^2}(\theta_{(x+\Delta x),t} - 2\theta_{x,t} + \theta_{(x-\Delta x),t})$$

where Δt and Δx are independent variables, i.e. the time interval Δt can be chosen to give

$$\alpha \frac{\Delta t}{(\Delta x)^2} = \frac{1}{2} \tag{10.6}$$

and the equation reduces to

$$\theta_{x,(t+\Delta t)} = \frac{1}{2}(\theta_{(x+\Delta x),t} + \theta_{(x-\Delta x),t}). \tag{10.7}$$

The right hand side of the equation contains only temperatures at time t, while the left hand side gives the temperature at time $t + \Delta t$. The use of this equation for a slab is shown in Figure 10.11.

When the separation between the temperature curves becomes small, it is

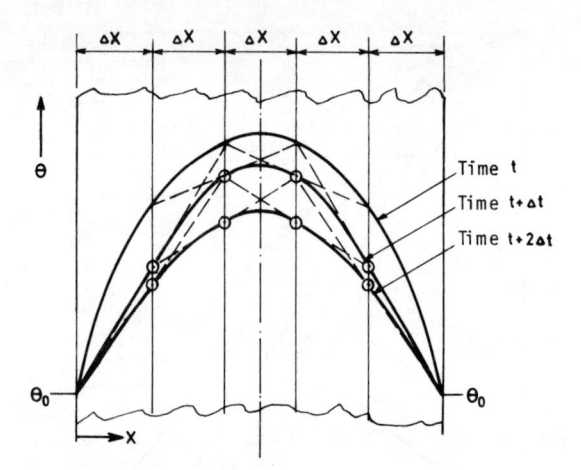

Figure 10.11 Ernst Schmidt's method for determination of thermal diffusion.

convenient to increase Δx, for example, by doubling. The corresponding new value of the time interval, Δt, is calculated by use of equation (10.6).

In Figure 10.11 the surface temperature is constant. With constant ambient temperature, θ_0, and constant heat transfer coefficient, h, the heat flow through the surface is

$$\left(\frac{dQ}{dt}\right)_{x=0} = hA(\theta_{x=0} - \theta_0) = kA\left(\frac{d\theta}{dx}\right)_{x=0}$$

or

$$\left(\frac{d\theta}{dx}\right)_{x=0} = \frac{\theta_{x=0} - \theta_0}{k/h}. \tag{10.8}$$

Equation (10.8) implies that an extension of the temperature curve from the surface will intercept the θ_0 line at a distance k/h from the surface. This is shown in Figure 10.12. In order to include the surface temperature, the first interval is located at a distance $\Delta x/2$ outside the surface (Figure 10.13).

Consider a cylinder with length L, inner radius r, and outer radius $r + dr$. The difference between the heat flow into and out of this cylinder is

$$\left(\frac{dQ}{dt}\right)_r - \left(\frac{dQ}{dt}\right)_{r+dr} = -k2\pi rL\left(\frac{\delta\theta}{\delta r}\right)_r + k2\pi(r+dr)L\left[\left(\frac{\delta\theta}{\delta r}\right)_r + \left(\frac{\delta^2\theta}{\delta r^2}\right)_r dr\right].$$

This difference is the heat accumulated per unit time,

$$\left(\frac{dQ}{dt}\right)_r - \left(\frac{dQ}{dt}\right)_{r+dr} = 2\pi rL\, dr\, \rho c_p \left(\frac{\delta\theta}{\delta t}\right)_r.$$

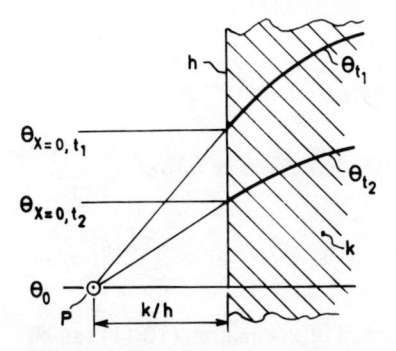

Figure 10.12 Plot of point P for a slab with thermal conductivity k and heat transfer coefficient h.

Combining the two equations and omitting the term involving dr^2 gives

$$\frac{\delta \theta}{\delta t} = \alpha \left(\frac{\delta^2 \theta}{\delta r^2} + \frac{1}{r} \frac{\delta \theta}{\delta r} \right) \qquad (10.9)$$

where $\alpha = k/\rho c_p$ is the thermal diffusivity of the cylinder, m^2/s.

As for a slab, the differentials can be replaced by differences and t selected to give

$$\alpha \frac{\Delta t}{(\Delta r)^2} = \frac{1}{2}. \qquad (10.10)$$

Also

$$\frac{\delta \theta}{\delta r} = \frac{\theta_{(r+\Delta r),t} - \theta_{r,t}}{\Delta r}.$$

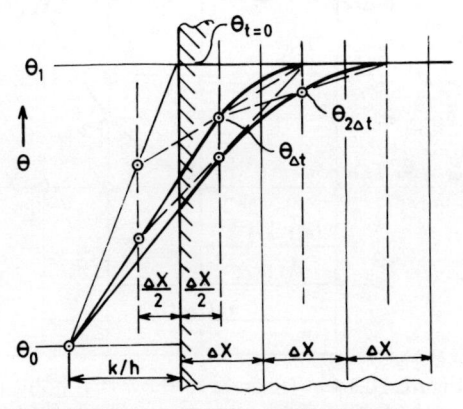

Figure 10.13 Graphical solution of thermal diffusion in a slab with constant initial temperature $\theta_{t=0} = \theta_1$.

can be replaced by

$$\frac{\theta_{(r+\Delta r),t} - \theta_{(r-\Delta r),t}}{2\Delta r}$$

This gives equation (10.9) in difference form,

$$\theta_{r,(t+\Delta t)} = \frac{1}{2}\left[\theta_{(r+\Delta r),t} + \theta_{(r-\Delta r),t} + \frac{\Delta r}{2r}(\theta_{(r+\Delta r),t} - \theta_{(r-\Delta r),t})\right]. \quad (10.11)$$

With ln r as abscissa instead of x, equation (10.11) can be solved graphically by the same procedure as for a slab. This can be seen from the following reasoning. In Figure 10.14 the curve 1–2–3 represents the temperature at time t. The two similar triangles in Figure 10.14 give,

$$\frac{\theta_B - \theta_{(r-\Delta r),t}}{\theta_{(r+\Delta r),t} - \theta_{(r-\Delta r),t}} = \frac{\ln r - \ln(r-\Delta r)}{\ln(r+\Delta r) - \ln(r-\Delta r)} = \frac{\ln\dfrac{1}{1-\Delta r/r}}{\ln\dfrac{1+\Delta r/r}{1-\Delta r/r}}$$

$$\approx \frac{1}{2} + \frac{1}{4}\frac{\Delta r}{r}.$$

The last expression is the Taylor approximation at the point $\Delta r/r = 0$. Rearranging the equation gives

$$\theta_B = \frac{1}{2}\left[\theta_{(r+\Delta r),t} + \theta_{(r-\Delta r),t} + \frac{1}{2}\frac{\Delta r}{r}(\theta_{(r+\Delta r),t} - \theta_{(r-\Delta r),t})\right]. \quad (10.12)$$

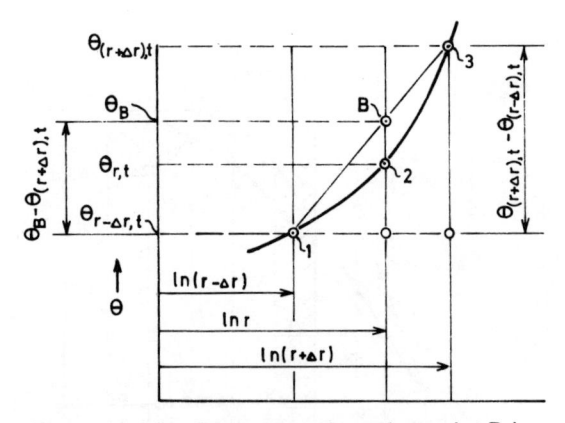

Figure 10.14 Diagram to show that point B is on the temperature curve for a cylinder at time $t + \Delta t$ when 1–2–3 is the temperature curve at time t.

The right hand sides of equations (10.11) and (10.12) are identical. Hence $\theta_B = \theta_{r,(t+\Delta t)}$, i.e. point B in Figure 10.14 is the temperature at radius r and time $t + \Delta t$.

The differential equation for a sphere is

$$\frac{\delta\theta}{\delta t} = \alpha\left(\frac{\delta^2\theta}{\delta r^2} + \frac{2}{r}\frac{\delta\theta}{\delta r}\right) \tag{10.13}$$

The graphical solution for a sphere is carried out as for a cylinder with $1/r$ as abscissa instead of $\ln r$.[7]

EXAMPLE 10.1. FREEZING OF ICE

Figure 10.15 shows an ice making machine ('Flake-Ice') in which a refrigerated metal band, a, slides on a drum and passes through water, b, where a layer of ice, c, is formed. The smaller roller, d, reduces the radius of the metal band and the ice loosens from the metal, as indicated in the figure, and breaks off. The temperature of the metal band is $-12°C$ and that of the water is $+10°C$.

Estimate the rotational speed of the metal band if the ice flakes are to be 3 mm thick, and 70% of the metal band is submerged in water.

Data for water:

latent heat of freezing,	$\lambda = 335$ kJ/kg
specific heat capacity,	$c = 4.2$ kJ/(kg K)
density of ice,	$\rho = 920$ kg/m^3
thermal conductivity of ice,	$k = 2.33$ W/(m K).

Solution

The heat removed from the water is the sum of the latent heat and the sensible heat,

$$\lambda' = 335 + (4.2)(10) = 377 \text{ kJ/kg}.$$

With $x_1 = 0$, equation (10.3) becomes

$$t = \frac{\lambda'\rho}{k(-\theta_0)}\frac{x_2^2}{2} = \frac{(377)(10^3)(920)}{(2.33)(12)}\frac{0.003^2}{2} = 56 \text{ seconds}$$

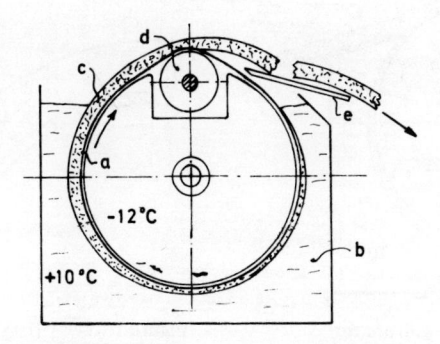

Figure 10.15 'Flake-Ice' machine.

which gives the time for one revolution as,

56/0.7 = 80 s

or 0.75 rev./min.

EXAMPLE 10.2. PRODUCTION OF FLAKES OF FATTY ACIDS

Figure 10.16 shows a rotating water cooled drum partly submerged in a mixture of palmetic and stearic acids that crystalize on the drum. The solid acids are removed by a knife with the edge approximately 0.5 mm from the drum as shown in Figure 10.17.

Estimate the speed in rev./min, n, and the flake thickness, if the production is to be 6000 kg/24 hours.

Given:

drum diameter	1.0 m	melting point of acids	54°C
effective drum length	3.0 m	heat removed	270 kJ/kg
drum temperature	25°C	thermal conductivity	
temperature of liquid acid	70°C	of solid acid,	$k = 0.21$ W/(m K)
		density of solid acid,	$\rho = 880$ kg/m^3.

Solution

Equation (10.3) gives the freezing time

$$t = \frac{(880)(270)(1000)}{(0.21)(3600)(54 - 25)} \left[\frac{(0.0005 + x)^2}{2} - \frac{0.0005^2}{2} \right]$$

$$= 5.42x\,(1000x + 1) \text{ hours.} \tag{a}$$

This is also the time during which a point on the surface of the drum is submerged,

$$t = \frac{1}{60n} \frac{(360 - 150)}{360} = \frac{1}{103n} \tag{b}$$

$$5.42x\,(1000x + 1) = \frac{1}{103n}. \tag{c}$$

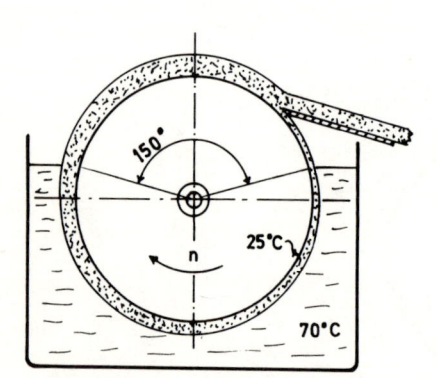

Figure 10.16 Rotating drum for production of fatty acid flakes.

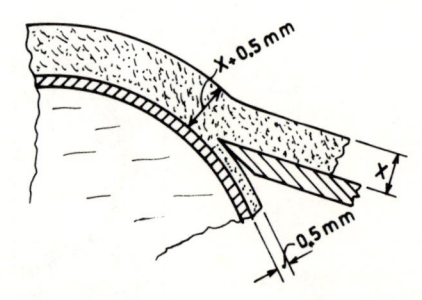

Figure 10.17 Detail of drum and knife in Figure 10.16.

The production is

$$\frac{6000}{24} = 250 = 3\pi 60 nx\ 880 = 498{,}000 nx \text{ kg/h}$$

or

$$n = \frac{1}{1992x} \qquad (d)$$

which is inserted into equation (c).

$$5.42(1000x + 1) = \frac{1992}{103},$$

$$x = 0.0026 \text{ m}$$
$$= 2.6 \text{ mm},$$

$$n = \frac{1}{(1992)(0.0026)}$$
$$= 0.19 \text{ rev./min.}$$

EXAMPLE 10.3. FREEZING IN A CYLINDER

Water at $0°C$ is frozen inside vertical steel tubes with inner diameter $D_1 = 34$ mm and outer diameter $D_2 = 42$ mm as shown in Figure 10.18. The refrigerant outside the tubes is at $\theta_2 = -20°C$ and its heat transfer coefficient is $h_2 = 300$ W/(m² K). Excess water is drained from the core when the liquid core has $D_0 = 10$ mm. The tubes are then heated and the ice cylinders removed by gravity.

Calculate the time required to freeze the cylinder if the heat capacities of frozen ice and tube walls are negligible compared to the latent heat of freezing. You are given that:

latent heat of freezing,	$\lambda = 335$ kJ/kg
density of ice,	$\rho = 920$ kg/m³
thermal conductivity of ice,	$k = 2.33$ W/(m K).

The thermal conductivity of steel is assumed to be infinite.

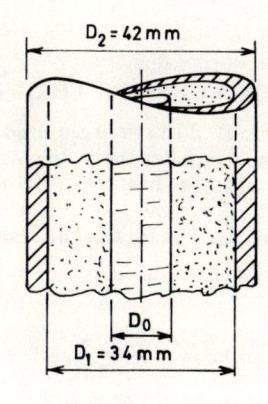

Figure 10.18 Section of a cylinder of frozen ice.

Solution

A layer dr is frozen in the time interval dt. The latent heat removed is

$$dp = 2\pi r dr \rho \lambda \text{ J/m tube} \tag{a}$$

This amount of heat is removed by conduction through a cylinder with inner radius r and outer radius $r_1 = D_1/2 = 0.017$ m. Equation (9.3) for heat conduction through a cylinder with temperature θ_0 at the inner and temperature θ_1 at the outer surface gives, per unit length of tube,

$$\frac{dq}{dt} = \frac{2\pi k (\theta_1 - \theta_0)}{\ln(r_1/r)} \quad \text{or} \quad \theta_1 - \theta_0 = \frac{dq}{dt} \frac{\ln(r_1/r)}{2\pi k}.$$

Equation (9.14) for heat transfer at the outer surface of the tubes gives

$$\theta_2 - \theta_1 = \frac{dq}{dt} \frac{1}{2\pi h_2 r_2}.$$

Sum of the two temperature differences

$$\theta_2 - \theta_0 = \frac{dq}{dt} \frac{1}{2\pi} \left(\frac{\ln \frac{r_1}{r}}{k} + \frac{1}{h_2 r_2} \right).$$

or

$$dq = \frac{-2\pi(\theta_0 - \theta_2)dt}{\ln(r_1/r)/k + 1/h_2 r_2}. \tag{b}$$

Equation (a) and (b) give

$$\int_0^{t_0} dt = \frac{-\rho\lambda}{k(\theta_0 - \theta_2)} \left[\ln r_1 \int_{r_1}^{r_0} r\, dr - \int_{r_1}^{r_0} r \ln r\, dr + \frac{k}{h_2 r_2} \int_{r_1}^{r_0} r\, dr \right]$$

and when integrated,

$$t_0 = \frac{-\rho\lambda}{2k(\theta_0 - \theta_2)} \left[r_0^2 \ln \frac{r_1}{r_0} + \frac{k}{h_2 r_2}(r_0^2 - r_1^2) + \frac{r_0^2}{2} - \frac{r_1^2}{2} \right]$$

or

$$t_0 = \frac{(-920)(335)(1000)}{(2)(2.33)(20)} \left[0.005^2 \ln 3.4 + \frac{2.33}{(300)(0.021)}(0.005^2 - 0.017^2) + \frac{0.005^2}{2} - \frac{0.017^2}{2} \right]$$

= 658 seconds or 11 minutes.

EXAMPLE 10.4. APPLICATION OF A GURNEY–LURIE DIAGRAM

Estimate for the ice cylinder in Example 10.3 the time required to heat the ice until its outer surface temperature reaches 0°C under the following conditions: the ice has a constant temperature of -14°C when the heating starts. The heating is carried out by circulation of a liquid at $+15$°C around the outside of the tubes with a heat transfer coefficient $h_2' = 135$ W/(m^2 K). The thermal diffusivity of ice is

$$\alpha = k/\rho c = 2.33/(920)(2000) = (1.27)(10^{-6}) \text{ m}^2/\text{s}.$$

The thermal diffusivity of steel is

$$\alpha_{\text{steel}} = 55/(8000)(460) = (1.5)(10^{-5}) \text{ m}^2/\text{s}$$

Solution

As an approximation, the steel tube is considered to be an ice cylinder with a wall thickness

$$s = \frac{(42-34)}{2} \frac{(1.27)(10^{-6})}{(1.5)(10^{-5})} = 0.3 \text{ mm}.$$

This gives as the outer diameter of the ice, 34.6 mm, and the heat transfer coefficient is corrected to this diameter,

$$h = 110 \frac{42}{34.6} = 136 \text{ W/(m}^2\text{ K)}.$$

The numbers to be inserted in the Gurney–Lurie diagram Figure 10-8 are

$$Y = \frac{15-0}{15+14} = 0.52,$$

$$m = \frac{2.33}{(135)(0.0173)} = 1.0,$$

$$n = 1.0.$$

The diagram gives

$$X = \frac{(1.27)(10^{-6})}{0.0173^2} t = 0.25,$$

$t = 59$ seconds or one minute.

Note Strictly, Figure 10.8 is only valid for a cylindrical solid rod. However, the 10 mm hole will make very little difference.

EXAMPLE 10.5. THERMAL DIFFUSION, ANALYTICAL SOLUTION

On watching the 'Flake-Ice' machine in Example 10.1, it is noticed that it takes 4 seconds from the time that the ice surface turns dull until the ice loosens from the metal band as shown in Figure 10.19. The temperature of the metal band is constant at $-12°C$. The temperature curve through the ice layer is approximately linear as shown in the right half of Figure 10.20 at the time when the surface turns dull. Heat exchange with the ambient air can be neglected. A customer wants to know:

(*a*) the surface temperature;
(*b*) the average temperature of the ice when it falls down into a storage bin below.

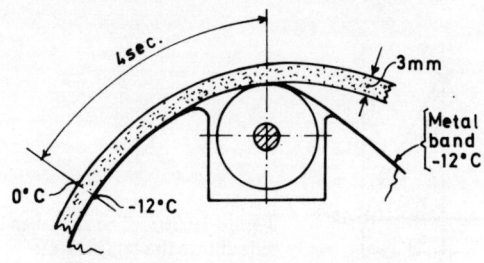

Figure 10.19 Top of the machine shown in Figure 10.15.

The thermal diffusivity of ice is

$$\alpha = \frac{k}{\rho c} = \frac{2.33}{(920)(2000)} = (1.27)(10^{-6}) \text{ m}^2/\text{s}.$$

Solutions

The problem is solved by means of the equations

$$\frac{\partial \theta}{\partial t} = \alpha \frac{\partial^2 \theta}{\partial x^2} \qquad (10.4)$$

$$\theta = f(t) g(x). \qquad (10.5)$$

The simplest analytical solution is obtained by

(1) assuming a symmetrical initial temperature curve as shown in Figure 10.20,
(2) choosing a convenient temperature scale, in this case adding 12°C to all temperatures so that the constant surface temperature is $\theta = 0$, and
(3) introducing boundary conditions (written below as BC) as soon as possible.

Combination of equations (10.4) and (10.5) gives

$$f'(t) g(x) = \alpha f(t) g''(x) \qquad (a)$$

or

$$\frac{f'(t)}{\alpha f(t)} = \frac{g''(x)}{g(x)}.$$

This equation is valid for all values of t and is independent of the value of x. This condition is only possible if the ratio is a constant, i.e.

$$\frac{f'(t)}{\alpha f(t)} = \frac{g''(x)}{g(x)} = \pm c$$

or

$$f'(t) = \pm c\, \alpha f(t) \qquad (b)$$

and

$$g''(x) = \pm c g(x). \qquad (c)$$

Equation (b) is satisfied by the exponential function

$$f(t) = c_1 e^{\pm c \alpha t}. \qquad (d)$$

BC: $\theta = 0$ when $t = \infty$. This boundary condition can only be fulfilled if the minus-sign is used, i.e.

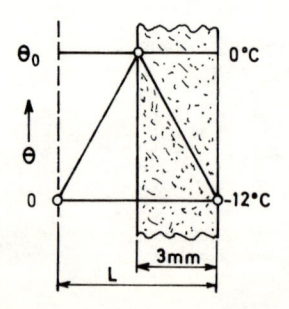

$$f(t) = c_1 e^{-c \alpha t}. \qquad (e)$$

Figure 10.20 The right hand side shows the temperature when the ice turns dull (see Example 10.5).

The minus sign must also be used in equation (c). With the minus sign, equation (c) is satisfied by a sine and cosine function,

$$g(x) = A \sin(x\sqrt{c}) + B \cos(x\sqrt{c}).$$

BC: $g(x) = 0$ when $x = 0$. This will not be the case for the cosine function, Thus,

$$g(x) = A \sin(x\sqrt{c}). \tag{f}$$

BC: $g(x) = 0$ when $x = L$. This will be the case when $L\sqrt{c} = n\pi$ where n is an integer,

$$c = \left(\frac{n\pi}{L}\right)^2.$$

This value of c is inserted in equation (e) and (f), and we get the solution of equation (10.5) in the general form

$$\theta = \sum_{n=1}^{\infty} b_n e^{-(n\pi/L)^2 \alpha t} \sin\left(\frac{n\pi}{L} x\right). \tag{g}$$

A time $t = 0$,

$$\theta(x, 0) = \sum_{n=1}^{\infty} b_n \sin\left(\frac{n\pi}{L} x\right). \tag{h}$$

This is the equation for a Fourier series for the odd function in the interval from $-L$ to L and it coincides with $\theta(x, 0)$ in the interval from 0 to L. Hence, the coefficient b_n is

$$b_n = \frac{2}{L} \int_0^L \theta(x, 0) \sin\left(\frac{n\pi x}{L}\right) dx \tag{i}$$

where $\theta(x, 0) = 2\theta_0 x/L$ for $0 \leq x \leq L/2$

$\theta(x, 0) = 2\theta_0 - 2\theta_0 x/L$ for $L/2 \leq x \leq L$.

These values of $\theta(x, 0)$ inserted into equation (i) give

$$b_n = \frac{2}{L} \int_0^{L/2} \frac{2\theta_0}{L} x \sin\left(\frac{n\pi x}{L}\right) dx + \frac{2}{L} \int_{L/2}^L \left(2\theta_0 - \frac{2\theta_0 x}{L}\right) \sin\left(\frac{n\pi x}{L}\right) dx$$

$$= \frac{8\theta_0}{n^2 \pi^2} \sin\left(\frac{n\pi}{2}\right).$$

Here, $\sin(n\pi/2)$ is 0 when n is even, $+1$ when $n = 1, 5, 9, \ldots$, and -1 when $n = 3, 7, 11, \ldots$. This is taken into account if b_n is written as

$$b_n = \frac{-(-1)^n 8\theta_0}{\pi^2 (2n-1)^2}.$$

Also, in equation (g), n must be replaced by $2n - 1$ and we get the solution

$$\theta = \frac{8\theta_0}{\pi^2} \sum_{n=1}^{\infty} \frac{-(-1)^n}{(2n-1)^2} e^{-[(2n-1)\pi/L]^2 \alpha t} \sin\left[\frac{(2n-1)\pi}{L} x\right]. \tag{j}$$

This is the solution with a temperature scale chosen to give a surface temperature 0°. Using the Celsius temperature scale with surface temperature θ_1, θ in equation (j) must be replaced by $\theta - \theta_1$ and θ_0 by $\theta_0 - \theta_1$. With the values of n

inserted, equation (j) with temperature in °C becomes

$$\frac{\theta - \theta_1}{\theta_0 - \theta_1} = \frac{8}{\pi^2} \left[e^{-(\pi/L)^2 \alpha t} \sin \frac{\pi x}{L} - \frac{1}{9} e^{-9(\pi/L)^2 \alpha t} \sin \frac{3\pi x}{L} + \ldots \right]. \quad (k)$$

In our case $L = (2)(0.003) = 0.006$ m and $t = 4$ s, and

$$\left(\frac{\pi}{L}\right)^2 \alpha t = \left(\frac{\pi}{0.006}\right)^2 (1.27)(10^{-6})(4) = 1.39.$$

This value, with $x = L/2$ inserted into equation (k), gives

$$\frac{\theta + 12}{12} = \frac{8}{\pi^2} \left(0.249 - \frac{1}{9} 0.0000037 + \ldots \right).$$

In this case it is sufficient to take only the first term into account. This gives the surface temperature of the ice

$$\theta = \frac{(8)(12)}{\pi^2} 0.249 - 12$$

$$= -9.6°\text{C}.$$

The average temperature θ_m is determined by integration over the cross-section,

$$\frac{\theta_m - \theta_1}{\theta_0 - \theta_1} = \frac{\int_0^L \frac{\theta - \theta_1}{\theta_0 - \theta_1} dx}{L}.$$

When $\left(\frac{\pi}{L}\right)^2 dt = 1.39$ is inserted in equation (k),

$$\frac{\theta_m - \theta_1}{\theta_0 - \theta_1} = \frac{8}{\pi^3 L} \int_0^L \left[e^{-1.39} \sin \frac{\pi x}{L} - \frac{1}{9} e^{-9(1.39)} \sin \frac{3\pi x}{L} + \ldots \right] dx.$$

It is sufficient to take account of only the first term

$$\frac{\theta_m - \theta_1}{\theta_0 - \theta_1} = \frac{16}{\pi^3} e^{-1.39} = 0.129$$

$$\theta_m = (0.129)(12) - 12$$

$$= -10.5°\text{C}.$$

EXAMPLE 10.6. THERMAL DIFFUSION, GRAPHICAL SOLUTION

Find answers for Example 10.5 by Ernst Schmidt's method.

Solution

The symmetrical temperature profile in Figure 10.20 is repeated in Figure 10.21 on a larger scale and with chosen values of Δx. The chosen values of Δx in Figure 10.21 and the corresponding values of Δt [equation (10.6)] are given below.

$$\Delta t = \frac{1}{2} \frac{\Delta x^2}{\alpha} = \frac{1}{2} \frac{\Delta x^2}{(1.27)(10^{-6})} = (3.94)(10^5) \Delta x^2 \text{ s}$$

Interval, Δx m	0.0005	0.001	0.0015
Time interval, Δt s	0.099	0.394	0.887

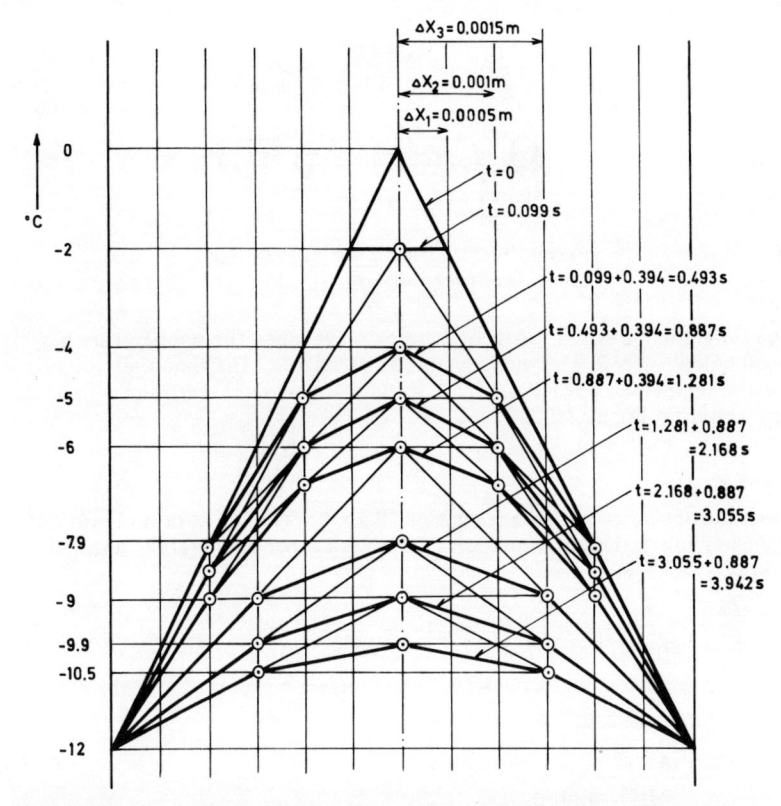

Figure 10.21 Graphical determination of the temperature as a function of the cooling time t in seconds for a $(2)(0.003) = 0.006$ m thick slab of ice with constant surface temperature $-12°C$. $\Delta x_1 = 0.0005$ m ($\Delta t = 0.099$ s) is only used once.

The temperature curve in Figure 10.21 at time $t = 3.94 \approx 4$ gives the temperature in the centre as $-9.9°C$ and the average temperature as

$$[(-12 - 10.5) + (-10.5 - 9.9)]/4 = -10.7°C.$$

EXAMPLE 10.7. GRAPHICAL SOLUTION OF THERMAL DIFFUSION IN A CYLINDER

Figure 10.22 shows a refrigerated storage room blasted into rock. The outer section is insulated and has two insulated doors. In heat loss calculations, the uninsulated part of the tunnel can be approximated by a cylinder of diameter 5.5 m. The influence of the ends can be neglected. Without the refrigerated tunnel the temperature of the rock is almost constant at $+6°C$ all year around. Thermal properties of the rock are:

thermal conductivity, $\quad k = 4$ W/(m K)
specific heat capacity per unit volume, $\rho c = (2660)(0.7) = 1860$ kJ/(m^3 K)
thermal diffusivity, $\alpha = k/\rho c = 4/(1860)(10^3) = (2.15)(10^{-6})$ m^2/s.

A supplier has offered refrigeration machinery with refrigeration capacity given in Figure 10.23. The heat transfer coefficient between the air and a cylindrical tunnel is estimated to be $h = 14$ W/(m^2 K). This estimate includes the influence of the rough blasted surface and other deviations from a cylindrical shape.

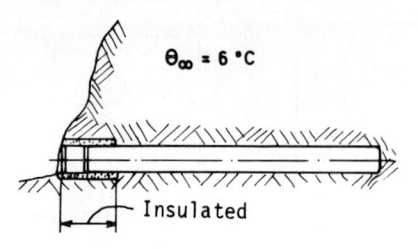

Figure 10.22 Tunnel serving as storage room for frozen fish.

The tunnel will be used for refrigerated storage when the temperature has dropped below $-22°C$. The owner wants an estimate of the time that will elapse before the temperature reaches $-22°C$ using the machinery with the refrigeration capacity given in Figure 10.23.

Solution

Heat transfer by convection [equation (9.14)] is treated as heat conduction through a cylindrical layer [equation (9.3)] with inner diameter D_1 and outer diameter $D_2 = 5.5$ m,

$$\frac{2\pi L k(\theta_s - \theta_a)}{\ln(D_2/D_1)} = \pi D_2 L h(\theta_s - \theta_a)$$

where θ_a is the air temperature and θ_s is the surface temperature of the tunnel.

$$\frac{(2)(4)}{\ln 5.5 - \ln D_1} = (5.5)(14),$$

$$D_1 = 4.96 \text{ m}$$

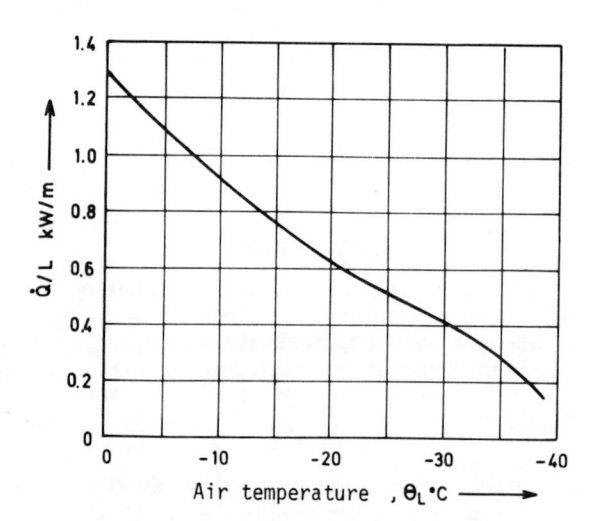

Figure 10.23 Refrigeration capacity per metre of uninsulated tunnel plotted as a function of the temperature in the tunnel.

and the radius of the superficial cylinder is,

$r_1 = 4.96/2 = 2.48$ m.

The heat transferred by convection per metre of the tunnel is

$$\dot{Q}/L = hA(\theta_s - \theta_a) = 14\pi 5.5(\theta_s - \theta_a) = 242(\theta_s - \theta_a) \text{ W/m}$$

or

$\theta_s = \theta_a + \dfrac{\dot{Q}/L}{242}$ as given below.

θ_a	°C	0	−5	−10	−15	−20	−25	Notes
$\dot{Q}/L$	W/m	1300	1080	920	760	620	520	from Figure 10.23
θ_s	°C	5.4	−0.5	−6.2	−11.9	−17.4	−22.9	$\theta_s = \theta_a + \dfrac{\dot{Q}/L}{242}$

In Figure 10.24 the air temperature θ_a is plotted as a function of the surface temperature of the tunnel θ_s.

Figure 10.25 shows the graphical determination of temperature as a function of time. Intervals, $\Delta r = 0.4$ m, have been chosen. The radius of the superficial layer is 2.48 m making ln 2.48 the abscissa. The first interval is located symmetrically about the surface, giving abscissae at $\ln(2.75 - 0.2) = \ln 2.55$ and $\ln(2.75 + 0.2) = \ln 2.95$. The time intervals are given by equation (10.10),

$$\alpha \frac{\Delta t}{\Delta r^2} = (2.15)(10^{-6}) \frac{\Delta t}{0.4^2} = \frac{1}{2},$$

$\Delta t = 37{,}200$

or

$37{,}200/3600 = 10.33$ h.

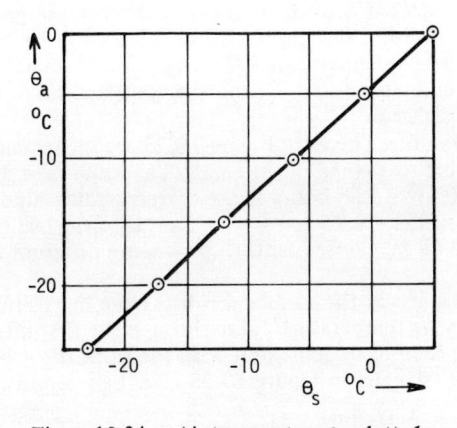

Figure 10.24 Air temperature θ_a plotted as a function of the surface temperature of the tunnel θ_s.

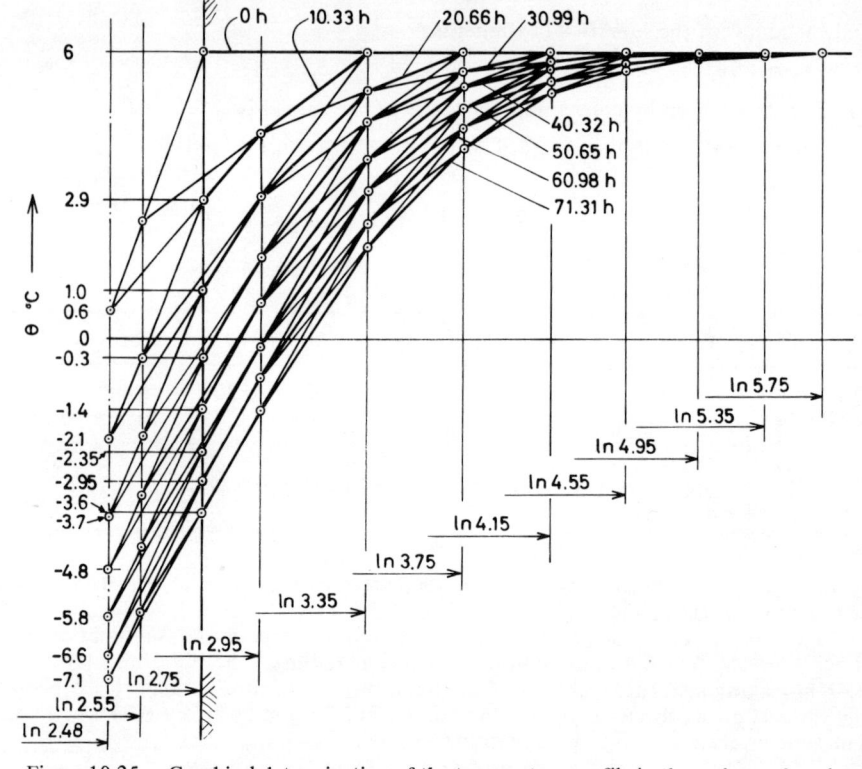

Figure 10.25 Graphical determination of the temperature profile in the rock as a function of time for the tunnel shown in Figure 10.22. The small circles to the far left give the corresponding air temperatures.

The initial rock temperature is +6°C. Figure 10.24 gives the air temperature, $\theta_a = 0.6°C$. This value is plotted at ln 2.48 in Figure 10.25 and connected by a straight line with the temperature +6°C at the surface. The point where this line crosses ln 2.55 is connected with the point (ln 3.35, 6°C) to give the temperature at ln 2.95 and at time 10.33 h. This point is connected with the air temperature point 0.6°C to give the surface temperature 2.9°C. $\theta_s = 2.9°C$ inserted in Figure 10.24 gives the new air temperature $-2.1°C$, and the construction of further temperature lines is seen on the diagram.

The lowest temperature curves in Figure 10.25 are inconveniently close together. Hence, it is reasonable to change the intervals. The value $\Delta r = 2.0$ m has been chosen with $\Delta t = 10.33(2/0.4)^2 = 258$ hours as the corresponding time interval. However, the inner radius, $r - \Delta r/2 = 2.75 - 1.0 = 1.75$ m, is to the left of the radius for the air temperature at 2.48 m. Consequently a procedure different from the previous one has to be used.

The first interval starts at the surface, i.e. it is from ln r to ln($r + \Delta r$) Figure 10.26. Furthermore the temperature curve through the first interval is replaced by a straight line in the semilogarithmic plot with the slope $(\theta_s - \theta)/[\ln r - \ln(r - s)]$. The resulting similar triangles in Figure 10.26 give the relationship,

$$\frac{\theta_1 - \theta_a}{\theta_s - \theta_a} = \frac{\ln(r + \Delta r) - \ln(r - s)}{\ln r - \ln(r - s)}$$

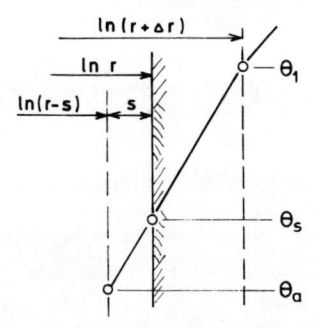

Figure 10.26 Diagram for derivation of equation (a) in Example 10.7.

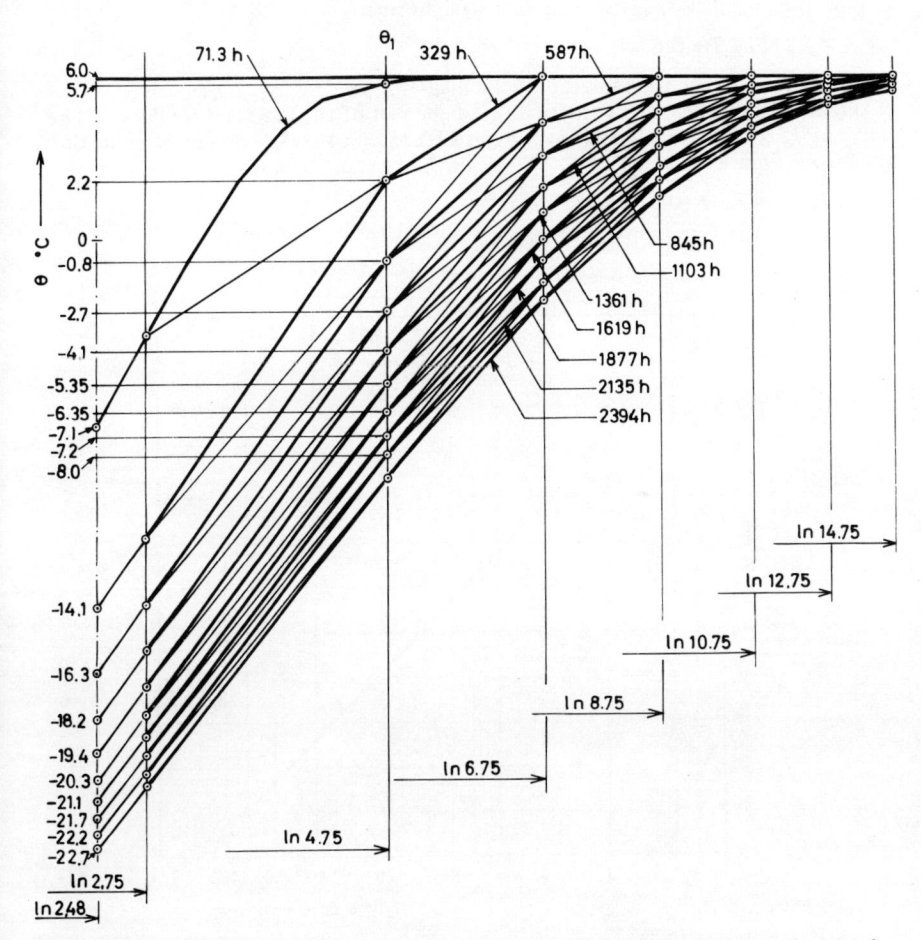

Figure 10.27 Graphical determination of the temperature in the rock adjacent to the tunnel shown in Figure 10.22 as a function of time. The small circles to the far left give the air temperatures.

or

$$\theta_1 = \theta_a + \frac{\ln\dfrac{r+\Delta r}{r-s}}{\ln\dfrac{r}{r-s}} (\theta_s - \theta_a) = \theta_a + \frac{\ln\dfrac{4.75}{2.48}}{\ln\dfrac{2.75}{2.48}} (\theta_s - \theta_a) = \theta_a + 6.29\,(\theta_s - \theta_a) \quad (a)$$

where s is the thickness of the layer that corresponds to the heat transfer coefficient, in our case $2.75 - 2.48 = 0.27$ m.

Using the values of θ_a from the previous table, equation (a) gives

θ_a	°C	−15	−20	−25	Notes
$6.29(\theta_s - \theta_a)$	°C	19.5	16.4	13.2	$\theta_s - \theta_a$ from previous table
θ_1	°C	4.5	−3.6	−11.8	from equation (a)

A plot of θ_a against θ_1 gives an almost straight line,

$$\theta_a \approx -17.7 + 0.63\,\theta_1. \quad (b)$$

This equation can be used instead of the diagram.

Figure 10.27 shows the construction of the rest of the temperature curves. The curve at 71.31 h is transferred from Figure 10.25 and the corresponding value of θ_1 read from Figure 10.27 is 5.7°C. This gives the next air temperature,

$$\theta_a = -17.7 + (0.63)(5.7)$$
$$= -14.1°C$$

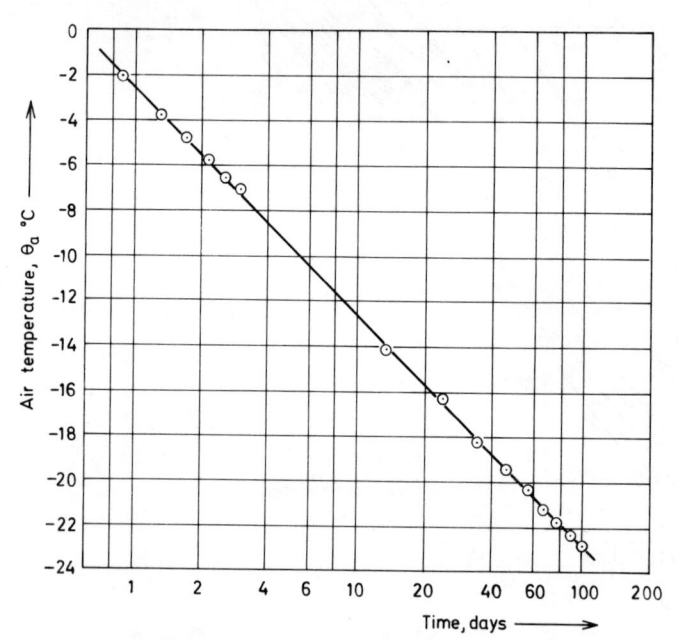

Figure 10.28 Air temperatures from Figure 10.25 and 10.27 plotted against time.

which is plotted at ln 2.48. The lines for next time interval give $\theta_1 = 2.2°C$ and

$\theta_a = -17.7 + (0.63)(2.2)$
$= -16.3°C$.

Figure 10.28 shows air temperatures from Figure 10.25 and 10.27 plotted against time in days. According to Figure 10.28 the desired air temperature of $-22°C$ is obtained after 85 days.

This estimate is based on two major approximations: a cyndrical tunnel and negligible influence of the ends. Hence it is advisable to add a safety factor, say 20%, i.e. be prepared to wait $(1.2)(85) = 102$ days or 3½ months before the desired temperature of $-22°C$ is reached.

Problems

10.1. Drums containing a viscous, aqueous solution are placed on edge. The diameter of the drums is 450 mm. Heat transfer through the ends and convection in the solution are neglected.

(a) The drums, temperature $30°C$, are placed in a cold storage room kept at $4°C$. Determine the time elapsed before the temperature in the centre is $12°C$.

(b) Calculate the time elapsed before the solution is frozen in the centre, when drums with solution at the freezing point are placed in a refrigerated room kept at $-22°C$.

Given:

	Density kg/m^3	Heat capacity kJ/(kg °C)	Thermal conductivity W/(m °C)
Liquid solution	1010	4.05	0.51
Frozen solution	960	1.8	2.1

The latent of freezing is 323 kJ/kg and the freezing temperature $0°C$. Overall heat transfer coefficient between the air and the inside of the drums, $U = 4.7$ W/(m^2 °C).

10.2. (a) Determine by Ernst Schmidt's method the time in seconds that it takes for the temperature in the centre of a 10 mm thick plate to reach $30°C$. The plate has thermal diffusivity $\alpha = (1.4)(10^{-7})$ m^2/s and a temperature $\theta_0 = 20°C$ when the surface temperature is suddenly changed to $100°C$.

(b) The time you determine before the temperature in the centre starts to increase, t_c, depends on the number of intervals chosen, $m = L/\Delta x$ where $L = 0.01$ m. Derive a function $t_c = f(m, L, \alpha)$ and comment on it.

10.3. A cold storage room has a 0.125 m thick outer concrete wall and an inner 0.25 m thick brick wall with 0.15 m insulation between the two walls. The wall temperature equals the ambient air temperature, $30°C$, when the air temperature in the room is lowered to and kept at $-20°C$.

(a) Using the Gurney–Lurie diagram, estimate the inner surface temperature of the brick wall after 10 hours, neglecting the heat transfer through the insulation.

(b) Solve the same problem by Ernst Schmidt's method.

(c) Solve the same problem, assuming the temperature of the concrete wall to remains constant at 30°C but the insulation is now substituted by a heat transfer coefficient,

$$h_{ins} = k_{ins}/L_{ins}.$$

Given:

brick wall, thermal diffusivity, $\alpha = (8.68)(10^{-7})$ m²/s
heat transfer coefficient (inner surface), $h = 6$ W/(m² °C)
insulation, thermal conductivity, $k_{ins} = 0.07$ W/(m °C).

References

1 Plank, R.: Beiträge zur Berechnung und Bewertung der Gefriergeschwindigkeit von Lebensmitteln, *Z. ges. Kälte-Ind.*, Beihefte, Reihe 3, H. 10 (1941).
2 Watzinger, A., A. Lydersen and H. Watzinger: *Freezing of Fish Fillets* (in Norwegian), Fiskeridirektoratet, Bergen, Norway, 1949.
3 Carslwaw, H. and J. C. Jaeger: *Conduction of Heat in Solids*, Clarendon Press, Oxford, 2nd Edn., 1959.
4 Gurney, H. P. and J. Lurie: Charts for estimating temperature distributions in heating and cooling solid shapes, *Ind. Eng. Chem.*, 15, 1170–1172 (1923).
5 Schneider, P. J.: *Conduction Heat Transfer*, Addison-Wesley, Cambridge, Mass. 1955.
6 Schmidt, E.: *Flöppl-Festschrift*, Springer-Verlag, Berlin, 1924.
7 Schmidt, E.: Das Differenzverfahren zur Lösung von Differentialgleichungen der nichtstationären Wärmeleitung, Diffusion und Impulsausbreitung, *Forsch. Ing.-Wes.*, 13, No. 5, 177–185 (1942).

CHAPTER 11
Energy Economy

In many industrial processes energy is a major expense. Energy consumption can be reduced if a process approaches more closely a reversible process. Means of obtaining this can be by

>   recovery of waste heat
>   operation at reduced temperature
>   multiple-effect vaporization
>   thermocompression ('heat pump')
>   use of a different process.

The savings in energy must be tempered by other considerations, such as maintenance, reliability, initial capital expenditure, and labour costs.[1]

Minimum energy requirement

The theoretical minimum energy required for a separation or concentration process is independent of the type of process used. Real processes require more energy than the theoretical minimum. Nevertheless, calculation of the theoretical minimum energy tells us whether or not there are possibilities for reducing the energy consumption. It may also give ideas about possible alternative processes.

There is no general equation for calculation of minimum energy in a separation process, but the calculations are correct if they are based on a *reversible* process.

Figure 11.1 shows schematically reversible, isothermal separation of a volatile component from a solution with a non-volatile solute, such as fresh water produced from sea water. Water vapour from sea water (A) at a temperature T_0 is compressed isothermally to the saturation pressure of water at temperature T_0. The corresponding minimum energy of separation is given by equation (4.8),

$$(-w)_T = RT_0 \ln(p_1/p_0) \text{ J/kmol water} \qquad (11.1)$$

where p_1 and p_0 are the vapour pressures of the pure component and of the solution at temperature T_0.

The saturation pressure of a 3.5 weight % aqueous NaCl solution is approximately 0.98 times the saturation pressure of water, giving minimum energy of separation,

$$(-w)_T \approx 8314 \, T_0 \, \ln(1/0.98) = 168 \, T_0 \text{ J/kmol water,}$$

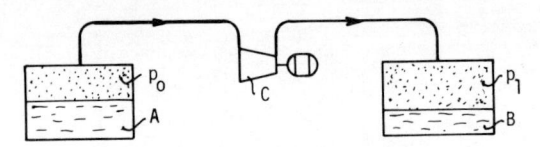

Figure 11.1 Reversible separation of a volatile component (B) from a solution (A) with a non-volatile solute.

The separation can also be carried out by reversed osmosis with osmotic pressure p_{os}. This gives a theoretical energy consumption

$$(-w)_T = p_{os}(18/1000) \text{ N m/kmol water}$$

and the osmotic pressure of the 3.5 weight % NaCl solution is

$$p_{os} \approx 168\, T_0/0.018 = 9333\, T_0 \text{ N/m}^2$$

Figure 11.2 shows schematically the reversible separation of a binary gas mixture by means of semipermeable membranes and isothermal compression. Ideal, semipermeable membranes do not exist, but the assumption that they do does not offend any law of nature, and they provide a simple means for calculating the minimum energy required.

Assuming ideal gases and an ideal gas mixture, the pressures behind the semipermeable membranes in Figure 11.2 are $p_A = y_A p_1$ and $p_B = y_B p_1$. For each of the components the isothermal work of compression is

$$(-w)_T = yRT_0 \int_{yp_1}^{p_1} \frac{dp}{p} = -RT_0 y \ln y \text{ J/kmol mixture.}$$

The total energy required is the sum of the work of isothermal compression for both components,

$$(-w)_T = -RT_0(y_A \ln y_A + y_B \ln y_B) \text{ J/kmol mixture}$$

or in a general form with n components,

$$(-w)_T = -RT_0 \sum_{i=1}^{n} y_i \ln y_i \text{ J/kmol mixture.} \qquad (11.2)$$

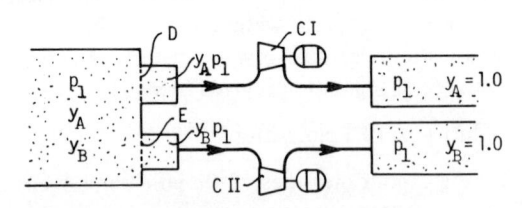

Figure 11.2 Reversible separation of a binary gas mixture with mole fraction y_A of component A and $y_B = 1 - y_A$ of component B. D and E are semipermeable membranes.

323

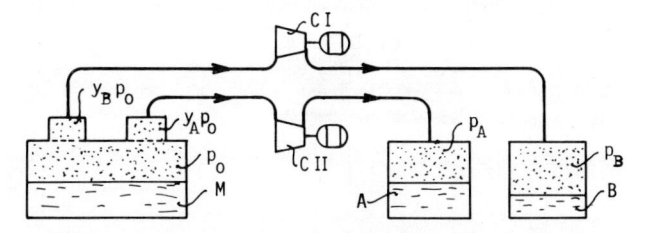

Figure 11.3 Reversible, isothermal separation of a binary liquid mixture by means of semipermeable membranes in the vapour phase.

Reversible separation of a binary liquid mixture is shown schematically in Figure 11.3. The mixture M in Figure 11.3 is boiling at temperature T_0. Assuming ideal gas behaviour and mole fractions y_A and y_B in the vapour, the pressures behind the membranes are $y_A p_0$ and $y_B p_0$. The isothermal compressors CI and CII compress the vapours to their saturation pressures at temperature T_0, p_A and p_B, and the vapours are condensed. The corresponding minimum energy requirement per kmole vapour is

$$(-w)_T = y_A R T_0 \int_{y_A p_0}^{p_A} \frac{dp}{p} + y_B R T_0 \int_{y_B p_0}^{p_B} \frac{dp}{p}$$

$$= R T_0 \left(y_A \ln \frac{p_A}{y_A p_0} + y_B \ln \frac{p_B}{y_B p_0} \right). \tag{11.3}$$

Recovery of waste heat

Reversible processes may be approached by recovering waste heat in heat exchangers. To a large extent this is done in many process industries, such as oil refining where heat recovery is of great importance to the economy.

The waste heat is often available at a temperature which is too low for direct use in the process. However, it may be utilized in conjunction with thermocompression, or in operating part of the plant at reduced temperature.

Operation at reduced temperature

There are cases where the heat requirement can be reduced by reducing the operating temperature below what has been common previously. An example is the concentration of alcohol from fermenting liquor from sulphite paper mills. Figure 11.4 shows the conventional concentration process with fractional distillation of cold, fermented liquor at atmospheric pressure. The feed F contains only about 2 mole % ethanol. Steam S is injected into the liquid at the bottom of the column. The vapour enriched in alcohol at the top is condensed in the condenser CI. CII is a smaller heat exchanger that reduces alcohol losses through the vent. R is the reflux liquid and D the distillate rich in alcohol. W is mainly water at about 100°C.

By this arrangement approximately half of the steam S is used to heat the water

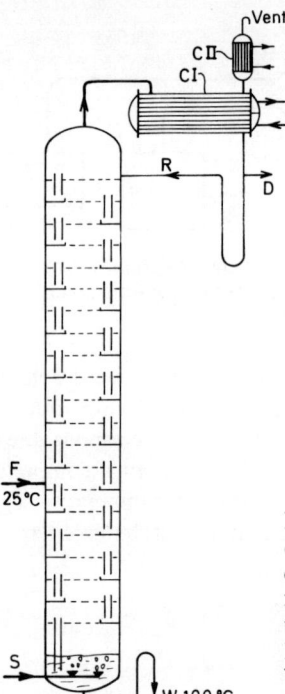

Figure 11.4 Continuous distillation column for concentration of alcohol from sulphite liquor. F is the feed with 2% ethanol, D distillate, R reflux, and W alcohol-free water. S is steam.

to 100°C. Most of this heat could be recovered by countercurrent heat exchange between the bottom product W and the feed F. But some liquors give prohibitive heavy deposits on heat transfer surfaces. If, however, the column is operated under vacuum, say 0.2 bar instead of 1 bar, the boiling temperature is reduced to 60°C instead of 100°C, and steam consumption is reduced by more than 25%.

Multiple-effect evaporation

Figure 11.5 shows a simple evaporator with no heat recovery. F is the solution to be concentrated, S steam from a steam boiler, and A the exhaust vapour. C is the thickened liquor.

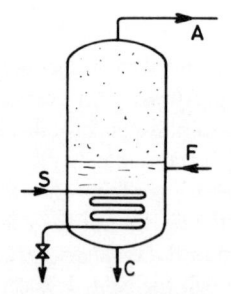

Figure 11.5 Simple evaporator without heat recovery.

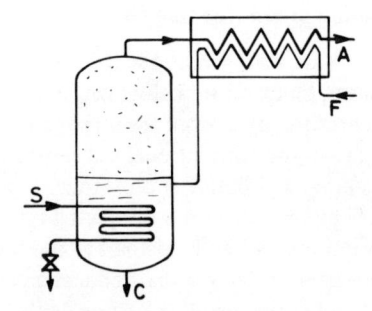

Figure 11.6 Evaporator with perheated feed.

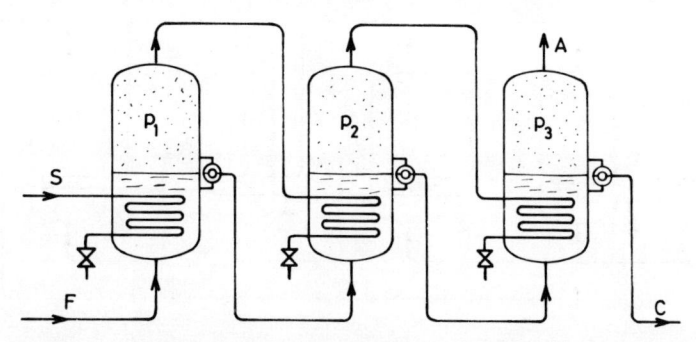

Figure 11.7 Triple-effect evaporator with forward feed. F is feed (solution), A exhaust vapour from the last stage, C the concentred liquor, and S is steam as heat source for the effect. Pressure $p_1 > p_2 > p_3$ give the necessary temperature differences between boiling liquid and condensing vapour.

Steam consumption in concentration of aqueous solutions can be reduced by up to 15 to 20% by heat exchange between the exhaust vapour and the feed (Figure 11.6).

In multiple-effect evaporators the vapour from one effect is the heating medium for another effect in which boiling takes place at a lower temperature and pressure. Figure 11.7 shows a triple-effect evaporator with forward feed, Figure 11.8 with backward feed, and Figure 11.9 a quadruple-effect evaporator with mixed feed.

In forward feed operation, the feed is introduced in the first effect and the liquor is transferred to succeeding effects by means of the pressure difference without the use of pumps. The method is advantageous if the concentrated liquor deposits scale on the heat transfer surfaces with the highest temperatures.

Backward feed requires pumps to transfer liquor between the effects, but has the advantage of better heat economy for cold feed and less viscous liquid, and therefore a higher heat transfer coefficient for the concentrated solution. In the quadruple-effect evaporator Figure 11.10, the feed is introduced to the first effect,

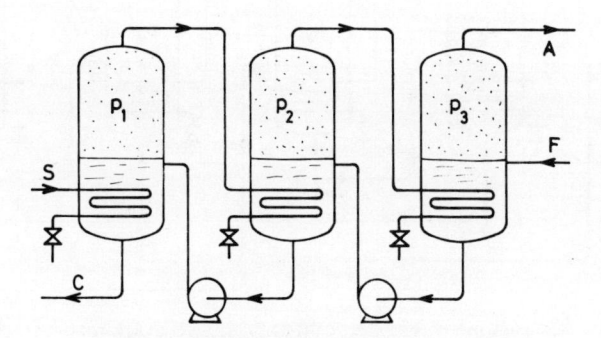

Figure 11.8 Triple-effect evaporator with backward feed: symbols as in Figure 11.7. Pressure $p_1 > p_2 > p_3$.

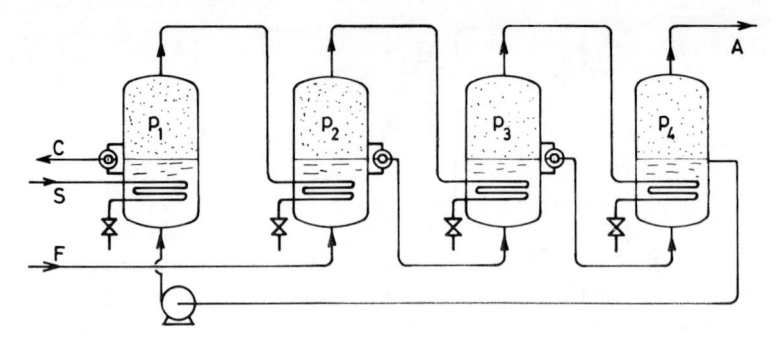

Figure 11.9 Quadruple-effect evaporator with forward feed in the three last effects and backward feed to the first. Pressure $p_1 > p_2 > p_3 > p_4$.

transferred forward to the third and fourth effect and pumped from the fourth effect back to the second effect, consisting of two evaporators. The two evaporators in this effect do reduce the amount of heat transferred to the liquor with maximum concentration, which gives the lowest coefficient of heat transfer.

Figures 11.7 to 11.10 are only schematic drawings, and additional heat recovery may be included. In some cases the condensate from one effect is released into the vapour supply for the following effect, where vapour flashing from the condensate will be an additional heat source. In total, the steam consumption in multiple-effect evaporation can be approximately proportional to the reciprocal of the number of effects.

Examples of its application are concentration of sugar solutions, of brine, of spent liquor in cellulose factories, of stick water in herring meal factories, etc.

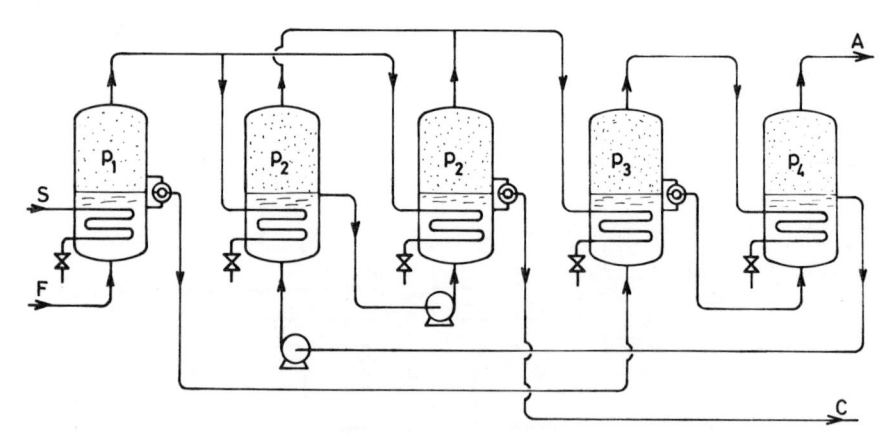

Figure 11.10 Quadruple-effect evaporator with two evaporators in the second effect where the liquor has maximum concentration. Pressure $p_1 > p_2 > p_3 > p_4$.

Heat pumps

Thermocompression, also called the heat pump, can reduce the energy requirement considerably. However, most of the energy is mechanical, and is more valuable than heat. Vapour recompression is shown schematically in Figure 11.11. The exhaust vapour at A is compressed to a pressure p_1 high enough to give a sufficient temperature difference in the heat exchanger E. Depending on heat losses to the surroundings and the temperature of the feed, some extra steam may be required or there may be some excess vapour, S (Figure 11.11).

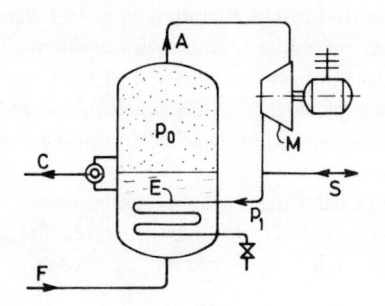

Figure 11.11 Vapour recompression: exhaust vapour from A at pressure p_0 is compressed to pressure p_1. F is feed and C concentrate. E is a heat exchanger, S steam supplied or rejected, M mechanical compressor.

Turbocompressors are used for vapour recompression of large vapour volumes (Figure 11.11), while steam ejectors are used in many smaller installations. Figure 11.12 shows schematically an installation for evaporating milk under vacuum. Part of the exhaust vapour is recompressed in the steam ejector to the right, while excess vapour is condensed in the barometric condenser to the left.

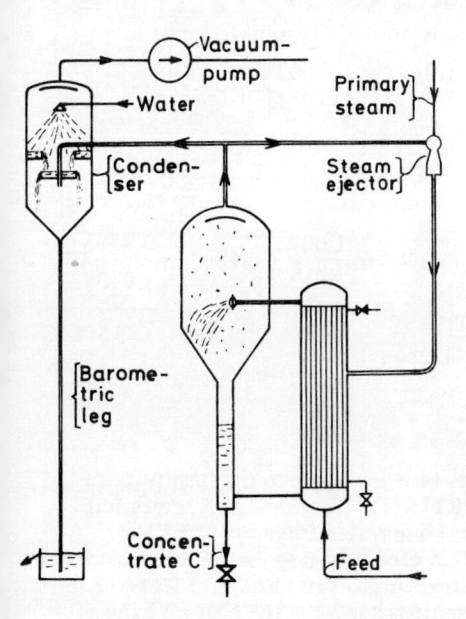

Figure 11.12 Vacuum evaporator with vapour recompression and steam ejector as compressor. Excess vapour is condensed in the barometric condenser to the left.

Other processes

Minimum energy requirement is obtained in reversible processes. This makes it worth considering operating possibilities closer to reversible processes. For example, the irreversible parts of the process in Figure 11.11 are heat transfer with a finite temperature difference, compression with losses, and irreversible mixing of the dilute feed F with the concentrated solution C. The result is a lower vapour pressure (and also a lower heat transfer coefficient) than for the dilute feed. This is avoided in a batch process where the pressure ratio p_1/p_0 increases gradually and only at the end reaches the value of the continuous process.

Liquid-liquid extraction is another process that can be operated close to reversibility, and the total energy requirement may be reduced. Penicillin was originally produced from dilute aqueous solutions by freezing and sublimation of the ice. Today penicillin is extracted from the aqueous phase into an organic liquid with a smaller volume, and the energy required for the separation is a fraction of what was needed previously.

Reversed osmosis is another process with possibilities for approaching reversibility. It was first used for production of fresh water from brackish water. Better membranes have opened up new applications, such as in dairies and cellulose factories.

EXAMPLE 11.1. SEPARATION OF A LIQUID MIXTURE

Calculate the minimum energy requirement for production of absolute ethanol from the azeotrope with 96 vol. % (89.43 mole % or 96.46 weight %) ethanol in water at a temperature of 78.15°C.

	96 vol. % ethanol	Ethanol	Water
Vapour pressure at 78.15°C, bar	1.013	1.007	0.439

Solution

The calculation is carried out for the reversible process in Figure 11.3, and equation (11.3) gives minimum energy per kmole mixture,

$$(-w)_T = (8314)(351.3)\left(0.8943 \ln \frac{1.007}{1.013} + 0.1057 \ln \frac{0.439}{1.013}\right)$$

$$= -274{,}000 \text{ J/kmol}$$

or 274,000/0.8943 = 306,000 J/kmol ethanol.

EXAMPLE 11.2. HEAT PUMP

Figure 11.13 shows a heat pump used to heat tap water to 95°C with waste water at 55°C as heat source. Refrigerant R114 (1,2-dichloro-1,1,2,2-tetrafluoroethane) evaporates in HE III. The vapour is compressed from p_7 = 2.52 bar to p_8 = 14.22 bar in the compressor SC. (Oil injected helical screw compressors can be used with R114 for condensation temperatures up to 110–120°C.[2]) R114 under pressure p_8 condenses in HE I. The condensate is cooled in the heat exchanger

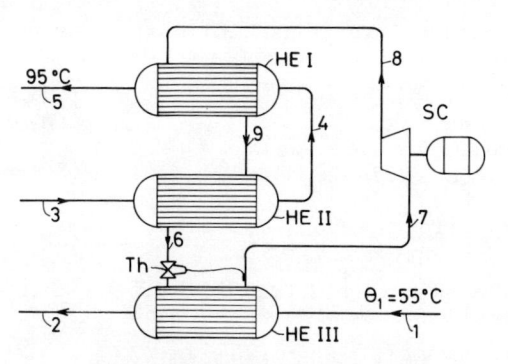

Figure 11.13 Heat pump with R114.

HE II and throttled from 14.2 bar to 2.5 bar in the throttle valve Th before it enters the evaporator HE III. The hot waste water at 1 (55°C) is cooled to a temperature θ_2 giving a mean temperature difference $\Delta\theta_{III} = 12°C$ in the R114 evaporator (HE III). Cold tap water at 3 (15°C) is heated in the counter current heat exchanger HE II and the R114 condenser HE I to give an outlet temperature $\theta_5 = 95°C$. The temperature of R114 from HE I is $\theta_9 = 96°C$ and from HE II $\theta_6 = 20°C$.
The adiabatic efficiency of the compressor is $\eta_{ad} = 0.70$.

Physical data for R114 (1,2-dichloro-1,1,2,2-tetrafluoroethane):

Temperature °C	Saturation pressure bar	Enthalpy Liquid h' kJ/kg	Enthalpy Vapour h'' kJ/kg	Entropy Liquid s' kJ/(kg K)	Entropy Vapour s'' kJ/(kg K)
20	1.82	439.1	569.7	4.259	4.705
30	2.52	449.7	576.2	4.294	4.711
96	13.12	524.2	615.4	4.514	4.761
100	14.22	529.1	617.4	4.527	4.763

Calculate per kg waste water

(a) kg R114 to be compressed,

(b) kg hot water produced,

(c) the heat transferred and the mean temperature differences in heat exchangers HE I and HE II.

(d) If the electric motor, M, has 90% efficiency, calculate the energy consumption of the motor in kWs = kJ per kJ energy utilized for hot water.

Solution

(a) R114 boils at 30°C in the evaporator. The logarithmic mean temperature difference, 12°C, is obtained in the evaporator with waste water temperature $\theta_2 = 34.6°C$. Heat transfer in the evaporator,

$$\dot{Q}_{III} = (55 - 34.6)4.19 = 85.5 \text{ kJ/kg waste water.}$$

Amount of R114 evaporated, $85.5/(576.2 - 439.1) = 0.62$ kg/kg waste water.

(b) Isentropic compression of saturated R114 vapour from $p_7 = 2.52$ bar to $p_8 = 14.22$ bar will result in vapour quality $x < 1.0$:

$4.763x + 4.527(1 - x) = 4.711$, $\qquad x = 0.78$.
Enthalpy before compression, $\qquad h_7 = 576.2$ kJ/kg R114.
Enthalpy after isentropic compression to $p_8 = 14.22$ bar,
$(0.78)(617.4) + (1 - 0.78)(529.1)$ $\qquad = 598.0$ kJ/kg R114.

Theoretical energy of compression, $\qquad 21.8$ kJ/kg R114.
Enthalpy after real compression, $h_8 = 576.2 + 21.8/0.7$ $\qquad = 607.3$ kJ/kg R114.

Heat transferred to the water in HE I and HE II, $607.3 - 439.1 = 168.2$ kJ/kg R114, or $(0.62)(168.2) = 104.3$ kJ/kg waste water. Amount of tap water heated,

$$104.3/[(95 - 15)(4.19)] = 0.31 \text{ kg/kg waste water.}$$

(c) Heat transfer in HE I,

$(0.62)(607.3 - 524.2) = 51.5$ kJ/kg waste water,

∴ $(0.31)(95 - \theta_4)(4.19) = 51.5$,

$$\theta_4 = 55.4°C.$$

$$\Delta\theta_m = \frac{(100 - 55.4) - (100 - 95)}{\ln \dfrac{100 - 55.4}{100 - 95}}$$

$= 18.1°C$.

Heat transfer in HE II,

$(0.31)(55.4 - 15)(4.19) = 52.4$ kJ/kg waste water.

$$\Delta\theta_m = \frac{(96 - 55.4) - (20 - 15)}{\ln \dfrac{96 - 55.4}{20 - 15}}$$

$= 17°C$.

(d) Energy consumption of the electric motor per unit energy absorbed by the heated water,

$(0.62)(607.3 - 576.2)/[(0.9)(0.31)(95 - 15)(4.19)] = 0.206$ kJ/kJ,

i.e. the amount of energy recovered is $1/0.206 = 4.85$ times the energy spent for compression.

Note Some additional energy from the waste water can be recovered in an extra counter current heat exchanger inserted between stream 2 and stream 3. It is also worth noting that more than 50% of the energy transferred to the tap water is transferred in heat exchanger HE II.

EXAMPLE 11.3. MULTIPLE-EFFECT EVAPORATION

Distilled water is to be produced by vaporization of tap water and condensation of the vapour. The heat source is steam with pressure 5.5 bar and quality $x = 0.98$. Condensate from the steam is returned to the steam boiler as feed water.

Calculate the steam consumption per kg evaporated water, S/A, for different arrangements, assuming that the tap water temperature is 10°C, the feed of tap water, F, is 18% more than the evaporated water, the temperature of the condensate from evaporators is the arithmetic average of the temperature of the boiling water in the evaporator and the condensing vapour. Heat exchange with the surroundings can be neglected.

(a) Calculate S/A for the single effect evaporator Figure 11.5 with pressure in the evaporator, $p_0 = 1$ bar absolute.

(b) Carry out the same calculation with heat exchange between the feed and the exhaust vapour (Figure 11.6), assuming average temperature difference $\Delta\theta_m = 25°C$.

(c) Calculate the amount of distilled water produced per kg steam with the triple-effect evaporator of Figure 11.7, assuming atmospheric pressure in the last effect.

(d) Carry out the same calculation with the feed preheated to $94°C$ by the exhaust vapour from the last effect.

Solution

(a) Enthalpy of steam, 5.5 bar and $x = 0.98$ (from diagram, page 348), $\qquad h_1 = 2711$ kJ/kg.
The saturation temperature at 5.5 bar is $155.5°C$, and condensate temperature, $(155.5 + 100)/2 = 127.8°C$
Enthalpy of water at $127.8°C = (127)(4.19)$ $\qquad h_2 = 536$ kJ/kg

Enthalpy used in evaporator, $\qquad h_1 - h_2 = 2175$ kJ/kg.
Feed per kg evaporated water, $F = 1/0.85 = 1.18$ kg/kg
Enthalpy of evaporated water (from steam table), $\qquad h_0'' = 2675$ kJ/kg.
Enthalpy of feed, $h_0 = (4.19)(10) = 42$ kJ/kg, or per kg
  evaporated, $Fh_0 = (1.18)(42) = $ $\qquad$ 50 kJ/kg.
Enthalpy of rejected water at C, $h_0' = (4.19)(100) = 419$ kJ/kg
  or per kg evaporated, $(F - 1)h_0' = (1.18 - 1)(419) = $ $\qquad$ 75 kJ/kg.
Enthalpy required per kg evaporated,
  $[h_0'' + (F - 1)h_0'] - Fh_0 = (2675 + 75) - 50 = $ $\qquad$ 2700 kJ/kg.
Steam consumption per kg distilled, $S/A = 2700/2175 = $ $\qquad$ 1.24 kg/kg.

(b) The temperature difference at the inlet of the heat exchanger is $\Delta\theta_i = 100 - \theta_i$, and at the outlet $\Delta\theta_o = 100 - 10 = 90°C$. Equation (9.43) gives

$$\Delta\theta_m = \frac{90 - (100 - \theta_i)}{\ln\dfrac{90}{100 - \theta_i}} = 25°C.$$

Trial and error gives $\theta_i = 97.3°C$. The corresponding enthalpy of the water is,

$h_0 = (4.19)(97.3) = 408$ kJ/kg.

With $F = 1.18$ kg feed per kg evaporated, the enthalpy per kg evaporated is

$[h_0'' + (F - 1)h_0'] - Fh_0 = (2675 + 75) - (1.18)(408) = 2269$ kJ/kg

or steam consumption,

$S/A = 2269/2175$
$\quad = 1.04$ kg/kg.

(c) Assuming the same temperature difference in each effect,

$(155.5 - 100)/3 = 18.5°C$, gives the following data:

	Effect		First	Second	Third	
Temp.	Condensing vapour,	θ_v °C	155.5	137	118.5	
	Boiling water,	θ_w °C	137	118.5	100	$\theta_w = \theta_v - 18.5$
	Condensate from heat exchanger,	θ_c °C	146.3	127.8	109.3	$\theta_c = (\theta_v + \theta_w)/2$
Enthalpies	Vapour to each effect,	kJ/kg	2711	2729	2704	from diagram
	Condensate from heat exchanger,	kJ/kg	616	537	458	from steam tables
	Vapour from each effect, h''	kJ/kg	2729	2704	2676	from steam tables
	Liquid from each effect, h'	kJ/kg	576	497	419	from steam tables

The enthalpies in this table are shown in Figure 11.14. The symbols F, S_1, S_2, and S_3 refer to one kg distillate which is the sum of the vapour from all three effects.

Enthalpy balances around the effects Figure 11.14 give, for 1st effect,

$$S_1\,2711 + (1.18)(42) = S_1\,616 + (1.18 - S_2)\,576 + S_2\,2729 \tag{a}$$

$$S_1 = 1.028\,S_2 + 0.308, \tag{b}$$

2nd effect,

$$S_2\,2729 + (1.18 - S_2)\,576 = S_2\,537 + (1.18 - S_2 - S_3)\,497 + S_3\,2704$$

$$S_2 = 1.044\,S_3 - 0.044 \tag{c}$$

and 3rd effect,

$$S_3\,2704 + (1.18 - S_2 - S_3)\,497 = S_3\,458 + (0.18)(419) + (1 - S_2 - S_3)\,2676$$

$$S_2 = -2.031\,S_3 + 0.994. \tag{d}$$

Equations (b), (c), and (d) give

$S_3 = 0.338$, $S_2 = 0.309$, and $S_1 = 0.626$ kg/kg.

(d) The enthalpy of water at 94°C is $(4.19)(94) = 394$ kJ/kg. This value can be inserted into equation (a) in place of 42. With the rest unchanged,

$$S_1\,2711 + (1.18)(394) = S_1\,616 + (1.18 - 0.309)(576) + (0.309)(2729)$$

$$S_1 = 0.42 \text{ kg/kg}.$$

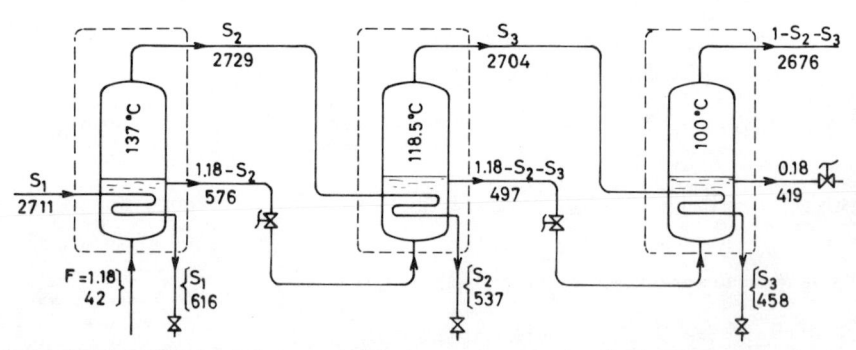

Figure 11.14 Enthalpy balance around each evaporator (broken line). The upper numbers or letters are kg per kg evaporated from all three evaporators. The numbers underneath are enthalpies in kJ/kg.

Comment The saving in heat going from the simple evaporator Figure 11.4 to a triple-effect evaporator, using most of the heat from condensation of the vapour from the last effect, is $(100)(1.24 - 0.42)/1.24 = 66\%$. This saving in heat has to be compared with the extra expense for equipment. Slightly more heat can be recovered if the condensate from the heat exchanger in the second effect is discharged into the vapour to the third effect.

Problems

11.1. Calculate the reduction in amount of steam to the first evaporator in Example 11.3 if the condensate from the second evaporator is throttled into the vapour line to the third evaporator.

11.2. Figure 11.15 shows the distillation column in Figure 11.4 supplied with a turbocompressor T for vapour recompression. The steam consumption S without vapour recompression is 215 kg per kmole distillate D. The feed F is mainly water with 2 mole % ethanol, specific heat capacity 76 kJ/(kmol °C). The distillate is 85 mole % ethanol and 15 mole % water. The heat of vaporization is assumed constant 41,000 kJ/kmol both for ethanol and for water. Heat of solution and heat losses to the surroundings can be neglected. Water to the condenser/evaporator CI is at a temperature 25°C, and it evaporates under a pressure 0.326 bar absolute. The pressure in the bottom of the column is 1.2 bar absolute.

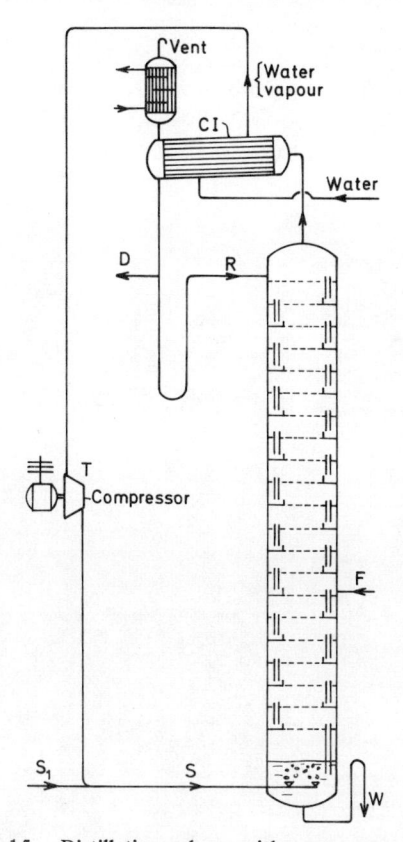

Figure 11.15   Distillation column with vapour recompression.

(*a*) Estimate the power supply to the turbocompressor in kJ/kmol feed if the motor has an efficiency $\eta = 0.90$ and the compressor an adiabatic efficiency $\eta_{ad} = 0.70$.

(*b*) Estimate the additional steam supply S_1 required.

References

1. Rozycki, J.: Energy conservation via recompression evaporation, *Chem. Eng. Progress*, **72** (May), 69–72 (1977).
2. Schibbye, H.: Helical screw compressors as refrigeration- and heat pump-compressors (in Swedish), *Scandinavian Refrigeration*, **6**, No. 1, 15–19 (1977).

APPENDIX 1
Units, Conversion Factors, and Symbols

The calculations in this book are carried out in the SI-system (*Systeme International d'Units*) as given in the International Standard ISO 1000–1973 (E) and with symbols as given in ISO Recommendations 31.

The base and supplementary units of this system are listed in Table A1.1.

Units

The metre is the length equal to 1,650,763.73 wavelengths in vacuum of the radiation corresponding to the transition between the levels $2\,p_{10}$ and $5\,d_5$ of the krypton-86 atom.

The kilogram is the mass of the international prototype.

The second is the duration of 9,192,631,770 periods of the radiation corresponding to the transition between the two hyperfine levels of the ground state of the caesium-133 atom.

The ampere is that constant electric current which, if maintained in two straight parallel conductors of infinite length, of negligible circular cross-section, and placed one metre apart in vacuum, would produce between these conductors a force equal to 2×10^{-7} newton per metre of length.

Table A1.1 Base and supplementary units in the SI-system.

Quantity	Symbol	SI-unit Name	Symbol
Length	l	metre	m
Mass	m	kilogram	kg
Time	t	second	s
Electric current	I	ampere	A
Thermodynamic temperature	T	kelvin	K
Amount of substance	n	mole	mol
Luminous intensity	I_v	candela	cd
Plane angle	α	radian	rad
Solid angle	ω	steradian	sr

The kelvin, unit of thermodynamic temperature, is the fraction 1/273.16 of the thermodynamic temperature of the triple point of water.

The mole is that amount of substance of a system which contains as many elementary entities as there are atoms in 0.012 kilogram of carbon 12.

The candela is the luminous intensity perpendicular to a surface of 1/600,000 square metres of a black body at the temperature of freezing platinum under a pressure of 101,325 newtons per square metre.

The radian is the plane angle between two radii of a circle which cut off an arc on the circumference which equal in length to the radius.

The steradian is the solid angle which, having its vertex in the centre of a sphere, cuts off an area of the surface of the sphere equal to that of a square with sides of length equal to the radius of the sphere.

Prefixes

The multiples of SI-units, symbols and prefixes, are given in Table A1.2. Values in parentheses should not be introduced except where they are already in common use.

Note Multiplying prefixes are printed adjacent to the SI unit symbol with which they are associated. The multiplication of units is usually indicated by leaving a gap between them, i.e. mN means millinewton, m N means metre times newton.

To avoid errors in calculations, it is recommended that coherent SI units rather than their multiples and sub-multiples be used, i.e. 10^6 N/m^2 instead of MN/m^2 or N/mm^2.

A prefix applied to a unit is part of that unit when applied to a power, i.e. mm^2 = (mm)^2 = 10^{-6} m^2.

Conversion factors

There is often a need for conversion between the different systems, SI, cgs, British, and technical metric system. Table A1.3 gives a selection of conversion factors for quantities used in this text. The kilocalories and the calories used in the table are the IT-calories (International Table calories). The two other calories not used in this text are the 15°C calorie equal to 4.1855 Joule and the thermochemical calorie exactly equal to 4.184 Joule. Joule in this text is N m and not the international Joule of 1948 which is 1.00017 N m.

Table A1.2 Prefixes and symbols for multiples of SI-units.

Prefix	Symbol	Value	Prefix	Symbol	Value
exa	E	10^{18}	(deci	d	10^{-1})
peta	P	10^{15}	(centi	c	10^{-2})
tera	T	10^{12}	milli	m	10^{-3}
giga	G	10^{9}	micro	μ	10^{-6}
mega	M	10^{6}	nano	n	10^{-9}
kilo	k	10^{3}	pico	p	10^{-12}
(hecto	h	10^{2})	femto	f	10^{-15}
(deca	da	10^{1})	atto	a	10^{-18}

Table A1.3 Conversion factors. Bold figures are exact. kp = kg force

Quantity	given in	multiplied by	gives
Absorption, coefficient of (see mass transfer)			
Acceleration	ft/s^2	0.3048	m/s^2
Area	in^2	0.00064516	m^2
	ft^2	0.09290304	m^2
	yd^2	0.83612736	m^2
	acre	4046.9	m^2
	da	**1000**	m^2
Diffusivity [see viscosity (kinematic)]			
Elasticity, modulus of (see pressure)			
Energy	erg	**10^{-7}**	N m = J = W s
	kp m	9.80665	N m = J = W s
	kcal	**4186.8**	N m = J = W s
	cal	**4.1868**	N m = J = W s
	ft lb	1.3558	N m = J = W s
	ft pdl	0.042139	N m = J = W s
	Btu	1055.056	N m = J = W s
	therm	1.055 × 10^8	N m = J = W s
Enthalpy	kcal/kg	**4186.8**	$\dfrac{\text{N m}}{\text{kg}} = \dfrac{\text{J}}{\text{kg}} = \dfrac{\text{W s}}{\text{kg}}$
	cal/g	**4186.8**	$\dfrac{\text{N m}}{\text{kg}} = \dfrac{\text{J}}{\text{kg}} = \dfrac{\text{W s}}{\text{kg}}$
	Btu/lb	2326	$\dfrac{\text{N m}}{\text{kg}} = \dfrac{\text{J}}{\text{kg}} = \dfrac{\text{W s}}{\text{kg}}$
Entropy (see heat capacity)			
Force	dyn	**0.00001**	N
	kp = kg$_f$	9.80665	N
	lb$_f$	4.44822	N
	pdl	0.138255	N
Heat capacity	kcal/(kg °C)	**4186.8**	J/(K kg)
	cal/(g °C)	**4186.8**	J/(K kg)
	Btu/(lb °F)	**4186.8**	J/(K kg)
Heat flux	kcal/(m^2 h)	**1.163**	W/m^2 = J/(m^2 s)
	cal/(cm^2 s)	41868	W/m^2 = J/(m^2 s)
	Btu/(ft^2 hr)	3.1546	W/m^2 = J/(m^2 s)
Heat transfer coefficient	kcal/(m^2 h °C)	**1.163**	W/(m^2 K) = J/(s m^2 K)
	cal/(cm^2 s °C)	41868	W/(m^2 K) = J/(s m^2 K)
	Btu/(ft^2 hr °F)	5.6784	W/(m^2 K) = J/(s m^2 K)
Henry's law constant	atm/(g/cm^3)	**101.325**	N m/kg
	bar/(kg/cm^3)	**101,000**	N m/kg

Table A1.3 (*continued*)

Quantity	given in	multiplied by	gives
	atm/(kg/m^3)	**101,325**	N m/kg
	atm/(lb/ft^3)	6325.8	N m/kg
	psi/(lb/ft^3)	430.43	N m/kg
Length	Å	**0.1**	nm
	mil	**25.4**	μm
	in	**0.0254**	m
	ft	**0.3048**	m
	yd	**0.9144**	m
	mile	1609.3	m
Mass	grain	0.0647989	g
	oz	28.3495	g
	troy oz	31.103486	g
	lb	**0.45359237**	kg
	slug	14.594	kg
	cwt	50.802	kg
	tonne	1000	kg
	long ton	1016.05	kg
	short ton	907.18	kg
Mass flow rate	kg/h	0.00027778	kg/s
	lb/hr	0.0001260	kg/s
	lb/s	**0.45359237**	kg/s
Mass flux	kg/(m^2 h)	0.00027778	kg/(m^2 s)
	lb/(ft^2 hr)	0.0013562	kg/(m^2 s)
	lb/(ft^2 s)	4.8824	kg/(m^2 s)
Mass transfer coefficient			
(a) dimensionless driving force	kg/(m^2 h)	0.0002778	kg/(m^2 s)
	g/(cm^2 s)	**10**	kg/(m^2 s)
	lb/(ft^2 hr)	0.0013562	kg/(m^2 s)
(b) concentration as driving force	kg/[m^2 h (kg/m^3)]	0.0002778	m/s
	g/[cm^2 s (g/cm^3)]	**0.01**	m/s
	lb/[ft^2 hr (lb/ft^3)]	84.667 $\times 10^{-6}$	m/s
(c) Pressure as driving force	kg/(m^2 h atm)	0.0027413	kg/(s MN)
	kg/(m^2 h bar)	0.0027778	kg/(s MN)
	g/(cm^2 s atm)	98.687	kg/(s MN)
	lb/(ft^2 hr atm)	0.013384	kg/(s MN)
	lb/[ft^2 hr (lb$_f$/in^2)]	0.0009351	kg/(s MN)
Power	erg/s	**10^{-7}**	N m/s = J/s = W
	kp m/s	9.8065	N m/s = J/s = W
	Hp (metric)	735.5	N m/s = J/s = W
	Hp (British)	745.7	N m/s = J/s = W
	cal/s	**4.1868**	N m/s = J/s = W
	kcal/h	1.163	N m/s = J/s = W
	ft lb/min	0.022597	N m/s = J/s = W
	ft lb/s	1.3558	N m/s = J/s = W
	Btu/hr	0.2931	N m/s = J/s = W

Table A1.3   (continued)

Quantity	given in	multiplied by	gives
	Btu/s	1055.1	N m/s = J/s = W
	tons refr.	3516.85	N m/s = J/s = W
Pressure	dyn/cm^2	**0.1**	N/m^2 = Pa
	Pa	**1.0**	N/m^2 = Pa
	kp/m^2	9.8067	N/m^2 = Pa
	mm water	9.8067	N/m^2 = Pa
	mmHg = torr	133.32	N/m^2 = Pa
	bar	**0.1**	MN/m^2 = MPa
	kp/cm^2	0.0980665	MN/m^2 = MPa
	at(techn.)[1]	0.0980665	MN/m^2 = MPa
	atm	0.101325	MN/m^2 = MPa
	pdl/ft^2	1.4881	N/m^2 = Pa
	lb$_f$/ft^2	47.88	N/m^2 = Pa
	inches water	249.09	N/m^2 = Pa
	in Hg	3386.6	N/m^2 = Pa
	psi	6894.8	N/m^2 = Pa
Shear stress (see pressure)			
Stress (see pressure)			
Surface tension	dyn/cm	**0.001**	N/m
	kp/m	9.80665	N/m
	lb$_f$/ft	14.6005	N/m
	lb$_f$/in	1.2167	N/m
Temperature (see after the table)			
Thermal conductivity	cal/(cm s °C)	418.68	W/(m °C) = W/(m K)
	kcal/(m h °C)	**1.163**	W/(m °C) = W/(m K)
	Btu/(ft hr °F)	1.7308	W/(m °C) = W/(m K)
	Btu/[ft^2 hr (°F/in)]	0.1442	W/(m °C) = W/(m K)
	Btu/(ft s °F)	6232	W/(m °C) = W/(m K)
Thermal diffusivity	m^2/h	0.0002778	m^2/s
	ft^2/sec	0.092903	m^2/s
	ft^2/hr	0.00002581	m^2/s
Velocity	km/h	0.2778	m/s
	ft/hr	84.67 x 10^{-6}	m/s
	ft/min	**0.00508**	m/s
	knots	**0.51444**	m/s
Viscosity (absolute = dynamic)	cP	**0.001**	N s/m^2 = kg/(m s)
	P	**0.1**	N s/m^2 = kg/(m s)
	kp s/m^2	9.80665	N s/m^2 = kg/(m s)
	lb/(ft hr)	0.00041338	N s/m^2 = kg/(m s)
	lb/(ft sec)	1.4482	N s/m^2 = kg/(m s)
(kinematic)	cSt	**10^{-6}**	m^2/s
	St = cm^2/s	0.0001	m^2/s
	ft^2/hr	0.00002581	m^2/s
	ft^2/sec	0.092903	m^2/s

Table A1.3   (continued)

Quantity	given in	multiplied by	gives
Volume	ounces (Brit. fluid)	0.28413×10^{-6}	m^3
	ounces (US fluid)	0.29574×10^{-6}	m^3
	in^3	16.387×10^{-6}	m^3
	gallons (Brit.)	0.004546	m^3
	gallons (US)	0.0037853	m^3
	ft^3	0.028317	m^3
	bushels (Brit.)	0.036369	m^3
	bushels (US)	0.035239	m^3
	barrels (Brit.)	0.163659	m^3
	barrels (US petr.)	0.15898	m^3
	barrels (US dry)	0.11563	m^3
	barrels (US liq.)	0.11924	m^3
	yd^3	0.76455	m^3

[1] German and Russian publications use 'at' for technical atmospheres. In German 'ata' is technical atmospheres absolute and 'atü' technical atmospheres gauge pressure, while Atm with capital A means atmospheres equal to 760 mm mercury.

Temperature,

$1°R = \frac{5}{9} K$

$\theta°F = (459.67 + \theta)°R = [(5/9)(\theta - 32)] \;°C$

$\theta°C = (273.15 + \theta) K$

The gas constant, $R = 8314.3$ N m/(K kmol).
Volume of an ideal gas

at 1 atm and 0°C, 22.414 m^3/kmol
at 30 inches mercury and 60°F, saturated with water vapour (Brit. gas industry), 386.23 ft^3/lb-mole
at 1 atm and 60°F, dry gas (US), 379.34 ft^3/lb-mole.

Acceleration due to gravity, 'normal', $g = 9.80665$ m/s^2.

Symbols

A	Area, m^2
A_p	Area of particle projected on a plane perpendicular to the direction of motion, m^2
B	Width of baffles (Chapter 5), m
C	Constant
Ca	Cauchy number, $\rho V^2/E$
C_b	Stefan–Boltzmann constant, 5.7×10^{-8} W/(m^2 K^4)
c	Specific heat capacity, J/(kg °C)
c_p	Specific heat at constant pressure, J/(kg °C)
c_v	Specific heat at constant volume, J/(kg °C)

D	Diffusivity, m²/s
D	Diameter, m
	Coil diameter, m
D_c	Collector diameter, m
D_p	Particle diameter, m
D_{pc}	Particle 'cut size' (50% removed), m
D_h	Hydraulic diameter [equation (1.10)], m
D_s	Diameter of a sphere with the same volume as a particle, m
D_s	Shell diameter, m
E	Emitted energy, J/s = W
E	Modulus of elasticity, N/m² = Pa
E_H	Elevation factor for pressure drop, equation (1.24)
Eu	Euler number, $p/\rho V^2$
f	Friction factor
f_D	Drag factor, Figure 6.2
F	Force, N
Fr	Froude number, $V^2/(L\,g)$
F_T	Correction factor for temperature difference
Gr	Grashof number, $\rho^2 D^3 (\beta \Delta T) g/\mu^2$
g	Local acceleration due to gravity, m/s²
h	Enthalpy, J/kg
h	Individual heat-transfer coefficient, W/(m² °C) or J/(s m² °C)
k	Thermal conductivity W/(m °C) = J/(s m °C)
L	Length, m
L_e	Length of straight pipe giving the same pressure drop as valves, fittings, etc., m
M	Molecular weight
M	Mass, kg
m	Cross-sectional ratio of orifices and nozzles, $(D_2/D)^2$
N	Number of tube rows in the direction of flow
n	Number of lost velocity heads
n	Revolutions per second, s⁻¹
n	Exponent in empirical equations
Nu	Nusselt number, hD/k
P	Total pressure of a mixture, Pa = N/m²
P	Power, N m/s = J/s = W
P_t	Pitch = shortest centre-to-centre distance between adjacent tubes, m
Pr	Prandtl number, $c_p \mu/k$
p	Pressure, Pa = N/m²
p_c	Critical pressure, Pa = N/m²
Δp_f	Pressure drop due to friction, Pa = N/m²
Q	Volume, m³
Q	Energy, J = Ws
$\dot{Q}$	Volumetric rate of fluid flow, m³/s
q	Heat added, J/kg = N m/kg
$\dot{q}$	Heat flux, W/m²

R	Gas constant, 8314.3 J/(K kmol)
R	Agitation intensity, kW/m^3
r	Radius, m
Re	Reynolds number, $\rho VD/\mu = VD/\nu$
S	Cross section, m^2
s	Specific surface area of particles, m^2/m^3
s	Thickness of plate or fin, m
s	Entropy, J/(kg K)
s	Length, m
Sc	Schmidt number, $\mu/(\rho D)$
T	Absolute temperature, K
t	Time, s
U	Over-all heat transfer coefficient, W/($^\circ$C m^2) or J/(s $^\circ$C m^2)
U	Internal energy, J/kg = N m/kg
u	Fluid velocity, m/s
v	Specific volume, m^3/kg
V	Average velocity of fluid, m/s
V_0	Superficial velocity referred to total cross section, m/s
V_t	Terminal sedimentation velocity, m/s
W	Mass, kg
w	Rate of mass flow, kg/s
w	Mechanical work, N m/kg = J/kg or J/kmol
w	Sedimentation velocity in an electric field, m/s
We	Weber number, $\rho L V^2/\gamma$
x	Coordinate, m
x	Mole fraction of a component in a liquid mixture
y	Coordinate vertical to direction of flow, m
y	Mole fraction of a component in a vapour mixture
z	Height above reference plane, m
α	Angle, degrees or radians
α	Nozzle coefficient
α	Thermal diffusivity = $k/(\rho c_p)$, m^2/s
β	Thermal coefficient of cubic expansion, K^{-1} or $^\circ$C^{-1}
γ	Surface tension, N/m
Δ	Difference
ϵ	Emissivity
ϵ	Porosity = void fraction = (void volume)/(total volume)
ϵ	Height of unevennesses, m or mm
ϵ	Expansion coefficient (flow meters)
ϵ_M	Minimum porosity during fluidization (Figure 6.7)
η	Efficiency
θ	Temperature, $^\circ$C
$\Delta\theta_c$	Critical temperature difference in boiling, $^\circ$C
τ	Shear stress in fluid, Pa = N/m^2
κ	Ratio of specific heat capacities, c_p/c_v
λ	Latent heat of phase change, J/kg

λ_m	Mean free path of the molecules, m
μ	Dynamic or absolute viscosity, N s/m^2 or cP
ν	Kinematic viscosity = μ/ρ, m^2/s, or cSt
π	Dimensionless group
ρ	Fluid density, kg/m^3
ρ_s	Density of slurry, kg/m^3
Φ_s	Shape factor, equation (1.19)
ψ	Critical pressure ratio, equation (4.20)
ψ	Shape factor, equation (6.2)
Ψ	Inertia parameter, equation (6.25)
ω	Angular velocity, rad

APPENDIX 2
Physical Properties of Gases and Liquids

Main sources: Reid, Prausnitz, and Sherwood: 'Properties of Gases and Liquids', McGraw-Hill, New York, 3rd Edn., 1977, and 'VDI-Wärmeatlas', VDI-Verlag, Düsseldorf, 3rd Edn., 1977.

The data for gases are for low and moderate pressures. Constant α W/(m K^2) for thermal conductivity of liquids is for a limited temperature range. T is the absolute temperature in K.

Table A2.1 Dynamic viscosity and liquid heat capacity.

Dynamic viscosity in N s/m^2 = cP × 10^{-3}

Liquid: $\mu = A e^{B/T}$

Gas: $\mu \approx \dfrac{b\, T^{1.5}}{S + T}$

Compound	Temperature range liquid °C	$A \times 10^6$	B	$b \times 10^6$	S	Liquid heat capacity at 20°C c_{20} kJ/(kg K)
Acetic acid	20 to 100	13	1328			2.00
Acetone	−75 to 50	17.1	864	1.17	527	2.16
Air				1.46	110	
Ammonia	−50 to 50	8.6	813	1.56	492	4.61
Benzene	20 to 170	8.7	1264	0.99	366	1.73
n-Butanol	−50 to 100	1.06	2324			2.35
Carbon dioxide	−25 to 20	0.71	1345	1.50	220	2.28 (0°C)
Carbon tetrachloride	0 to 175	11.3	1302	1.17	300	0.85
Chlorine	−100 to 0	68	470	1.66	336	0.94
Diethyl ether	−100 to 70	14.5	826	0.94	343	2.34
Ethane	−150 to −50	22.8	358	0.97	229	4.31
Ethanol	−50 to 100	4.0	1660	1.32	514	2.41
Ethylene				1.04	225	2.41(−100°C)
n-Heptane	−20 to 100	13	1010	0.88	513	2.36
Hydrogen				0.65	67	
Methane				1.00	168	4.82(−100°C)
Methanol	−75 to 50	7.6	1275	1.44	477	2.50
Nitrogen				1.40	108	2.15(−180°C)
Oxygen				1.69	126	1.70(−180°C)
Propane	−40 to 40	7.8	746	0.86	241	
n-Propanol	−75 to 100	1.4	2154	1.13	479	2.35
Toluene	−25 to 200	13.3	1110	0.97	420	1.72
Water	10 to 115	2.6	1750	1.75	626	4.19

Table A2.2 Gas heat capacity and liquid and gas thermal conductivity.

Compound	\multicolumn{4}{c}{Gas heat capacity $c_p = A + BT + CT^2 + DT^3$ kJ/(kg K)}	\multicolumn{2}{c}{Liquid $k_2 = k_{2s} - \alpha(T_2 - 298)$}	Gas $k_g \approx k_{g2s}\left(\dfrac{T_2}{298}\right)^{1.8}$
	A	$B \times 10^3$	$C \times 10^6$	$D \times 10^{10}$	k_{2s}	$\alpha \times 10^4$	k_{g2s}
Acetic acid	0.081	4.24	−2.92	8.24	0.159	2.0	
Acetone	0.108	4.49	−2.16	3.51	0.160	5.0	0.013
Air	1.057	−0.37	0.86	2.42			0.026
Ammonia	1.604	1.40	1.00	−6.96	0.516	(10)	0.024
Benzene	−0.434	6.07	−3.86	9.13	0.145	3.3	0.013
n-Butanol	0.044	5.64	−3.02	6.32	0.153	2.4	0.013
Carbon dioxide	0.450	1.67	−1.27	3.90	0.08	(10)	0.016
Carbon tetrachloride	0.265	1.33	−1.48	5.75	0.104	1.8	0.0067
Chlorine	0.380	0.48	−5.46	2.18			0.0093
Diethyl ether	0.289	4.53	−1.40	−1.26	0.133	0.7	0.015
Ethane	0.180	5.92	−2.31	2.90			0.021
Ethanol	0.196	4.65	−1.82	0.30	0.167	5.6	0.015
Ethylene	0.136	5.58	−2.98	6.26	0.119	3.5	0.021
n-Heptane	−0.051	6.75	−3.64	7.64	0.123	3.3	0.012
Hydrogen	13.5	4.60	−6.85	37.9	0.118 (−253°C)		0.181
Methane	1.200	3.25	0.75	−7.05			0.034
Methanol	0.660	2.21	0.81	−8.90	0.201	3.0	0.016
Nitrogen	1.112	−0.48	0.96	4.17			0.026
Oxygen	0.878	0	0.55	−3.33			0.026
Propane	−0.096	6.95	−3.60	7.29	0.08		0.018
n-Propanol	0.041	5.53	−3.09	7.15	0.156	1.7	0.014
Toluene	−0.264	5.56	−3.00	5.33	0.140	1.6	0.015
Water	1.790	0.11	0.59	−2.00	0.605	1.5	0.019

Table A2.3 Saturation pressure p_s and specific volume v'' of saturated steam.

θ °C	p_s kN/m²	v'' m³/kg	θ °C	p_s MN/m²	v'' m³/kg	θ °C	p_s MN/m²	v'' m³/kg
10	1.228	106.4	90	0.0701	2.361	240	3.348	0.0597
15	1.705	77.96	100	0.1013	1.673	250	3.978	0.0500
20	2.337	57.84	110	0.1433	1.210	260	3.694	0.0421
25	3.166	43.41	120	0.1985	0.891	270	5.505	0.0355
30	4.241	32.94	130	0.2701	0.668	280	6.419	0.0301
35	5.621	25.26	140	0.3614	0.508	290	7.445	0.0255
40	7.374	19.56	150	0.4760	0.392	300	8.592	0.0216
45	9.581	15.28	160	0.6180	0.307	310	9.870	0.0183
50	12.33	12.05	170	0.7920	0.243	320	11.29	0.0155
55	15.74	9.583	180	1.003	0.194	330	12.87	0.0130
60	19.92	7.682	190	1.255	0.1564	340	14.61	0.0108
65	25.01	6.205	200	1.555	0.1273	350	16.54	0.0088
70	31.16	5.048	210	1.908	0.1043	360	18.67	0.0070
75	38.55	4.135	220	2.320	0.0861	370	21.05	0.0050
80	47.36	3.410	230	2.798	0.0715	374	22.09	0.0036

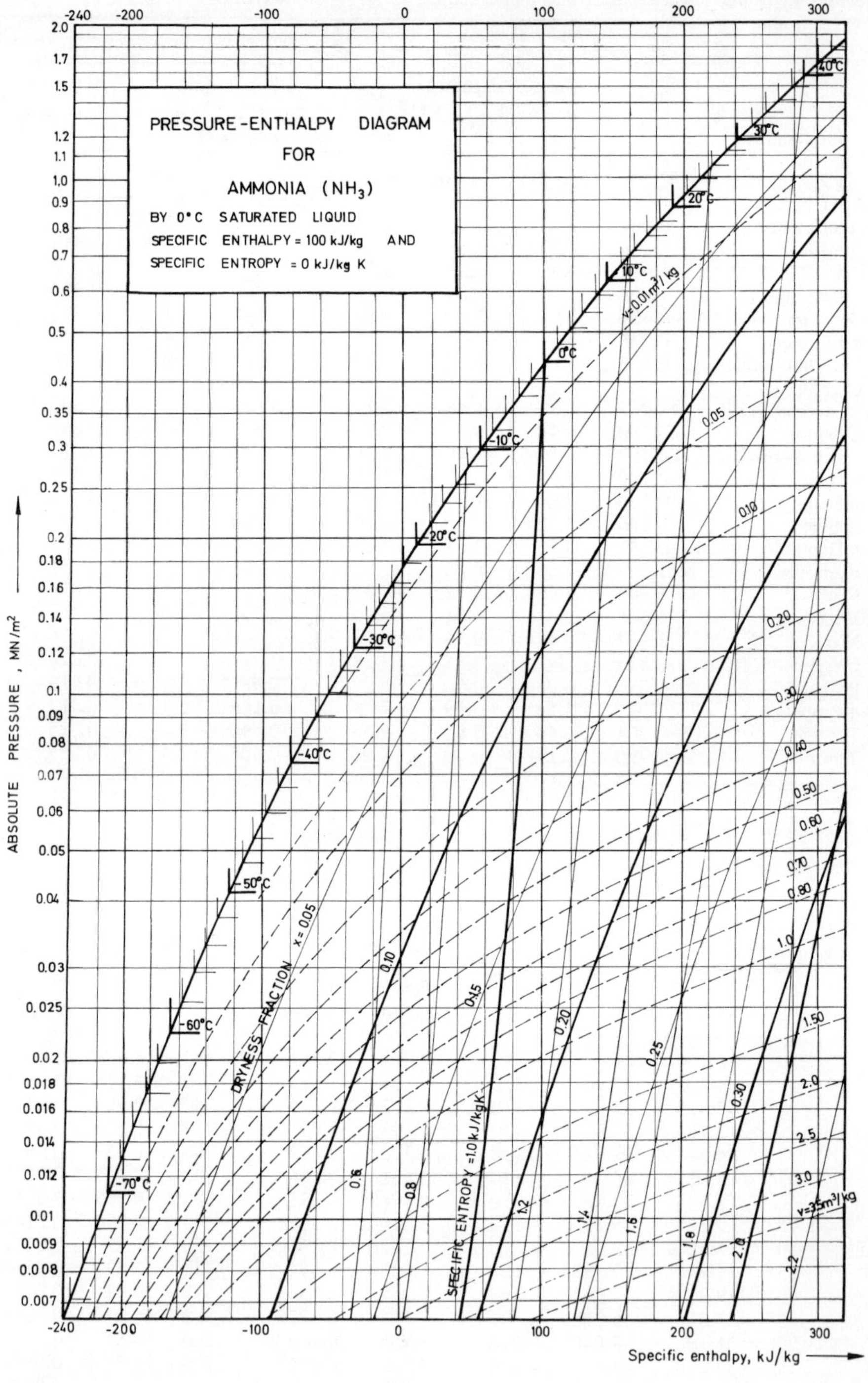
PRESSURE-ENTHALPY DIAGRAM
FOR
AMMONIA (NH_3)
BY 0°C SATURATED LIQUID
SPECIFIC ENTHALPY = 100 kJ/kg AND
SPECIFIC ENTROPY = 0 kJ/kg K
ABSOLUTE PRESSURE, MN/m²
Specific enthalpy, kJ/kg
DRYNESS FRACTION x = 0.05
SPECIFIC ENTROPY = 1.0 kJ/kg K
v = 0.01 m³/kg
v = 35 m³/kg
-70°C
-60°C
-50°C
-40°C
-30°C
-20°C
-10°C
0°C
10°C
20°C
30°C
40°C

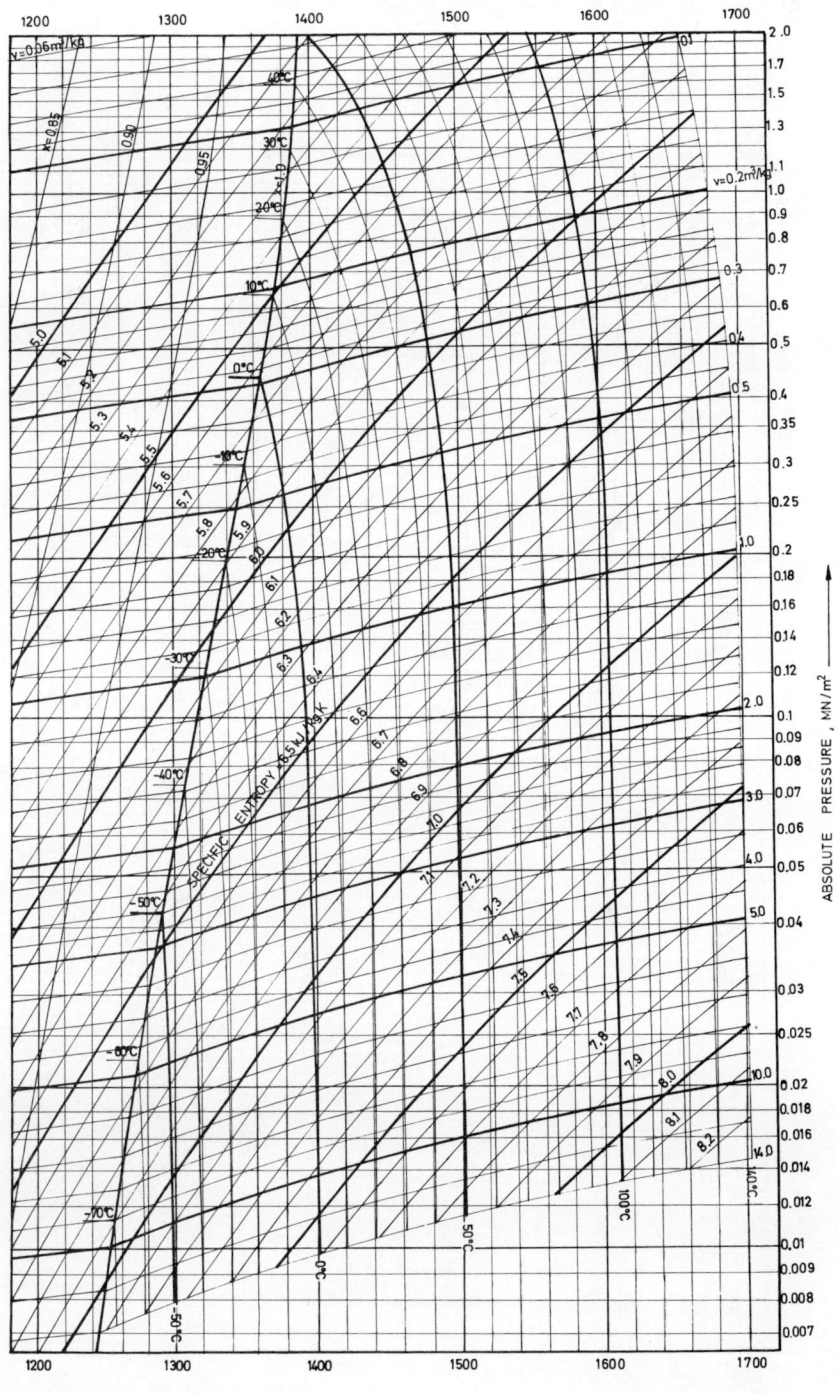

BASED ON MOLLIER- DIAGRAM FROM
KÄLTEMASCHINENREGELN 5th ED.

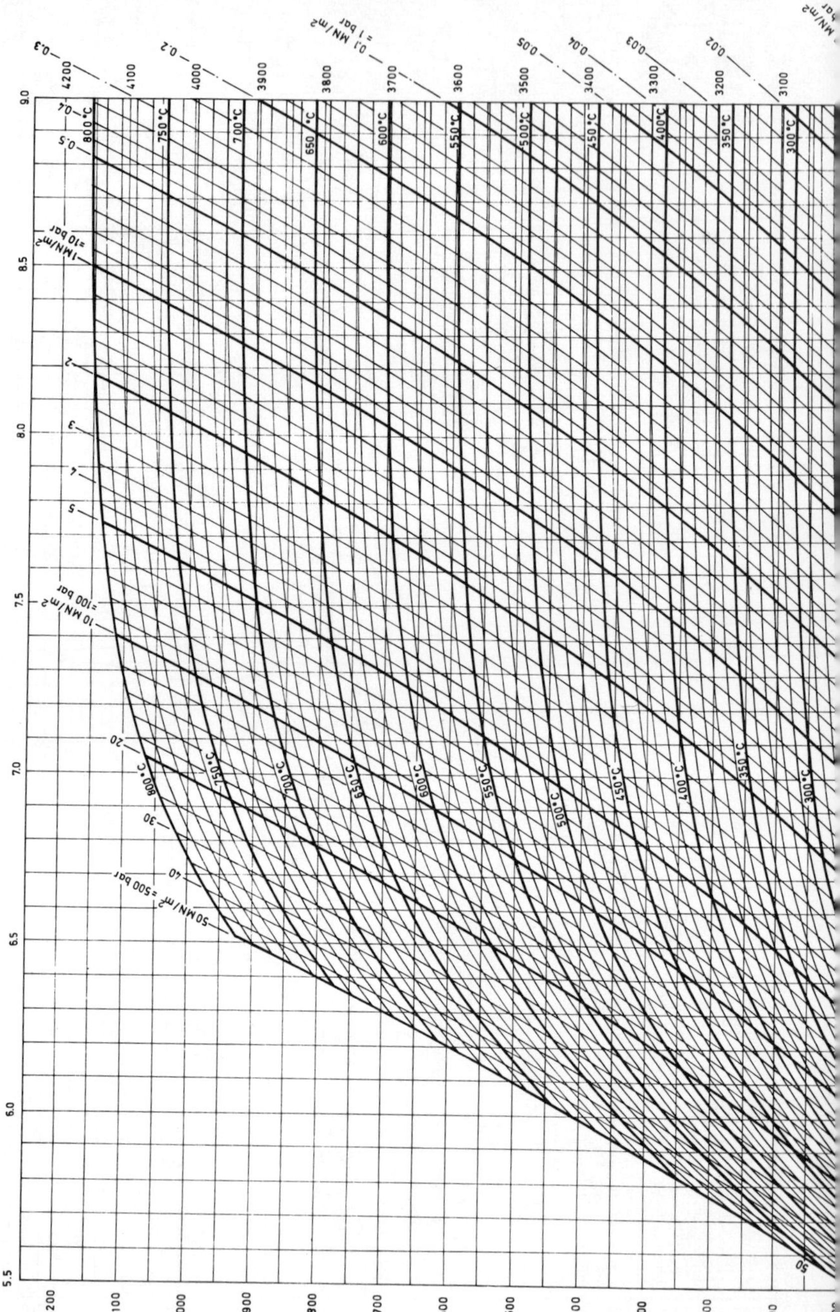

SPECIFIC ENTHALPY, kJ/kg
0.1 MN/m² = 1 bar
1 MN/m² = 10 bar
10 MN/m² = 100 bar
50 MN/m² = 500 bar

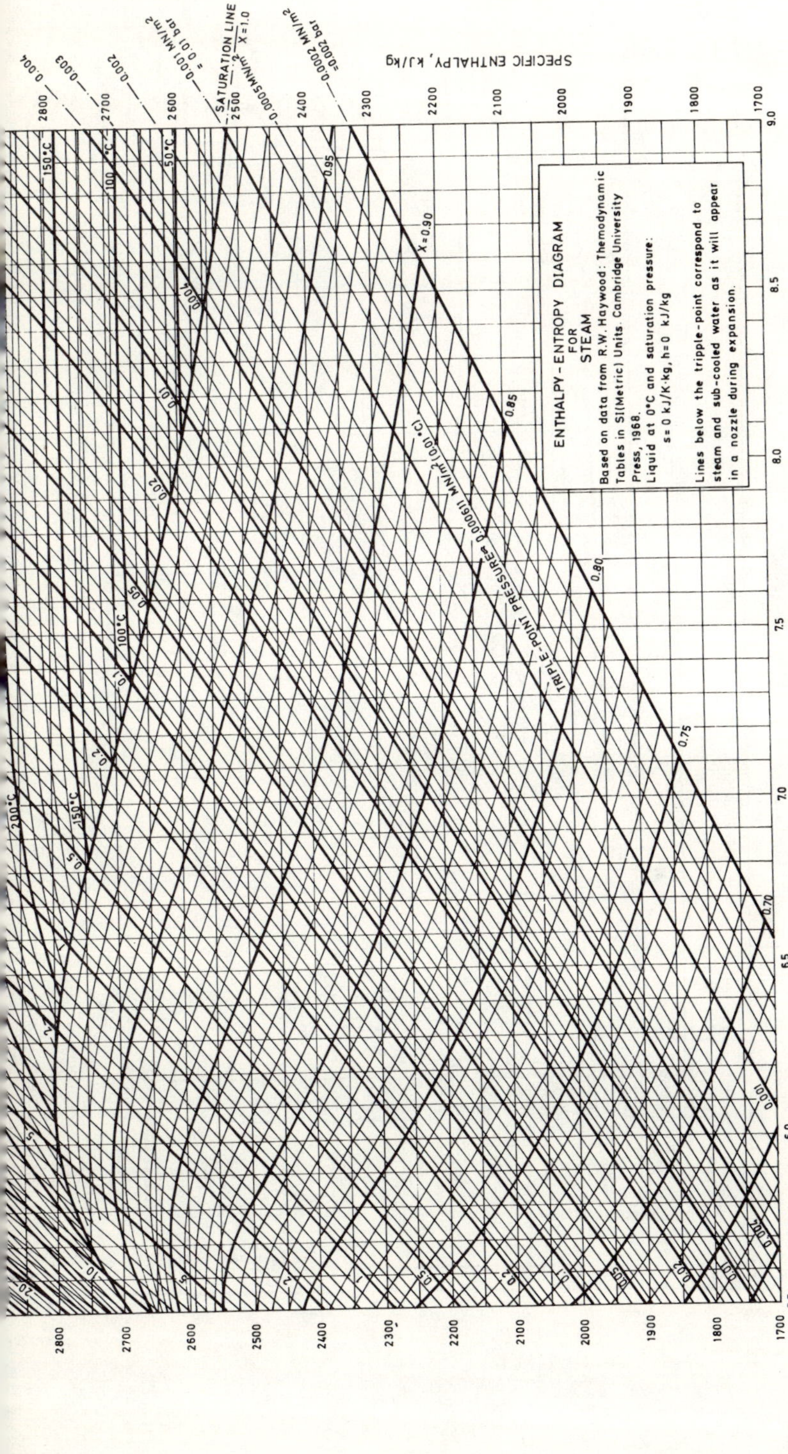
ENTHALPY - ENTROPY DIAGRAM
FOR
STEAM
Based on data from R.W. Haywood: Thermodynamic Tables in SI(Metric) Units. Cambridge University Press, 1968.
Liquid at 0°C and saturation pressure:
s = 0 kJ/K·kg, h = 0 kJ/kg
Lines below the tripple-point correspond to steam and sub-cooled water as it will appear in a nozzle during expansion.
SPECIFIC ENTROPY, k J/kg·K
SPECIFIC ENTHALPY, kJ/kg
SATURATION LINE
X = 1.0
X = 0.90
TRIPLE-POINT PRESSURE= 0.000611 MN/m² (0.01 °C)
0.001 MN/m² = 0.01 bar
0.0002 MN/m² =0.002 bar
0.0005MN/m²
0.95
0.85
0.80
0.75
0.70
5.5
6.0
6.5
7.0
7.5
8.0
8.5
9.0
1700
1800
1900
2000
2100
2200
2300
2400
2500
2600
2700
2800

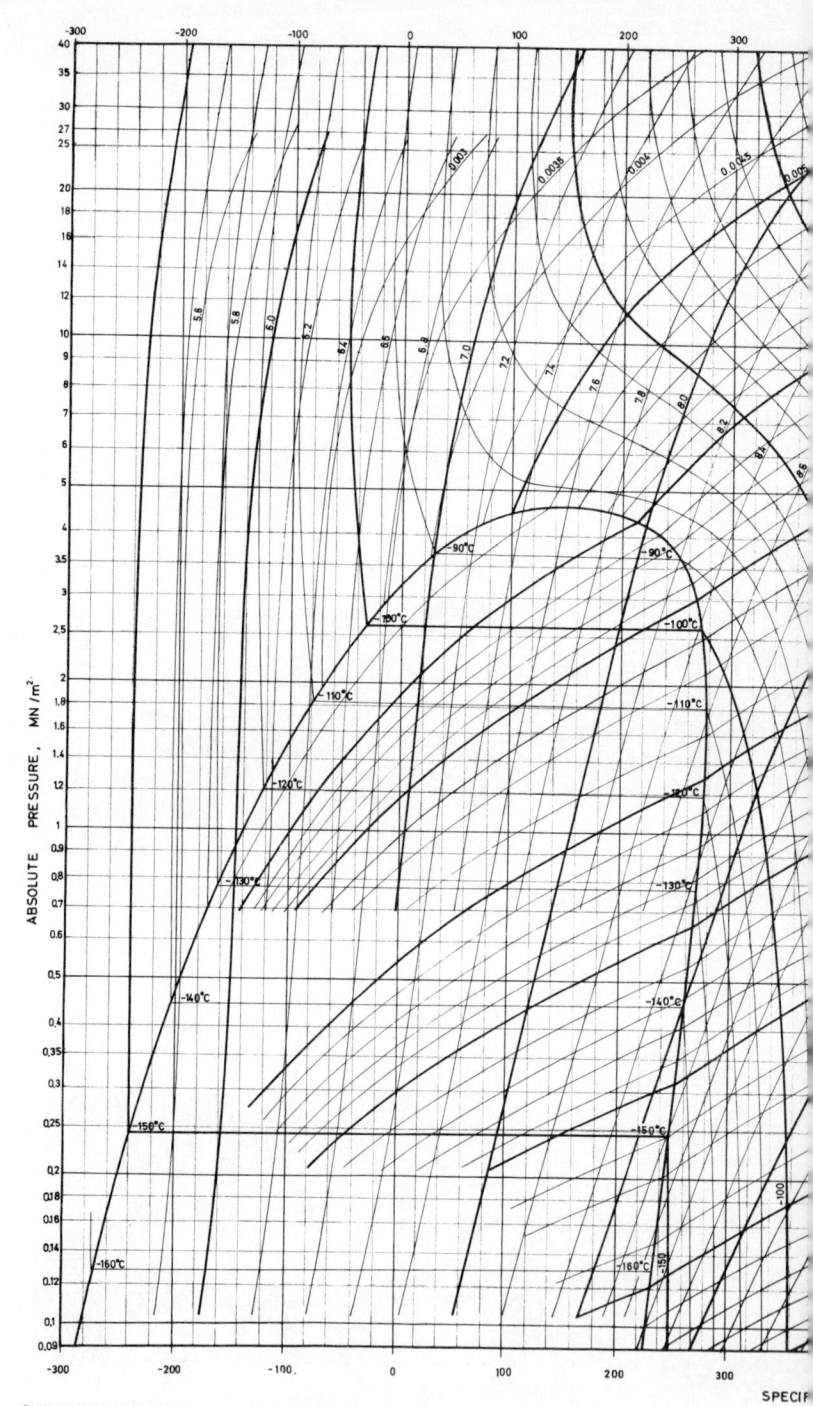

BASED ON GRAPH AND TABLES IN
W.C. EDMINSTER: APPLIED THERMODYNAMICS, Gulf PUBL.CO.
AND F. DIN: THERMODYNAMIC FUNCTIONS OF GASES, VOL. 3, BUTTERWORTHS.

351

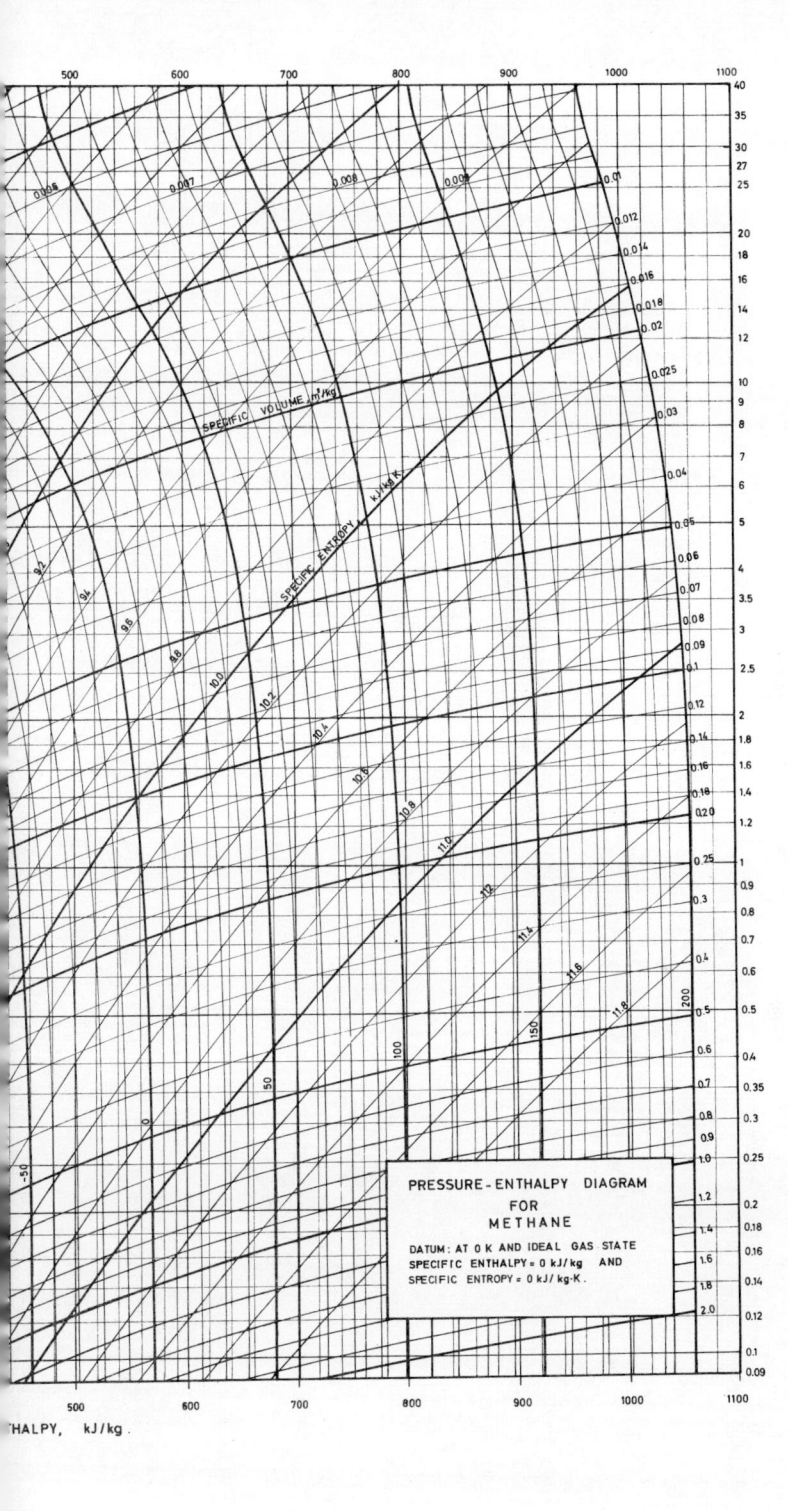

PRESSURE-ENTHALPY DIAGRAM
FOR
METHANE

DATUM: AT 0 K AND IDEAL GAS STATE
SPECIFIC ENTHALPY = 0 kJ/kg AND
SPECIFIC ENTROPY = 0 kJ/kg·K.

Index

Absorption 124
Adiabatic
   compression 74
   efficiency 77, 78
   expansion 85
Agglomeration 139, 142, 193
Agitated vessels 117
   heat transfer 244
Agitation 117
   dimensional analysis 41
   dispersed gas 120, 124, 220
   effects of 119
   intensity 41, 119
   power consumption 119
Agitators 117
Ammonia, enthalpy diagram 346, 347
Anemometers 61
Arithmetic mean temperature difference 252
Atomization of liquids 219

Backward feed 325
Bacteria filters 153, 184
Bag filters 153, 185
Barometric condenser 83, 103, 109, 262, 327
Berl saddles 11
Bernoulli's equation 4
Bingham plastic fluid 1
Blaine surface area 225
Blenders 50, 120
Blowers 77, 78
   see also Compressors
Boiling liquids 247
   binary mixtures 249
   critical temperature difference 248
Brownian motion 129, 153, 154
Bubbles 137, 219

Callandria evaporator 265
Cascade coolers 255

Cauchy number 39
Centrifugal
   blowers 77, 78
   compressors 78
   fans 77
   pumps 70, 90
   separators 137
Centrifuges 137
Classifiers
   double-cone 169
   rake 132
   screw 226
Claude process 98
Coalescence 221
Co-current flow 253
Colloid mill 221
Comminution 223
Compressible filter cakes 201, 203, 208
Compression 74
   adiabatic 75, 91, 92
   isentropic 75, 76
   isothermal 75, 76
   polytropic 75
Compressors
   capacity 78, 83
   centrifugal 78
   efficiency 78, 83
   helical screw 78, 79
   liquid piston 78
   oil diffusion pump 83
   reciprocating 78, 80
   sliding vane 83
   steam ejector 84, 87
   turbo 79
   two-impeller 77, 79
   vacuum pumps 81
   water jet ejector 84
Condensation 246, 289
Condensers 256
   direct contact 84

Conduction, *see* Heat conduction
Convection   238
   forced   240
   free   239
Conversion factors   337
Cooling coils   256
Counter-current flow   253
Critical pressure ratio   86
Critical temperature difference   248
Crushers   223
Crushing and grinding   223, 229
   closed circuit   227
   power consumption   225, 229
Crystallizers   244, 265
Cunningham correction factor   129
Cut size   141, 145
Cyclones   137, 139, 177, 178
Cylindrical filter cakes   203, 215

Decanters   137, 172, 175
Demisters   149, 150, 182
Diffusion pumps   83, 84
Dimensional analysis   37
Direct contact heat exchange   262
Dispersion
   gas in liquid   120, 219, 227
   liquid in liquid   221, 229
Dissolution of crystals   121
Distributor   12, 24
Dowtherm   262
Drag factor   128
Drops   45, 137
Dynamic similarity   39

Ejectors
   steam   83, 87, 102
   water   83, 96
Electrostatic precipitators   161, 186, 188
Emissivity   236
Energy balances   4, 69
Energy economy   321
Enthalpy   76
   ammonia   346, 347
   methane   350, 351
   steam   348, 349
Entrainment   150, 263
Entropy   76
   ammonia   346, 347
   methane   350, 351
   steam   348, 349
Euler number   39
Evaporation   263
   multiple-effect   324, 330

Evaporators   256, 263, 287
Expansion   85
   coefficient   59
   machines   79, 85
Extended surfaces   258

Fall velocity equivalent diameter   132
Fans   77
Fibre filters   153, 184, 186
Fibre mist eliminators   152
Filter
   aids   193
   cake   193
   calculations   200, 208, 213, 215
   centrifuges   198
   laboratory tests   204, 208
   presses   194
   rotating   196
Filters for
   gases   153
   liquids   193
Filtration   193
   compressible filter cakes   201, 203, 208
   incompressible filter cakes   201
Fin efficiency   261, 282
Finned tube heat exchangers   28, 258, 279, 284
Finned tubes   10, 258
Fittings   7, 9
Flotation   205
Flow distributors   12, 135
Flow measurements   55
Flow work   69
Fluid friction,
   see Pressure drop
Fluidization   134, 170
Fluids
   Newtonian   1
   non-Newtonian   1
   suspensions   12, 30, 133
Foaming   263
Forced convection   240
Forced vortex   144
Forward feed   325
Fouling resistance   254
Fourier's law   231
Free convection   239
Free vortex   143
Freezing   294, 305, 306, 307
Friction factor   6
   see also Drag factor; Pressure drop
Froude number   39, 120

Gas absorption 124
Gas constant 340
Gas distributor 136
Gas scrubbers 155
Granular beds 11, 245
Grashof number 239
Grinding 223
Gurney—Lurie diagrams 299

Head
  pressure 4
  velocity 4
Heat conduction 231
  graphical solution 234, 266, 267
  steady-state 231
  unsteady-state 294
Heat exchangers 254
Heat flux 249
Heat pipes 284
Heat pump 106, 327, 328
Heat radiation, see Radiation
Heat transfer 231, 294
Heat transfer coefficients
  agitated vessels 244
  boiling 247
  coils 243
  condensation 246
  direct contact 262
  finned tubes 260
  forced convection 240
  free convection 239
  granular beds 245
  jacketed vessels 244
  liquid films 245
  natural convection 239
  overall coefficients 250
  radiation 235
  scraped tubes 244
  shell-and-tube heat exchangers 242
  tube banks 243
  tube coils 243
Hindered settling 133, 168
Homogenization 221
Hydraulic diameter 6, 21
Hydrocyclones 137, 142, 180
Hyperfiltration 200

Ideal gas 75
Impellers 117
Impingement separators 148
Internal energy 69
Isentropic compression 75, 76
Isothermal compression 75, 76

Joule—Thomson expansion 97

Kinetic energy 4
Kozeny—Carman equation 135, 200

Labyrinth seal 80
Laminar flow 2
Laval nozzle 86
Leakage, vacuum systems
  flanges 81
  rotating shaft seals 81
Liquid films 245
Liquid—piston blowers 78
Logarithmic mean temperature
  difference 252
  correction factor 254

Magnetic flowmeters 61
Mean free path 129
Mechanical work 69
Mesh 227
Mills 223
Minimum energy requirement 321
Mixed feed 324
Mixers
  drum mixer 50
  fluid jet 120
Modal laws 38
Mono pump 74
Multiple-effect vaporization 324, 330

Natural convection 239
Net pressure suction head 19
Newtonian fluids 1
Non-ideal gas 76
Non-Newtonian fluids 1
Nozzle coefficient 57
Nozzle expansion 86
Nozzles for
  ejectors 86, 104
  metering 55, 61
Nusselt number 239

Optimum diameter 7
Orifice coefficient 57
Orifice meters 55, 63
Overall heat transfer 250

Paddles 117
Parallel flow 253
Particle and drop mechanics 128
Particles
  Brownian motion 129
  characteristics of 130, 225

classification of   132
drag factor   128
fall velocity equivalent diameter   132
fluidization of   134
pneumatic conveying of   12, 136
sieve analysis   227
terminal velocity   129
   in electric field   164
   in hindered settling   133
Physical properties   344
Piezometer ring   56
Pipe roughness   6
Pitot tubes   58, 66
Pneumatic conveying   12, 136
Polytropic compression   75
Porosity   11, 135, 200
Potential energy   4, 69
Power consumption in agitation   119
Prandtl number   239
Prefixes   336
Pressure drop   1
  distributors   12, 135
  filter cakes   195, 200
  finned tubes   10, 28
  fittings   7
  heat exchangers   14, 28
  helices   8
  spirals   21
  stack   26
  suspensions   12, 30
  tube banks   8
  two-phase flow   12
  valves   7
Pressure head   4
Propellers   117
Pumping   69
Pumps   70

Radiation   235, 268, 270
  exchange coefficient   237
  from flames   238
  from gases   238
  graphical solution   237, 269
Raschig rings   11
Reciprocating
  compressors   78, 80
  pumps   71
Recovery of waste heat   323
Refrigeration   94, 102
Reversed osmosis   200
Reversible separation   321
  gas mixture   322
  liquid mixture   323

Reynolds number   39, 239
Rheology   2
Rotameter   60
Rotating
  disc atomizers   219
  filters   196
Roughness in pipes   6

Sauter mean diameter   221
Scaling   254, 263
Schmidt number   247
Schmidt's method   300
Scraped tubes   244
Screw
  compressors   78, 79
  presses   198
  pumps   73
Scrubbers   155
  efficiency   157
Secondary heat transfer media   262
Shape factor   11, 128
Shell and tube heat exchangers   242, 256, 273, 274, 287, 289
Sieve analysis   227
Siloes, flow in   12
Specific heats of
  gases   345
  liquids   344
  metals   232
Spiral heat exchangers   21, 256
Spray
  dryers   167
  nozzles   219
  towers   155
Stack   26
Static head   265
Steam ejectors   83, 87, 102, 109, 327
Steam table   345
Stefan–Boltzmann's law   235
Stokes' law   129
Streamline flow   2
Supersonic nozzle   87
Surface roughness   6
Surface volume mean diameter   221
Suspensions   12, 30, 133
Symbols   340

Temperature differences   251, 289
Terminal velocity   42, 129, 167
Thermal conductivity   231
  gases   345
  liquid   345
  metals   232
Thermal diffusion   297, 309, 312, 313

Thermal diffusivity 297
Thermocompression 327
Thickeners 132
Throttle calorimeter 97
Tube banks 8, 243
Tubes, finned 10, 260
Turbines 85
Turbocompressors 79
Turbulent flow 3
Two-phase flow 12, 31
Tyler standard screen 227

Ultrafiltration 199
Units 335
Unsteady state heat transfer 294
   graphical solution 300, 312, 313
   with phase change 294, 305, 306, 307
   without phase change 297, 309

Vacuum pumps 81
Valves 7
Vapour recompression 327
Velocity head 4, 22
Velocity of sound 86
Venturi
   meters 58
   scrubbers 156, 186
Viscosity 1
   gases 344
   liquids 344
Viscous flow 2

Water jet ejectors 83, 96
Weber number 39
Wire mesh demisters 150, 182
Work index of crushing 225
Work of expansion 85

Zeppeliners 63